Natural Environment Research Council
BRITISH GEOLOGIC

Geology of th
between Dudley and Bridgnorth

One-inch geological sheet 167, New Series)

T. H. Whitehead, MSc, ARCS and R. W. Pocock, DSc

LONDON HER MAJESTY'S STATIONERY OFFICE 1947

HER MAJESTY'S STATIONERY OFFICE

HMSO publications are available from:

HMSO Publications Centre
(Mail and telephone orders)
PO Box 276, London SW8 5DT
Telephone orders (01) 622 3316
General enquiries (01) 211 5656
Queueing system in operation for both numbers

HMSO Bookshops
49 High Holborn, London WC1V 6HB
(01) 211 5656 (Counter service only)
258 Broad Street, Birmingham B1 2HE
(021) 643 3740
Southey House, 33 Wine Street, Bristol
BS1 2BQ   (0272) 264306
9 Princess Street, Manchester M60 8AS
(061) 834 7201
80 Chichester Street, Belfast BT1 4JY
(0232) 238451
71–73 Lothian Road, Edinburgh
EH3 9AZ   (031) 228 4181

HMSO's Accredited Agents
(see Yellow Pages)

And through good booksellers

BRITISH GEOLOGICAL SURVEY
Keyworth, Nottinghamshire NG12 5GG
Plumtree (060 77) 6111
Murchison House, West Mains Road,
Edinburgh EH9 3LA   (031) 667 1000

The full range of Survey publications is available through the Sales Desks at Keyworth and Murchison House. Selected items are stocked by the Geological Museum Bookshop, Exhibition Road, London SW7 2DE; all other items may be obtained through the BGS London Information Office in the Geological Museum ((01) 589 4090). All the books are listed in HMSO's Sectional List 45. Maps are listed in the BGS Map Catalogue and Ordnance Survey's Trade Catalogue. They can be bought from Ordnance Survey Agents as well as from BGS.

The British Geological Survey carries out the geological survey of Great Britain and Northern Ireland (the latter as an agency service for the government of Northern Ireland), and of the surrounding continental shelf, as well as its basic research projects. It also undertakes programmes of British technical aid in geology in developing countries as arranged by the Overseas Development Administration.

The British Geological Survey is a component body of the Natural Environment Research Council.

© *Crown copyright 1947*
First published 1947
Second impression 1961
Third impression 1987

ISBN 0 11 884241 2

Maps and diagrams in this book use topography based on Ordnance Survey mapping

PREFACE

The district represented on Sheet 167 (Dudley) of the New Series one-inch map includes parts of the Old Series one-inch quarter-sheets 61 S.E., 62 S.W., 55 N.E., and 54 N.W. These were surveyed on the one-inch scale by W. T. Aveline, H. H. Howell, E. Hull, J. B. Jukes, J. Phillips and A. C. Ramsay, and were published in 1852 and 1853, with some revisions in 1855, 1858, 1868, 1885, and 1898. Horizontal sections on the scale of six inches to the mile, crossing parts of the district, were issued between 1853 and 1859, some of which were revised in 1865. Those on sheets 53 and 54 were accompanied by brief explanations. Vertical Sections of many of the pits in the South Staffordshire Coalfield area were published in 1853 and 1860.

The portion of the South Staffordshire Coalfield lying within Sheet 167 has been described in a memoir on the southern part of that coalfield, published in 1927, and consequently has received only brief attention in the present volume. Details of several pits in the same area will be found on Sheet 94 of Vertical Sections published in 1931.

The six-inch survey of Sheet 167 has been mainly the work of Mr. T. H. Whitehead and Dr. R. W. Pocock, who are responsible for the present memoir. Minor portions of the area were mapped by W. Gibson, C. H. Cunnington, Dr. T. Robertson and Messrs. H. Dewey and T. Eastwood. The new one-inch map was issued in 1939.

The first draft of the memoir was completed in 1937 and it was ready for the press in 1939, but publication was prevented by the outbreak of war. Some details of the Coal Measures have since been brought up to date by Dr. G. H. Mitchell and certain alterations have been made in other parts of the text as a result of recent research.

Of the fossils the fish remains have been identified by Prof. D. M. S. Watson, the corals by Dr. Stanley Smith, the plant remains by Dr. R. Crookall, and the remainder by Drs. J. Pringle and C. J. Stubblefield. Dr. J. Phemister has contributed petrographical notes on the igneous rocks. The memoir has been edited partly by Mr. C. H. Dinham and partly by Mr. T. H. Whitehead and Dr. R. W. Pocock.

Assistance rendered during the survey by colliery officials, managers of quarries, waterworks etc., landowners, agents and others is gratefully acknowledged.

W. F. P. McLintock,
Director.

Geological Survey Office,
  Exhibition Road,
    South Kensington,
      London, S.W.7.

31*st March* 1947

CONTENTS

	PAGE
PREFACE BY THE DIRECTOR	ii
LIST OF ILLUSTRATIONS	viii
EXPLANATION OF PLATES	ix
LIST OF SIX-INCH MAPS	x

CHAPTER I.—INTRODUCTION 1

    Geographical Range, 1. Physiography, 2. Geological Sequence, 2.

CHAPTER II.—SILURIAN ROCKS AND OLD RED SANDSTONE 7

    General Account 7

        South Staffordshire Area, 8. Shropshire Area, 11. Worcestershire Area—The Trimpley Inlier, 12.

    Details 14

        South Staffordshire Area 14

            Dudley and Sedgley, 14. Ellowes Park and Turner's Hill, 17. Coalbournbrook, 17. Netherton, 17. Lye and Wollescote, 20.

        Shropshire Area 20

            Nordley and Morville, 20. The Lye, Wallsbatch and Oldfield, 22. North of Glazeley, 23. Overwood and Shutley, 23. Eudon George and Deuxhill, 23. Sidbury and Stottesdon, 23.

        Worcestershire Area—The Trimpley Inlier 24

            Beds below the *Psammosteus* Limestones, 24. The *Psammosteus* Limestones, 25. Beds above the *Psammosteus* Limestones, 26.

CHAPTER III.—CARBONIFEROUS ROCKS: INTRODUCTION 32

    Lower Carboniferous 32

    Upper Carboniferous 32

CHAPTER IV.—CARBONIFEROUS ROCKS: PRODUCTIVE COAL MEASURES... 35

    Middle Coal Measures (South Staffordshire Coalfield).

        General Account 35

            Sequence, 35. Coals, ironstones, fireclays, etc., 36. Fossils, 37. Marine bands, 38.

        Details 38

            Smestow and Claverley Borings 39

    Kinlet Group (Wyre Forest Coalfield) 40

        General Account 40

            Introduction, 40 Sequence, 41. Comparison with South Staffordshire and Coalbrookdale; marine bands, 41. Base of the Group, 42. Upper limit and variation in thickness, 44. Coals, 45. Correlation with the Clee Hills Coalfield, 45. Red beds, 47.

	PAGE
Details	48

Outliers near Middleton Scriven, 48. Billingsley and Chorley, 48. Kinlet, 50. Buttonbridge and Pound Green, 52. Highley, 53. Seckley Wood, etc., 54. Eymore Wood and Shatterford, 54. Witnells End, Arley Wood, etc., 56. The Compton Sinking, 57. Trimpley, 57.

CHAPTER V.—CARBONIFEROUS ROCKS: UPPER COAL MEASURES ... 58

Etruria (Old Hill) Marl Group (South Staffordshire Coalfield) ... 58

General Account 58
Introduction, 58. Lower limit, 58. Lithological characters, 59. Coal seams, 61. Thickness and upper limit, 61. Distribution, 61.

Details 61
  Smestow and Claverley 62

Halesowen Group (South Staffordshire Coalfield) 62

General Account 62
Introduction, 62. Lithological characters and sequence, 62. Relation to underlying beds, 63. Upper limit and thickness, 64. Fossils, 64. Economic products, 64.

Details 65
  Smestow and Claverley 65

Highley Group (Forest of Wyre Coalfield) 65

General Account 65
Introduction, 65. Lithological characters and sequence, 65. Associated unconformities, 68. Upper limit, 69. Fossils, 69. Economic products, 71.

Details 71
Severn Lodge, Tiphouse and Highley, 71. Billingsley, 73. Borle Brook and Chelmarsh Ridge, 73. Eardington Pit and Boring, 74. Deuxhill Outlier, 74. Tasley, 75. Upper Arley (west of R. Severn), 76. Buttonbridge, 77. Bradley and Meaton, 77. Upper Arley (east of R. Severn) and Shatterford, 77. Witnells End, High Trees, Compton, etc., 79.

Keele Group 81

General Account 81
Introduction, 81. Lithological characters, 81. *Spirorbis* limestones, 81. Fossils, 83. Thickness, 84. Distribution, 84.

Details 84
Smestow and Claverley, 84. Wooton and Shatterford, 85. Arley and Alveley, 86. West of R. Severn, 89.

Enville Beds 91

General Account 91
Introduction, 91. Lithological characters, Calcareous Conglomerate Group, 91. Lithological characters, Breccia Group, 92. Relation of Breccia Group to underlying beds, 93. Nomenclature and classification, 94. Distribution, 96.

	PAGE
Details	97
Calcareous Conglomerate or Bowhills Group...	97

      Enville, Four Ashes, Gatacre, etc., 97. Kingsnordley, Bowhills and Romsley, 98. Quatt, 100. Trimpley, 101.

Breccia or Clent Group	101

      Enville, Gatacre and Bobbington, 101. Compton, Kingsford, etc., 103. Trimpley, 103. Romsley, Aston and Coton Hall, 104.

CHAPTER VI.—TRIASSIC ROCKS	105
General Account	105

      Lower Mottled Sandstone, 105. Bunter Pebble Beds, 106. Upper Mottled Sandstone, 108. Lower Keuper Sandstone, 108.

Details	109
Lower Mottled Sandstone	109

         Wordesley, Wombourne, Himley and Wollaston, 109. Abbot's Castle Hill, Kinver, etc., 110. Bridgnorth, Quatford and Quatt, 110.

Bunter Pebble Beds	111

         Clent and Hagley, 111. Upper Penn and Wombourne, 111. Kingswinford, Wordesley, Wollaston, etc., 111. Abbot's Castle Hill, Kinver, Wolverley, etc., 111. Bridgnorth, 113.

Upper Mottled Sandstone	114

         Wollaston, Stourbridge and Churchill, 114. Lower Penn, Wombourne, Prestwood, etc., 114. Rudge and Claverley, 114. Worfield and Roughton, 115.

Lower Keuper Sandstone	115

         Stourbridge, Hagley and Clent, 115. Lower Penn and Wombourne, 115. Wollaston, 116. Rudge and Claverley, 116. Chesterton, Hopstone, Hilton and Farmcote, 116.

CHAPTER VII.—FOLDS AND FAULTS; EXTENSIONS OF THE COALFIELDS; IGNEOUS ROCKS	118
Folds and Faults	118
General Account	118

         Introduction, 118. Tectonic History, 118. The South Staffordshire Coalfield, 123. The Main Syncline, 123. The Stourbridge Syncline, 124. The Trimpley Anticline, 124. The Western Area, 126.

Details	126

         The Western Boundary Fault, 126. The Lloyd House Fault, etc., 127. The Stapenhill Fault, 127. The Enville Fault, 128. The Trimpley Anticline and associated Faults, 128. The Romsley Fault, 129. The Pattingham and Patshull Faults, 129. The Arley Park Fault, 129. The Station Fault, 129. The Highley-Kinlet Fault, 130. The Tiphouse Fault, 130. The Kinlet Hall Fault, 130. The Brock Hall or Billingsley Fault, 130. The Deuxhill Fault, 131.

Extensions of the Coalfields...	131

		PAGE
Igneous Rocks		134
South Staffordshire Coalfield		135
Shatterford		138
Kinlet and Highley		142

CHAPTER VIII.—GLACIAL AND POST-GLACIAL DEPOSITS 148

Glacial Deposits 148
    Introduction 148
    Older Glacial Gravels 150
    The Kingswinford Gravel Ridge 151
    Earlier River Terraces 153
        Stour Basin 153
        Severn Valley 154
            Bridgnorth and Eardington, 154. Quatt, 155.
    Boulder clay and associated Sand and Gravel 155
        Wolverhampton, Penn and Wombourne, 155. Trescott, Seisdon, Trysull, etc., 156. Bobbington area, 157. Rudge Heath, Claverley and Farmcote, 158. Allscott, Barnsley and Mose, 158. Rhodes Farm and Morville, 159. Astley Abbots, 159. Eardington, 159.
    Fluvio-glacial Deposits and Main Terrace 160
        Smestow Brook, 161. R. Stour, 161. Rudge Heath and Claverley, 163. Dalicote and Worfe Valley, 163. Wooton and Quatt, 163. R. Severn, 164.
    Breccia Gravels 165
    Pebble Spreads 168
    Glacial History 169

Post-glacial Deposits 175
    Later River Terraces 175
        Smestow Brook and R. Stour, 175. Worfe Valley, 175. Severn Valley, 176.
    Alluvium 177
        Smestow Brook and R. Stour, 177. R. Worfe, 177. R. Severn, 177.

CHAPTER IX.—MINERAL PRODUCTS, WATER SUPPLY AND AGRICULTURE 178
    Coal 178
    Ironstone 178
    Fireclays and Pottery Clays 179
    Brick and Tile Clays 180
    Lime and Cement 180
    Sand and Gravel 181
    Building Stone 181
    Road Metal 182
    Water Supply 182
        Silurian and Old Red Sandstone, 182. Coal Measures, 183. Trias, 183. Superficial Deposits, 184.
    Agriculture 184

	PAGE
APPENDIX I.—TABLE OF WELLS AND BOREHOLES FOR WATER	186
APPENDIX II.—DETAILS OF CERTAIN BOREHOLES	193
A. Smestow, near Wombourne	193
B. Kinlet	197
APPENDIX III.—LIST OF GEOLOGICAL PHOTOGRAPHS	202
INDEX	207

ILLUSTRATIONS

TEXT FIGURES

	PAGE
FIG. 1.—Sketch-map of the ' Solid ' Geology of the Dudley and Bridgnorth area	5
FIG. 2.—Section from Baggeridge Colliery to the north-east corner of Ellowes Park, Sedgley	16
FIG. 3.—Section in the north-eastern bank of the canal at Brewin's Bridge, Netherton	18
FIG. 4.—Section on the north side of the Halesowen road at The Hayes, Lye	21
FIG. 5.—Section in the north-east bank of the Kidderminster road near Birch Farms	27
FIG. 6.—Section along stream west of Holbeache	28
FIG. 7.—Section south-east of Mary Moors	30
FIG. 8.—Comparative Vertical Sections of Middle Coal Measures	43
FIG. 9.—Diagrammatic Section across part of the Wyre Forest Coalfield, Eardington to Stagbury (Stagborough) Hill, Sheet 182	46
FIG. 10.—Comparative Vertical Sections of the Principal Coals; Clee Hills and Wyre Forest Coalfields	47
FIG. 11.—Vertical Sections of Measures above the Thick Coal and of Etruria Marl in the part of the South Staffordshire Coalfield within Sheet 167	60
FIG. 12.—Section of Keuper basement beds in the road north-east of Ludstone	117
FIG. 13.—Map to illustrate the Geological Structure of the Dudley and Bridgnorth District in relation to surrounding areas	119
FIG. 14.—Hypothetical Section from the Claverley to the Smestow Borehole	125
FIG. 15.—Section of the Shatterford Basalt and Country Rock in dingle 250 yds. south of Witnell's End	139
FIG. 16.—Map of the Kinlet basalt outcrops	141
FIG. 17.—Section of a portion of the Kinlet Boring, 1929	145
FIG. 18.—Generalized Map of the Glacial Deposits between Bridgnorth and Birmingham	149
FIG. 19.—Gradient profiles of Breccia-gravels near Enville	167
FIG. 20.—Sketch-map showing Retreat Stages of the Irish Sea Ice-sheet near Bridgnorth and Wolverhampton	172

Plates

		FACING PAGE
PLATE	I.—The Severn Valley at Bridgnorth. (Survey photograph No. 6844)	1
PLATE	II.—A.—Escarpment of Meadowley Hill. (Survey photograph No. 6868) B.—Wren's Nest Hill, Dudley; from the west. (Survey photograph No. 2219)	9
PLATE	III.—A.—Downtonian sandstones overlain by Coal Measures sandstone and conglomerate, Brewin's Bridge, Netherton. (Survey photograph No. 1952) B.—View from Enville Sheepwalks, looking east towards the Clent Hills. (Survey photograph No. 6903)	20
PLATE	IV.—Comparative sections of the Highley Beds...	80
PLATE	V.—A.—Alveley grindstone and building stone quarry. (Survey photograph No. 6871) B.—Bunter Pebble Beds on Lower Mottled Sandstone; Ridge Sandpit near Stourbridge. (Survey photograph No. 2198)	88
PLATE	VI.—A.—Clent Hills, Clent; view looking west-south-west. (Survey photograph No. 2211) B.—Kinver Edge from near Shatterford. (Survey photograph No. 6889)	92
PLATE	VII.—A.—Queen's Parlor, Bridgnorth; Pebble Beds on Lower Mottled Sandstone. (Survey photograph No. 6841) B.—Lower Mottled Sandstone, false-bedded; Bridgnorth. (Survey photograph No. 6838)	106
PLATE VIII.—A.—Mass House Quarry, Kinlet; spheroidal weathering of basalt. (Survey photograph No. 6916) B.—Raggitts Quarry, Kinlet; 'Dykes' in decomposed basalt. (Survey photograph No. 6915)		142
PLATE	IX.—A.—Boulder Clay on Keele Beds near Eardington Mill. (Survey photograph No. 6861) B.—View looking down valley of R. Severn; showing terrace features. (Survey photograph No. 6885)	160
PLATE	X.—Vertical Sections of Measures below the Thick Coal, South Staffordshire Coalfield	206

Explanation of Plates

PLATE   I.—View from Castle Hill, looking upstream. On the skyline is the Bunter escarpment with "High Rock" to the left. The houses on the left bank are on the alluvium and the Lower Danesford Terrace. The feature of the Main Terrace is seen on the fields to the right above the road.

PLATE   II.—A.—From the south-east; an escarpment formed by the outcrop of the *Psammosteus* Limestones and overlying sandstone. In the middle distance marls of the Downton series covered by boulder clay, sand and gravel.

PLATE   II.—B.—Shows the quarries on the west side of the hill, the white surface being the dip-slope of the Nodular Beds from which the Upper Limestone has been stripped.

PLATE III.—B.—The foreground is on Clent Breccia, the base of which runs approximately along the hedge on the right. In middle distance: Stour valley with Kinver (above bushes in right foreground). Beyond: the eastern escarpment of the Bunter Pebble Beds from Bunker's Hill (right) to the Bells Mill gap; in front of it (left) part of the main escarpment can be seen. On skyline, from right to left: Walton Hill, Clent Hill, Wychbury Hill, Hodge Hill and the South Staffordshire Coalfield.

PLATE VI.—A.—In the foreground: Clent Breccia outcrop with dry valley. The lower ground beyond is on the Stourbridge syncline (p. 124) with the eastern ridge of the Bunter Pebble Beds beyond. The higher dark ridge beyond, to right, is the Enville Beds area of S.E. Shropshire. Beyond it Titterstone Clee Hill can be faintly seen.

PLATE VI.—B.—Coal Measures and Downtonian rocks of the Trimpley Anticline under wooded valley in foreground and middle distance. Beyond and to right: Castle Hill, a feature of Clent Breccia. In the distance the Kinver Edge escarpment. Far distance left: the South Staffordshire Coalfield, with the Wren's Nest and Castle Hill (Dudley).

PLATE IX.—B.—View from near Quatt. The wooded bank in the middle distance rises from a lower terrace and is capped by the Upper Danesford Terrace, beyond which the ground rises to the Main Terrace. On the extreme left is Lodge Farm, on the Main Terrace of the left bank, with a steep bluff towards the river.

LIST OF SIX-INCH MAPS

The following is a list of the six-inch geological maps included, wholly or in part, in the one-inch geological map, Sheet 167, with the names of the surveyors, and dates of survey. Maps with an asterisk prefixed have been published. The remainder are available for public reference in MS. form at the office of the Geological Survey, and copies can be supplied at the cost of drawing and colouring; those marked † have been surveyed only in part.

SHROPSHIRE

51 S.W.	Round Hill	T. Robertson	1923–24
51 S.E.	Nordley	T. Robertson	1923–24
52 S.W.	Hartlebury	T. Robertson	1923
52 S.E.	Chesterton	T. Robertson	1923
†58 N.W.	Morville	R. W. Pocock	1929
58 N.E.	Tasley and Bridgnorth (N.)	R. W. Pocock	1929
†58 S.W.	The Down	R. W. Pocock	1929
58 S.E.	Eardington and Bridgnorth (S.)	R. W. Pocock	1929
59 N.W.	Worfield	R. W. Pocock	1929
59 N.E.	Claverley	T. H. Whitehead	1929
59 S.W.	Quatford	R. W. Pocock	1929
59 S.E.	Gatacre	T. H. Whitehead	1929
†66 N.W.	Middleton Scriven	R. W. Pocock	1930
66 N.E.	Glazeley and Deuxhill	R. W. Pocock	1930
†66 S.W.	The Highlands	R. W. Pocock	1930
*66 S.E.	Billingsley	R. W. Pocock	1930–31
67 N.W.	Quatt	R. W. Pocock	1931
		T. H. Whitehead	1930

67 N.E.	Four Ashes	T. H. Whitehead	1929
*67 S.W.	Highley (N.) and Alveley	R. W. Pocock	1931–32
		T. H. Whitehead	1929
67 S.E.	Romsley	T. H. Whitehead	1930
†73 N.W.	Stottesdon	R. W. Pocock	1930
*73 N.E.	Kinlet	R. W. Pocock	1930–32
†73 S.W.	Nethercott	R. W. Pocock	1930–32
73 S.E.	Baveneywood	R. W. Pocock	1930–32
*74 N.W.	Highley (S.) and Upper Arley	T. H. Whitehead	1930–32
*74 N.E.	Shatterford	T. H. Whitehead	1930–32
74 S.W.	Buttonbridge and Eymore	T. H. Whitehead	1932

STAFFORDSHIRE

61 S.W.	Rudge and Shipley	T. Robertson	1922–23
61 S.E.	Trescott	T. Roberston	1922
62 S.W.	Upper Penn	T. Robertson	1922
		H. Dewey	1915
*62 S.E.	Bilston	W. Gibson	1911–12
		T. Robertson	1921–22
66 N.E.	Trysull and Seisdon	T. H. Whitehead	1929
66 S.E.	Swindon	T. H. Whitehead	1929
*67 N.W.	Wombourn and Baggeridge	T. H. Whitehead	1921
*67 N.E.	Sedgley and Coseley	T. H. Whitehead	1921
		W. Gibson	1912
*67 S.W.	Himley	T. H. Whitehead	1920
*67 (Worcs. 1) S.E.	Dudley	T. H. Whitehead	1920
70 N.E.	Enville	T. H. Whitehead	1920
70 (Worcs. 3) S.E.	Kinver	T. H. Whitehead	1920
*71 N.W.	Kingswinford and Brierley Hill	T. H. Whitehead	1920
*71 (Worcs. 4) N.E.	Brierley Hill and Netherton	T. H. Whitehead	1919
*71 (Worcs. 4) S.W.	Stourbridge	T. H. Whitehead	1922
*71 (Worcs. 4) S.E.	Lye, Cradley and Wollescote	T. H. Whitehead	1922
74 (Worcs. 8) N.W.	Shatterford	T. H. Whitehead	1930–32
74 (Worcs. 8) N.E.	Wolverley	T. H. Whitehead	1930
75 (Worcs. 9) N.W.	Churchill and Hagley	T. H. Whitehead	1922

WORCESTERSHIRE

8 S.W.	Trimpley	T. H. Whitehead	1931
8 S.E.	Heathfield	T. H. Whitehead	1930
*9 N.E.	Clent	C. H. Cunnington	1914
		T. Eastwood	1920
		T. H. Whitehead	1922
9 S.W.	Broom	T. H. Whitehead	1922
9 S.E.	Holy Cross	T. H. Whitehead	1922

(for other Worcestershire Sheets see equivalent Staffordshire Sheets).

PLATE I

Geology of Dudley and Bridgnorth (*Mem. Geol. Surv.*)

THE SEVERN VALLEY AT BRIDGNORTH.

A 6844

DUDLEY AND BRIDGNORTH

CHAPTER I

INTRODUCTION

GEOGRAPHICAL RANGE

OF the district described in the present memoir and represented on Sheet 167 (Dudley) of the New Series one-inch map, rather more than one-fifth lies in the County of Worcester, the remainder being divided nearly equally between Shropshire and Staffordshire. The most thickly populated portion lies in the east, where it includes part of the industrial region of the Black Country. The greater part, however, is agricultural country pleasantly diversified by heath and woodland. The northern end of the Forest of Wyre extends into the district in the south-west. A few scattered collieries in the west have done little to spoil the rural aspect of their surroundings.

All the large towns lie in the east, where, indeed, they are hardly separated from one another. Dudley, a county borough of Worcestershire detached from the main portion of that county, lies close to the eastern border. To the north of it are Tipton and Bilston, with the southern suburbs of Wolverhampton just entering the district. South and south-west of Dudley, Brierley Hill, Cradley, Lye and Stourbridge are other notable manufacturing towns.

In a strip of country fringing the Black Country on the west are to be found the seats of landowners, such as Himley Hall and Hagley Hall, mansions built by ironmasters and other industrial leaders of the past, and growing residential suburbs in such places as Penn, Kingswinford, Wollaston, Pedmore and Hagley. Kidderminster lies just south of the district, which, however, includes its northern outskirts.

Outside the industrial area the only town is Bridgnorth, a market town with the one road-bridge over the R. Severn in the district.

Of villages in the middle portion of the district, Claverley in the north and Enville near the centre may be mentioned, and also Kinver, which owing to its proximity to Kinver Edge, now a property of the National Trust, has long been a holiday resort for people from the Black Country towns. In the Severn valley Highley and Alveley are in part colliers' villages, whilst Hampton and Upper Arley have ferries over the river. Morville, Sidbury and Stottesdon are typical of the small villages in the area west of the Severn.

Physiography

The district is one of considerable relief; about half of it rises over 300 ft., and only in the Severn valley below Bridgnorth does any part lie less than 100 ft. above sea-level. The highest ground is in the south-east where the Clent Hills reach just over 1,000 ft. above Ordnance Datum.

Crossing the north-eastern corner an elevated ridge, rising to over 700 ft. for the greater part of its length and reaching 800 ft. on Kate's Hill, Dudley, forms part of the main watershed between the Humber and Bristol Channel drainages. A central tract of high ground separates the valleys of the Severn and Stour and rises to over 650 ft. on Enville Sheepwalks and near Trimpley. West of the Severn the ground, though dissected by deep valleys, rises generally westward, reaching over 600 ft. in places within the district, towards the Clee Hills which lie some four miles beyond the western boundary.

The small part of the district that lies north-east of the main watershed is drained by the head-waters or tributaries of the R. Tame.[1] Of the remainder more than half is drained by the Severn or by tributaries of that river which join it within the district, the most important of these being the R. Worfe, on the east, and Mor Brook and Borle Brook on the west. Most of the eastern part of the district lies in the basin of the R. Stour which, flowing at first nearly from east to west, some three miles west of Stourbridge turns almost at right angles and flows south-south-westward to join the Severn about $4\frac{1}{2}$ miles beyond the southern border of the district. The most noteworthy tributary of the Stour is Smestow Brook, which rises just north of the district near Wolverhampton and flows southward to join the Stour near the point where the latter itself turns to a southward direction.

Geological Sequence

The following table summarizes the geological formations that occur in the district, and are represented on the map and section :—

SUPERFICIAL FORMATIONS (DRIFT)

Recent and Post Glacial :—

Peat
Alluvium
River Terraces

Glacial :—

Fluvio-glacial gravels and breccia-gravels
Main or Fluvio-glacial Terrace
Higher River Terraces
Sand and Gravel
Boulder clay

---

[1] The true source of the Tame is a matter of uncertainty, see 'The Geology of the Southern Part of the South Staffordshire Coalfield' (*Mem. Geol. Surv.*), 1927, p. 2, footnote 3.

SOLID FORMATIONS

Triassic :—
 Keuper :—   Thickness
  Lower Keuper Sandstone : red and brown sandstones with pebbly beds and conglomerates ... 150 to 500 ft.
 Bunter :—
  Upper Mottled Sandstone : soft, red and mottled sandstones... ... ... 200 to 500 ft.
  Pebble Beds : dull red sandstones, pebbly sandstones and conglomerates ... ... 370 to 400 ft.
  Lower Mottled Sandstone : soft, red and mottled sandstones... ... ... 0 to 850 ft.

Unconformity

Carboniferous? :—
 Upper Coal Measures? :—
  Enville Beds :—
   Breccia, or Clent, Group : red marls and sandstones with breccias ... ... 120 to 450 ft.

Minor Unconformity

   Calcareous Conglomerate, or Bowhills, Group : red and brown sandstone with marl and calcareous conglomerate ... ... 175 to 400 ft.

Carboniferous :—
 Upper Coal Measures :—
  Keele Group : red marls and sandstones with *Spirorbis* limestones ... ... 500 to 1,000 ft.
 Upper Coal Measures (South Staffordshire) :—
  Halesowen Group : greenish and buff sandstones and grey clays with thin coals and *Spirorbis* limestones ... ... 200 to 500 ft.

Minor Unconformity

  Etruria Marl : purple and ochreous marls with greenish grits and conglomerates ... ... 200 to 680 ft.
 Upper Coal Measures (Wyre Forest) :—
  Highley Group : greenish and buff sandstones and grey and purple clays with thin coals and *Spirorbis* limestones ... ... 300 to 600 ft.

Minor Unconformity

  ? Kinlet Group, upper part (see below).
 Middle Coal Measures (South Staffordshire) :—
  Productive Measures : grey clays and brown sandstones with coals, ironstones and fireclays 350 to 650 ft.
 Middle Coal Measures (Wyre Forest) :—
  Kinlet Group (top portion Upper Coal Measures?) : grey and purple clays and marls with greenish grits and conglomerates, coals, ironstones and fireclays... ... ... 800 to 1,400 ft.

Unconformity — Thickness

Old Red Sandstone :—

Upper (*Farlow Series*) : sandstones, marls and cornstones... — 200 ft.

Unconformity

Lower (*Ditton Series*) : red marls with sandstones and cornstones ... — 1,120 ft.

Silurian :—

Downton Series :—

Red Downtonian beds : red marls with some sandstones and cornstones ... — 465 ft.

Temeside Group :

Temeside shales : grey purple and green shales and thin sandstones with bone beds.

Downton Castle Sandstone : buff and greenish sandstones ... — 100 ft.

Ludlow Series :—

Upper Ludlow Shales : grey shales and calcareous flags with thin nodular limestones. Ludlow Bone Bed at top ... — 40 ft.

Aymestry or Sedgley Limestone : grey argillaceous nodular limestone ... — 25 ft.

Lower Ludlow Shales : grey shales with thin limestones ... — 500 ft.

Wenlock Series :—

Wenlock or Dudley Limestone : flaggy grey limestones separated by grey shales with limestone nodules ... — 178 to 190 ft.

Wenlock Shales : calcareous grey shales. Thickness : total, 650 ft. ; exposed ... — 100 to 150 ft.

Igneous Rocks :—

Basalt and dolerite, contemporaneous in the Coal Measures and intrusive in the Coal Measures and Silurian.

The principal area of outcrop of the pre-Carboniferous formations lies to the west of the R. Severn, but there are small inliers near Trimpley, in the southern part of the district, and in the South Staffordshire Coalfield. The oldest rocks, the Wenlock Shales, appear only in the cores of periclinal folds near Dudley and Sedgley. Movements after the deposition of the Lower Old Red Sandstone brought Upper Old Red Sandstone to lie unconformably upon it, there being no representative of the Middle Old Red Sandstone of Scotland. Further movements may have prevented the deposition of Lower Carboniferous rocks in the district, though these are found a few miles to the west following conformably upon the Upper Old Red Sandstone. Silurian rocks thus form the floor of the South Staffordshire Coalfield and, with those of the Old Red Sandstone, that of the Forest of Wyre Coalfield also.

The 'Middle' Coal Measures crop out in the east, where the district includes part of the South Staffordshire Coalfield, and in the west, where

Fig. 1.—Sketch-map of the 'Solid' Geology of the Dudley and Bridgnorth area.

the Wyre Forest Coalfield extends along the Severn valley. The Productive Measures of South Staffordshire contain many workable seams of coal, ironstone and fireclay, including such well-known measures as the Thick or Ten Yard Coal and the Stourbridge Old Mine Clay. In the western coalfield the productive measures are the Kinlet Group, which around Highley includes workable coal-seams. The uppermost beds of the Kinlet Group may, however, belong to the Upper Coal Measures and represent the lowest part of the Etruria Marl.

The outcrops of Upper Coal Measures tend to fringe the coalfields, and these rocks form part of the barren cover to the Productive Measures. Some of the highest beds, however, occupy a considerable area near the middle of the district, where they are brought up by an anticlinal fold. Thin seams of coal occur in certain of these Upper Coal Measures, and the sulphur coals of the Highley Group were formerly worked. Gentle folding has caused the Halesowen and Highley Groups to rest with some degree of unconformity on the beds below, and later folding resulted in a slight unconformity at the base of the Clent Group, which includes the well-known " trappoid " breccia of the Clent Hills and Enville Sheep-walks.

More widespread folding led to a greater unconformity at the base of the Triassic rocks. These rocks occupy the area between the two coalfields, except where the Upper Coal Measures are brought up by the anticline referred to above. They are represented by all three divisions of the Bunter Series, important as a waterbearing formation, and by the Lower Keuper Sandstone. The Keuper Marl does not enter the district but is found only a short distance away, both to the north and south.

Upon all these " solid " formations lie superficial deposits of clay, sand and gravel, chiefly in the northern half of the district and along the larger river-courses. Glacial deposits, consisting of boulder clay and of sand and gravel that often form mounds and ridges, and hundreds of boulders and smaller fragments of rock from far distant sources, tell of the invasion of the district by an ice-sheet from the Irish Sea basin. The farthest limit reached by this ice-sheet can be traced across the district from west to east. Outside it other gravels with stones foreign to the district, and a few isolated boulders, seem to indicate another, and earlier, invasion by ice. Along the river-courses are terraces of gravel, some of which appear to be older than the deposits of the Irish Sea ice-sheet, others to have been carried down by the melt-waters of that sheet, and still others that belong to a later, more tranquil period when no ice lay anywhere near. Yet quieter conditions of river-flow are indicated by alluvial flats of sand, silt and mud, which are still being added to when floods carry the water beyond its normal channels.

T.H.W.

CHAPTER II

SILURIAN ROCKS AND OLD RED SANDSTONE

GENERAL ACCOUNT

The Silurian and Old Red Sandstone rocks found within the area of Sheet 167 include representatives of the Wenlock, Ludlow, and Downton Series, of the Ditton Series (Lower Old Red Sandstone) and of the Farlow Series (Upper Old Red Sandstone).

The question as to where the dividing line between the Silurian and Old Red Sandstone systems should be drawn is still a matter of controversy. Barrois, Pruvost and Dubois,[1] Stamp,[2] Robertson,[3] Straw[4] and others regard the Downtonian, including the Ludlow Bone Bed, as the lowest part of the Old Red Sandstone, while Wickham King[5] includes in the Silurian not only more than 2,000 ft. of rocks classed by him as Downtonian but also some 750 ft. of overlying beds, with the *Cephalaspis* Sandstone at their base, which he named Dittonian.

Murchison's original upper limit of the Silurian seems to have been the Ludlow Bone Bed.[6] He subsequently extended the Silurian System to include first the Downton Castle Sandstone and later the grey measures (Temeside Group) that occur above the Bone Bed at Ludlow, and also the Tilestones at Llandeilo. The Welsh Borderland was mapped on that classification and the top of the Silurian was drawn on the Old Series maps, and later on the Ammanford Sheet (230 New Series), in a position estimated to include the Temeside Shales in Shropshire and the Tilestones in South Wales.

On the New Series maps, Shrewsbury (152), Wolverhampton (153) and Birmingham (168), the Downtonian rocks were included in the Silurian System, but there the question of the upper limit of the Downtonian did not arise.

In the present sheet (167) it has been found that a boundary can be

---

[1] 'Description de la Faune Siluro-Dévonienne de Liévin,' *Mém. Soc. Géol. du Nord.*, tom. vi, fasc. 2, 1922 (for 1920), p. 163.

[2] 'The Base of the Devonian, with special reference to the Welsh Borderland,' *Geol. Mag.*, 1923, p. 276.

[3] 'The Siluro-Devonian Junction in England,' *ibid.*, 1928, p. 385.

[4] 'The Siluro-Devonian Boundary in South-central Wales,' *Journ. Manchester Geol. Assoc.*, vol. i, 1929, pp. 81–84.

[5] 'The Downtonian and Dittonian strata of Great Britain and North-western Europe,' *Quart. Journ. Geol. Soc.*, vol. xc, 1934, p. 526.

[6] *London and Edinb. Phil. Mag.*, new ser., vol. vii, 1835, p. 45.

drawn at the base of the " *Psammosteus* Limestones "[1] (these limestones forming a definite group which can be mapped) and it has been decided to classify all the beds between the top of the Ludlow Bone Bed and this boundary as " Downton Series " and the beds next above the boundary as " Ditton Series." These divisions will therefore not correspond strictly with King's Downtonian and Dittonian.

King has shown[2] that the Upper Old Red Sandstone rocks (Farlow Sandstones) of the Clee Hills area (Sheet 166) rest unconformably on the older rocks down to the Downtonian, but within the area of Sheet 167 there is no clear evidence of unconformity.

The oldest rocks to be found within the district are the Wenlock and Ludlow limestones and shales of the South Staffordshire inliers : Sedgley ; Turner's Hill ; Wren's Nest and Castle Hill, Dudley ; Saltwells ; and the Hayes near Lye. They are succeeded by rocks of the Downton Series.

Rocks of Silurian age have been proved beneath the Coal Measures in borings at Smestow near Wombourne (pp. 39, 195, 196) and near Claverley (p. 39).

Along the western margin of the district, in Shropshire, and in the Trimpley anticline, Worcestershire, near the southern margin, no beds older than those of the Downton Series are exposed.

The three areas of South Staffordshire, Shropshire, and Worcestershire will be described separately in the sequel.

R.W.P.

SOUTH STAFFORDSHIRE AREA

In that part of the South Staffordshire Coalfield within the present district Silurian rocks, ranging from the upper part of the Wenlock Shales to Red Downtonian Beds of the *Thyestes* (*Auchenaspis*) or Ledbury Group[3] a little above the horizon of the Ledbury fishband, crop out as inliers amongst the Coal Measures.

Wenlock.—The Wenlock Series consists of the following subdivisions :—

	Ft.
Wenlock or Dudley Limestone[4] :—	
Upper or Thin Limestone	23–28
Nodular Beds	120
Lower or Thick Limestone	35–42
Wenlock Shales (uppermost beds only)	about 100–150

[1] King, W. W., ' Notes on the " Old Red Sandstone " of Shropshire,' *Proc. Geol. Assoc.*, vol. xxxvi, 1925, p. 383 ; later " Psammosteid limestones," *idem*, *Quart. Journ. Geol. Soc.*, vol. xc, 1934, p. 527. Originally named from the occurrence in them of *Psammosteus anglicus* Traquair, which, however, has been shown to be conspecific with *Phialaspis symondsi* (Lankester) ; see L. J. Wills, ' Rare and New Ostracoderm Fishes from the Downtonian of Shropshire,' *Trans. Roy. Soc. Edinb.*, vol. lviii, 1935, p. 427 (p. 440). The term *Psammosteus* Limestones is thus strictly a misnomer, but is retained herein for convenience.—C. J. S., T. H. W.

[2] *Proc. Geol. Assoc.*, vol. xxxvi, 1925, p. 383.

[3] King, W. W., ' The Downtonian and Dittonian Strata of Great Britain and North-western Europe,' *Quart. Journ. Geol. Soc.*, vol. xc, 1934, p. 526 (see p. 527).

[4] See R. I. Murchison, ' Silurian System,' 1839, pp. 484-7, and A. J. Butler, ' The Stratigraphy of the Wenlock Limestone of Dudley,' *Quart. Journ. Geol. Soc.*, vol. xcv, 1939, p. 37.

Geology of Dudley and Bridgnorth (*Mem. Geol. Surv.*) PLATE II

A.—Escarpment of Meadowley Hill.

A 6868

B.—Wren's Nest Hill, Dudley; from the west.

A 2219

The Wenlock Shales, grey shales with thin limestones, crop out only in the cores of the periclines of the Castle Hill (Survey Photograph 2218) and the Wren's Nest, Dudley (Plate IIB) and of Hurst Hill, near Sedgley. The Wenlock Limestone appears in these same periclines. The two limestone bands have been much worked, the workings extending far underground. The lower band was until recently still worked in the Wren's Nest by means of a shaft (Survey Photograph 1966).

Ludlow.—The Ludlow Series is subdivided as follows :—

	Ft.
Ludlow Bone Bed : calcareous sandstone with fish remains, etc.	up to 2
Upper Ludlow Shales	40
Aymestry Limestone and *Dayia navicula* Beds (Sedgley Limestone)	20–25
Lower Ludlow Shales	about 500

The Lower Ludlow Shales differ little from the Wenlock Shales either in lithological character or in fossil content. They crop out around the Castle and Wren's Nest hills, near Sedgley and, the uppermost beds only, at the Hayes, Lye (Fig. 4, p. 21, Survey Photograph 1958). The Sedgley Limestone, formerly often called the " brown lime," is an impure, nodular, shaly or flaggy limestone not sharply marked off from the beds above and below. It is best seen at Sedgley, where it has been much quarried, but crops out also at Turner's Hill and at the Hayes, Lye (Fig. 4). The upper beds have yielded *Dayia navicula*[1] (J. de C. Sowerby) and appear to represent the Mocktree Shales of the Ludlow district.[2] The fauna of the lower portion seems to correspond with that of the true Aymestry Limestone of Shropshire. The Upper Ludlow Shales are sandy calcareous flags with limestone nodules. In South Staffordshire they cannot, as in Shropshire, be divided into upper *Chonetes* and lower *Rhynchonella* flags[3]; *Chonetes striatellus* (Dalman) and *Camarotoechia nucula* (J. de C. Sowerby) being about equally abundant throughout. They crop out at Sedgley ; Turner's Hill ; the Hayes, Lye (Fig. 4) and (the upper portion only) at Saltwells near Netherton.

Downton Series.—The local sequence of the Downton Series[4] is as follows :—

	Ft.
Red Downtonian Beds (Ledbury Group) :—	
Purple mudstones and sandstones	135
Hard sandstones (Ledbury fish-band)	6–10
Purple marls and purple and green sandstones	305
Purple and green sandstones	15
Temeside Group :—	
Temeside Shales : purple and green rubbly shales and thin sandstones ; ' bone beds ' at top	44
Downton Castle Sandstone : massive to flaggy sandstones	60

---

[1] See p. 20 below and W. W. King and W. J. Lewis, *Geol. Mag.*, 1912, pp. 440-1.

[2] Elles, G. L. and I. L. Slater, ' The Highest Silurian Rocks of the Ludlow District,' *Quart. Journ. Geol. Soc.*, vol. lxi, 1906, p. 195 (see p. 197).

[3] Elles, G. L. and I. L. Slater, *op. cit.*, p. 199.

[4] See W. W. King and W. J. Lewis, ' The Downtonian of South Staffordshire,' *Proc. Birm. Nat. Hist. and Phil. Soc.*, vol. xiv, 1917, p. 90, from which the thicknesses of the members of the Red Downtonian (with the exception of the lowest item) are extracted.

The Downton Castle Sandstone, with the underlying Ludlow Bone Bed, crops out at Turner's Hill, at Saltwells (Netherton) and at the Hayes, Lye (Fig. 4, p. 21). The Temeside Shales occur on Turner's Hill and are well exposed in the canal cutting at Brewin's Bridge (Fig. 3, p. 18, Survey Photographs 1952-4), 480 yds. S.S.W. of Netherton Church. Red Downtonian beds appear in Ellowes Park near Upper Gornal, at Turner's Hill, in the canal cutting at Netherton above mentioned, in Coalbournbrook near Brettell Lane, near the Hayes and in Ludgbridge Brook, Lye.

In addition to the outcrops above mentioned Silurian rocks have been met with in several places beneath the Coal Measures in colliery shafts and workings.[1] The evidence from such occurrences indicates that, broadly speaking, the Silurian rocks are disposed in two north-westward pitching synclines separated by an anticline represented by the Dudley and Sedgley ridge, with its subsidiary complications. South-east of Dudley the anticline separating the two synclines is replaced by a fault.

In the north-eastern syncline beds of Upper Ludlow or later age have been penetrated in shafts at Parkfield near Wolverhampton, Deepfields near Coseley, and the Foxyards about half a mile north-east of Mons Hill. Accordingly the Aymestry Limestone must meet the base of the Coal Measures along a line between these places and the Silurian outcrops of Sedgley and Dudley. Near the Foxyards, however, the line of basset must turn north-eastward, in conformity with the synclinal structure, to leave the district near Bilston ; for farther east and south-east there is no evidence of rocks later than the Lower Ludlow beneath the Coal Measures.

On the south-west side of the Dudley and Sedgley ridge the Aymestry Limestone must meet the Coal Measures along a line from Shavers End, where it seems to have been proved beneath Upper Ludlow beds in a trial pit, to its outcrop near Turlshill House. A short distance south-west of this line the floor of the Coal Measures must consist of Downtonian rocks.

In a pit near Sedgley Hall Farm (about 400 yds. N.W. of Sedgley church) the Coal Measures appear to rest on Red Downtonian beds ; and they certainly do at Baggeridge Colliery, where Mr. W. W. King has identified the Ledbury fish-band in a gate road[2] (Fig. 2, p. 16). Evidence of similar rocks below the Coal Measures has been found at Oak Farm (near Himley), Shut End, Corbyns Hall (between Shut End and Bromley), Brierley Hill, Nagersfield (near Buckpool) and Stambermill, to north-west and west of the Netherton anticline, and in Saltwells Colliery to the east of it. In the Wassel Grove Pits, between Lutley and Hagley, the Coal Measures rest upon Temeside Beds, so that a short distance farther south the pre-Carboniferous floor may consist of Ludlow rocks. It is

---

[1] For details see T. H. Whitehead in ' The Geology of the Southern Part of the South Staffordshire Coalfield ' (*Mem. Geol. Surv.*), 1927, pp. 20-22, and works therein cited.

[2] King, W. W. and W. J. Lewis, *Proc. Birm. Nat. Hist. and Phil. Soc.*, vol. xiv, 1917, p. 96, and W. W. King, *Trans. Inst. Min. Eng.*, vol. lxi, 1921, p. 154.

possible that near Wassel Grove the line at which the base of the Downtonian meets the Coal Measures may curve round westward and northwestward, for boulders of Wenlock Limestone found by Mr. W. W. King in the Coal Measures sandstones of Chawn Hill and numerous fragments of Wenlock Limestone and Upper Llandovery Sandstone in the Clent Breccia of Doctor's Hill,[1] Old Swinford, suggest that these Silurian formations may, in former times, have been exposed to denudation in this vicinity. Farther north, however, Downtonian rocks appear to form the floor of the coalfield in the whole of the south-western syncline, except where small inliers of Ludlow beds are brought up in the subsidiary anticlines of Netherton and Turner's Hill.

T.H.W.

SHROPSHIRE AREA

Along the western margin of the sheet, in Shropshire, the oldest beds exposed are Red Downtonian rocks which emerge from beneath the Coal Measures west of Bridgnorth and are overlain to the south and south-west by sandstones, marls and cornstones of the Ditton Series and these by the Upper Old Red Sandstone rocks of the Clee Hills (Sheet 166).

The Downton Castle Sandstone and Temeside Shales are not present at the surface and the beds exposed are mainly red marls with subordinate sandstones passing up into the *Psammosteus* Limestones, at the base of the overlying Ditton Series.

Above the *Psammosteus* Limestones, which comprise one or more bands of limestone or nodular cornstone separated by red and green marls, come grits and purple sandstones intercalated with red and grey-green marls yielding *Pteraspis leathensis* White and *Pachytheca*. It has not been found possible to recognize the *Cephalaspis* Sandstone with any certainty in the area and, as explained above (p. 8), the beds between the base of the *Psammosteus* Limestones and the *Cephalaspis* Sandstone, if present, are included in the Ditton Series.

The lower part of the Ditton Series comprises purple and green sandstones, red and green marls and cornstones and has yielded *Cephalaspis fletti* Stensiö, *Pteraspis rostrata* Agassiz, etc. These rocks are exposed in the ground north of Middleton Scriven round Oldfield and Wellsbach, north of Glazeley and again in the south-west corner of the sheet round Overwood and Shutley.

The upper part of the Ditton Series is exposed in the ground south of Middleton Scriven and Glazeley and round Sidbury and Stottesdon. It is composed of sandstones, marls and cornstones. *Pteraspis* has been found in beds near the top of this group at Farlow, 2 miles S.W. of Stottesdon.

R.W.P.

---

[1] King, W. W., *op. cit.*, pp. 155, 166.

WORCESTERSHIRE AREA—THE TRIMPLEY INLIER

The Trimpley inlier of Downtonian and Lower Old Red Sandstone strata[1] extends in a north-north-east to south-south-west direction from Arley Wood to North Wood (Sheet 182), its length being about 3 miles and its greatest breadth, near Trimpley, a little over a mile. The strata range from beds below the *Ischnacanthus* Zone of the Downtonian to approximately the top of the Lower Old Red Sandstone (Dittonian of Mr. King).[2]

The inlier is fairly equally divided by a cross fault running in a west-south-west direction near Park Attwood. The apparent vertical throw of this fault changes from side to side along its course, so that it would seem either that it represents a mainly horizontal displacement (see p. 128 below) or that it was formed subsequently to the folding of the rocks on either side. South of the fault the beds are disposed in a broken anticline with a steep western limb but with suggestions of a subsidiary syncline on the east side. North of the fault the structure is in the main synclinal, with traces of a subsidiary anticline in the north-east.

The lowest beds of the inlier crop out at the north-west end in Coldridge Wood. The first noteworthy horizon is a belt of greenish and purple sandy marls and micaceous sandstones, about 8 or 10 ft. thick, with a lenticular cornstone in the upper part. These beds have yielded fish-remains and lamellibranchs[3] and were placed by Mr. King[4] in his Trimpley Fish Zone. The beds below these consist mainly of red marls and are but little exposed. Their thickness is difficult to estimate, but if there is no repetition by faulting or folding it might amount to several hundred feet, in which case the lowest beds of the inlier would be on a horizon not far above the top of the Ledbury Group.

Above the fish-bearing beds mentioned in the preceding paragraph come 300 ft. or more of red marls, also not well exposed, amongst which are some thin sandstones and cornstones some of which have yielded fossils to Mr. King.[5] These beds belong in part to that author's "*Ischnacanthus* Zone" (I.6) and in part to the group above (I.7). They are followed by the most prominent group of strata of the northern part of the inlier—the *Psammosteus* or Birch Hill Limestones. These consist of lenticular cornstones and inorganic limestones, often relatively thick and massive, interbedded with purple and green flaggy sandstones, in

---

[1] King, W. W., 'The Geology of Trimpley,' *Trans. Worcs. Nat. Club*, vol. vii, 1921, p. 319, in which, however, only a brief account and an incomplete statement of the succession is given. The writer desires gratefully to acknowledge Mr. King's assistance in the study of the area, including the loan of his six-inch geological maps. Mr. King's agreement with the writer's interpretation of the structure and stratigraphy is, however, not necessarily implied.

[2] King, W. W., *Proc. Geol. Assoc.*, vol. xxxvi, 1925, p. 384, and *Quart. Journ. Geol. Soc.*, vol. xl, 1934, p. 527.

[3] King, W. W., *op. cit.* (1921), p. 320 and *op. cit.* (1934), p. 52.

[4] *Op. cit.* (1921), p. 320. Later, however, Mr. King appears to have removed the lower part of these beds from the Trimpley Fish Zone (or *Ischnacanthus* Zone, I.6) and placed them at the top of his I.5 division. See below, p. 24.

[5] *Loc. cit.* (1921).

part micaceous, and with marls. This belt of strata, as a whole, may be about 100 ft. thick ; but it is variable in detail and in consequence the thickness of beds included in it may differ from place to place. The group makes a strong feature in Birch Wood and the wood to the south-west, where the limestones have been quarried for lime-burning. South of the Kidderminster road its outcrop, on both sides of the syncline, forms part of a steep bluff and, near Park Attwood, overlooks a striking amphitheatre-like area excavated in the beds below.

South of the Park Attwood cross-fault the *Psammosteus* Limestones have a broad outcrop, owing to locally low dip, north-east of Holbeache. Quarries in this outcrop afford the best examples of massive limestone (see below, p. 26). Farther south faulted inliers of the *Psammosteus* Limestones are brought up in the core of the anticline west of Mary Moors.

The beds of the Trimpley inlier above the *Psammosteus* Limestones have a total thickness of over 1,100 ft. Though marls still predominate sandstones and cornstones are more numerous than in the beds below. The sandstones, few of which exceed 10 ft. in thickness, appear to be lenticular and impersistent and it is difficult to be sure of the identity of individual beds from place to place. Groups of lenticular sandstones forming in the aggregate comparatively thick belts can, however, in some cases be followed for considerable distances. Certain sandstones, with cornstones, lying about 240 ft. above the *Psammosteus* Limestones, represent the " *Cephalaspis* sandstone-cornstone " (II. 1) of Mr. King.[1] The beds between these and the *Psammosteus* Limestones include, at about the middle, purple and yellowish-green, flaggy sandstones with, in places, grey or greenish rubbly cornstones which represent the Eurypterid Sandstones (I. 9) of Mr. King and include the " Eastham Sandstone " which crops out on the crest of the steep bluff overlooking the amphitheatre of Park Attwood.

A prominent belt of sandstones and cornstones occurs with its base about 650 ft. above the *Psammosteus* Limestones. These include the " Hall Barn Cornstones " of Mr. King,[2] so called from their being exposed in a quarry at Hall's Farm (formerly Hall Barn),[3] south-west of Trimpley. The sandstones of this belt are prevalently purple in colour with some green beds. Many of them are calcareous and hard, others are micaceous. The cornstones are of the rubbly, breccia type, some purple and green, others of a greenish-grey colour. This belt of sandstones and cornstones, the total thickness of which is about 100 ft., appears to represent the greater part of Group II. 3 of Mr. King's classification. Above it the strata include a number of beds of purple and green sandstone, with several cornstones, mostly of the greenish-grey type. What are probably the highest beds of the inlier crop out at the northern edge of North Wood.

T.H.W.

---

[1] *Op. cit.* (1934), p. 527.

[2] *Op. cit.* (1921), p. 321.

[3] Shown as ' Allsbarn or Halls Farm ' on the Old Series geological map Sheet 55 N.E.

DETAILS

SOUTH STAFFORDSHIRE AREA

Dudley and Sedgley.—In the neighbourhood of these towns the Silurian rocks crop out as a broadly anticlinal tract, its axis directed north-west and south-east, with minor structures superimposed upon it. Of these minor structures two are the periclines of the Castle and Wren's Nest hills, Dudley, in the cores of which the Wenlock Shales crop out, but are indifferently exposed. The Wenlock Limestone is in two bands separated by rubbly and nodular limestones and calcareous flags. They are for the most part quarried away at their outcrop, but may still be seen in the pillars left to support the roofs of the old workings (Survey Photographs 1962-5, 1967).

Mr. A. J. Butler has recently made a detailed examination of the Wenlock Limestone of the Dudley area,[1] distinguishing a number of lithological subdivisions. In a measured section on the Wren's Nest he finds that the total thickness of the limestone group is 198 ft. There is little variation in the thickness of the group as a whole or of its major divisions, but some of the minor lithological subdivisions are variable in thickness, and some do not persist over the whole area. From his study of the " crog-balls," masses of unstratified hard limestone that occur chiefly in the unworked basement beds of the Lower Limestone and in the Nodular Beds, Mr. Butler concludes that these are similar in origin and constitution to the ballstones of Wenlock Edge.[2] As the result of his discovery of *Monograptus flemingii* (Salter) var. δ Elles in the shales just below the basement beds of the Lower Limestone and in those just above the passage beds from the Upper Limestone to the Lower Ludlow Shales, Mr. Butler considers that there can be little or no difference in age between the Dudley Limestone and the Wenlock Limestone of Shropshire.[3]

At Hurst Hill the Wenlock limestones again form a pericline, which is, however, much broken by an axial fault. On the east side of this fault both bands of limestone appear at the surface; on the west only the upper bed crops out. The limestones are almost completely quarried away; but the intermediate Nodular Beds are well displayed in the sides of the road leading from Sedgley to Hurst Hill, where they dip eastward at 70° (Survey Photograph 1972).

The Lower Ludlow Shales crop out around the Castle Hill and Wren's Nest periclines and for some distance to the south of the latter, where they are exposed in a small stream just south of the Old Priory Farm, and also in some ditches between the Priory and St. James's Church, Eve Hill, Dudley. In the first-mentioned locality they are very fossiliferous and yielded *Atrypa reticularis* (Linnaeus), *Strophonella euglypha* (Hisinger) and *Dalmanites sp.*[4] (Reg. Nos. T.W.302-6). They consist of grey clays, with calcareous sandy mudstones and thin limestones. Their outcrop extends a short distance up the valley south of Woodsetton. They again crop out round the broken pericline of Hurst Hill, and rise from beneath the Aymestry Limestone of Sedgley. Just north-west of Cinder Hill some crownings-in, resulting from the working of the Wenlock Limestone underground, afford good exposures in the Lower Ludlow Shales, which consist of grey clays, calcareous mudstones

---

[1] 'The Stratigraphy of the Wenlock Limestone of Dudley,' *Quart. Journ. Geol. Soc.*, vol. xcv, 1939, p. 37.

[2] Crosfield, M. C., and M. S. Johnston, 'A Study of Ballstone and the Associated Beds in the Wenlock Limestone of Shropshire,' *Proc. Geol. Assoc.*, vol. xxv, 1914, p. 193. See also T. Robertson in 'The Shrewsbury District,' (*Mem. Geol. Surv.*), 1938, p. 114.

[3] *Cf.* T. Robertson and C. J. Stubblefield in 'The Shrewsbury District' (*Mem. Geol. Surv.*), 1938, pp. 101-2.

[4] These and other fossils from the Silurian of South Staffordshire have been identified by Drs. J. Pringle and C. J. Stubblefield.

and thin limestones, with many fossils, including the following (Reg. Nos. T.W.315–36) :—

Atrypa reticularis (*Linn.*).
Chonetoidea (Aegirina) cf. grayi (*Davidson*).
Cyrtia exporrecta (*Wahlenberg*).
Leptaena rhomboidalis (*Wilckens*).
Lingula sp.
Orthis (Bilobites) biloba (*Linn.*).
Orthis (Parmorthis) elegantula *Dalman*.
Orthis (Skenidioides) lewisi *Davidson*.
Strophonella euglypha (*Hisinger*).
" Pleurotomaria " cf. lloydi (*J. de C. Sowerby*).
Dalmanites vulgaris (*Salter*).

North of the middle of Sedgley the Silurian rocks take the form of a syncline, pitching southward and well marked by the outcrop of the Aymestry (or Sedgley) Limestone. This last forms the crest of a bold escarpment, facing nearly east, of which Beacon Hill is the highest point. In the quarries here the limestone is well exposed and dips at about 20° W.S.W. From Beacon Hill the outcrop runs for about half a mile to north-north-west and then curves round and runs due south, again forming the crest of a bold escarpment, here facing west (Survey Photographs 1975–6) in which some quarries show the limestone dipping eastward at 10° to 20°. The outcrop is terminated, in part by a fault and in part by the overstep of the basal Coal Measures, about 300 yds. N.W. of Sedgley Church.

In the syncline above described the Upper Ludlow Shales follow upon the Limestone. They were well exposed in 1921 in some temporary excavations. They are calcareous flags with limestone nodules, distinctly more sandy than the beds below the limestone. The following fossils (Reg. Nos. T.W.337–47) were collected from them :—

Serpulites longissimus *Murchison*.
Camarotoechia nucula (*J. de C. Sowerby*).
Chonetes minimus (*J. de C. Sowerby*).
C. striatellus (*Dalman*).
Goniophora cymbaeformis (*J. de C. Sowerby*).
Orthonota sp.
Pterinea sp.
" Orthoceras " cf. bullatum *J. de C. Sowerby*.

At Park Hill, north of Sedgley, there is another curved outcrop of the Aymestry Limestone, apparently part of a dome-like structure of which the south-west portion is cut off by a fault.

About the middle of Sedgley an east and west fault crosses the outcrops; south of it an anticline replaces the syncline, the axes being nearly aligned. In the centre of this anticline the Aymestry Limestone has a broad outcrop. The opposing dips can be seen in two quarries respectively north and south of Catholic Lane, which leads from the Dudley road to Cotwall End. A little farther south the western limb of the anticline is replaced by a fault, for the Upper Ludlow Shales on the west, dipping westward at about 20°, abut against the Aymestry Limestone, dipping eastward at about the same amount.

To the east of the anticline a shallow syncline brings up the Aymestry Limestone again at Turlshill House. Some 300 to 400 yds. west of the axis of the anticline a narrow strip of Upper Ludlow Shales extends from a point 500 yds. N.N.W. of the Sedgley church to Cotwall End. This outcrop is faulted on the east against basal Coal Measures sandstones and on the west against

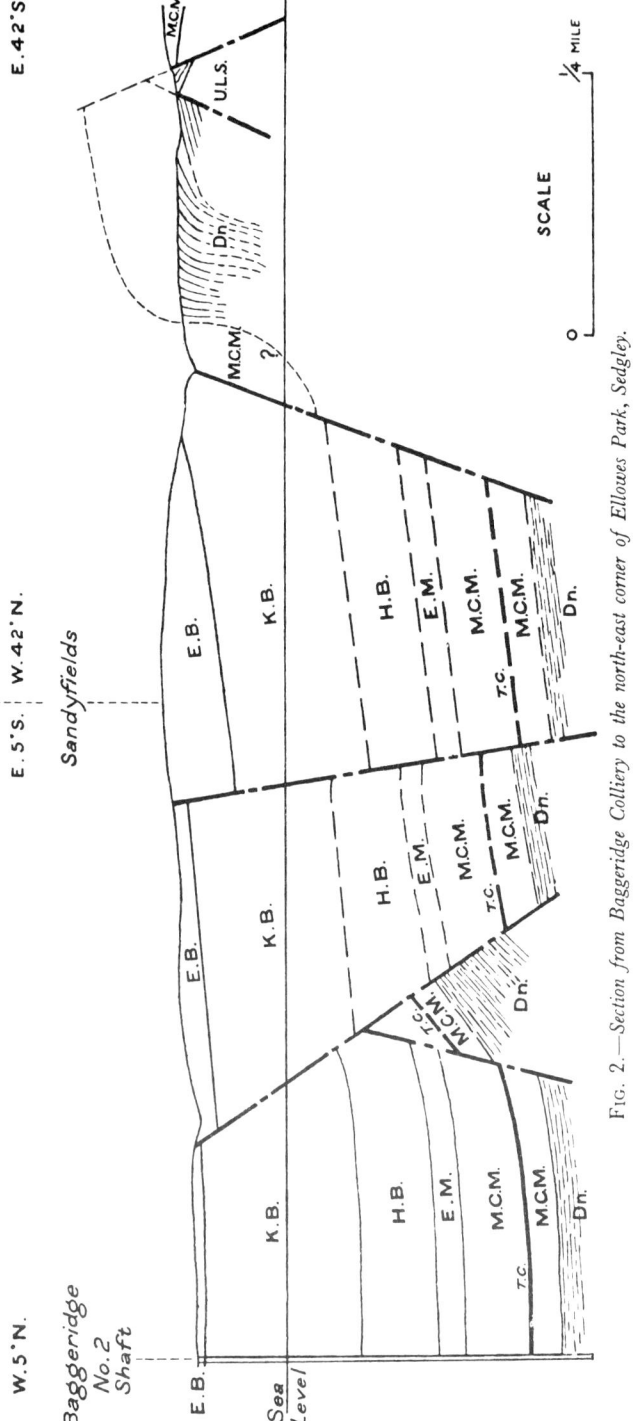

Fig. 2.—Section from Baggeridge Colliery to the north-east corner of Ellowes Park, Sedgley.

E.B. Enville Beds. K.B. Keele Beds. H.B. Halesowen Beds. E.M. Etruria Marl. M.C.M. Middle Coal Measures. T.C. Thick Coal. Dn. Downtonian. U.L.S. Upper Ludlow Shales.

higher Coal Measures. At the southern end of this strip some sandstones, identified by Messrs. King and Lewis[1] as the Downton Castle Sandstones, are exposed.

Ellowes Park and Turner's Hill.—Just south of Cotwall End a cross-fault running nearly east and west separates the Silurian outcrops of Sedgley from those of Ellowes Park and the fields to the north of it. Over this latter ground the purple and green marls and sandstones of the Red Downtonian Beds crop out, and are well exposed along two dip streams. The best section is afforded by the southern stream, which forms the northern boundary of Ellowes Park. Here purple marls and purple and green sandstones, the latter for the most part highly micaceous, are exposed to a total thickness estimated by Messrs. King and Lewis as 341 ft. Near the top of these beds a band of purple and white calcareous and micaceous sandstones yielded to Mr. King some fragments of *Hemicyclaspis* (*Cephalaspis*) *murchisoni* (Egerton) and *Didymaspis grindrodi* Lankester. He regards the fish-beds as the equivalent of the *Auchenaspis* band of Ledbury, and designates the horizon the Ledbury fish-band.[2] The rocks lie in the form of an anticline with a steep western limb and the eastern limb replaced by a fault[3] (Fig. 2). Some calcareous mudstones, with thin limestones, dipping eastward, in the tunnel under the drive to Ellowes Hall, about 60 yds. south of the line of the stream section, are probably Upper Ludlow Shales, on the upthrow (east) side of the fault. Hard greenish sandstones, in part micaceous, in which a spine of *Onchus* was found, were seen in the steep bank to the west of the drive at a point 450 yds. N.N.E. of the tunnel. These are probably Downton Castle Sandstones, and therefore are presumably also on the upthrow side of the fault. What is probably the same fault separates the outcrops of Ellowes Park from those of Turner's Hill.

At Turner's Hill the Aymestry Limestone is exposed in a quarry at the foot of the steep western scarp of the hill. Its outcrop, being cut off by faults, extends only a few yards north and south of the quarry. *Gypidula galeata* (Dalman), (Reg. Nos. T.W.310-11), was found in the limestone. The Upper Ludlow is poorly exposed in the steep slopes above the quarry, and fragments of the Ludlow Bone Bed can be found near the eastern edge of the wood that covers this slope. The Downton Castle Sandstones occupy the highest part of the hill. In the field immediately to the south-east olive mudstone and green sandstone, turned up by the plough, are assigned to the Temeside Shales. In the next field to the south, large slabs of purple and green highly micaceous sandstone were ploughed up. These are certainly Red Downtonian Beds, and are probably on much the same horizon as the beds lying above the Temeside Shales at Brewin's Bridge (see below, p. 19). Although the structure is complicated by faulting, the general dip in Turner's Hill is south-eastward, and the beds appear to form the eastern limb of an anticline that may be part of the same structure as that noted in Ellowes Park.

Coalbournbrook.—A small inlier of Red Downtonian rocks occurs between the base of the Coal Measures and the Western Boundary Fault at Dennis Park, about 700 yds. S.W. of Brettell Lane Station. The beds consist of highly micaceous purple and green sandstones and red and purple clays, to a thickness of about 200 ft.[4]

Netherton.—An inlier of Silurian rocks forms the core of the Netherton Anticline south of Netherton church.[5] The lowest beds are the Upper Ludlow

---

[1] *Proc. Birm. Nat. Hist. & Phil. Soc.*, vol. xiv, 1917, p. 92.

[2] *Trans. Inst. Min. Eng.*, vol. lxi, 1921, p. 154.

[3] For a somewhat different interpretation of the structure see W. W. King and W. J. Lewis, *op. cit.*, Fig. 2, p. 93.

[4] See T. H. Whitehead in ' The Geology of the Southern Part of the South Staffordshire Coalfield ' (*Mem. Geol. Surv.*), 1927, p. 15.

[5] King, W. W., and W. J. Lewis, ' The Uppermost Silurian and Old Red Sandstone of South Staffordshire,' *Geol. Mag.*, 1912, p. 437.

Shales of which a little over 15 ft. are exposed in the tramway leading from Messrs. Doulton's large clay pit northward to the canal. These beds consist of calcareous flags, some of them sandy, with limestone nodules. Fossils (Reg. Nos. T.W. 67–79 ; Pl. 4156–65) collected from them include :—

Serpulites longissimus *Murchison*.
Camarotoechia nucula (*J. de C. Sowerby*).
Chonetes striatellus (*Dalman*).
Orbiculoidea rugata (*J. de C. Sowerby*).
Orthis (Dalmanella) *cf*. lunata (*J. de C. Sowerby*).
Bythocypris *cf*. siliqua (*Jones*).
Orthoceras sp.

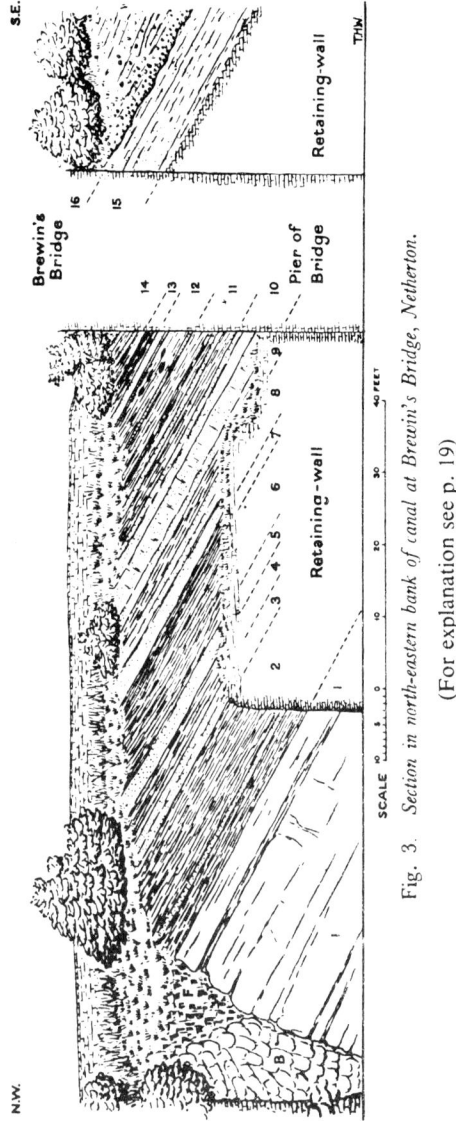

Fig. 3. *Section in north-eastern bank of canal at Brewin's Bridge, Netherton.*

(For explanation see p. 19)

Above the Upper Ludlow Shales the Ludlow Bone Bed crops out to the north and south of the footbridge over the tramway cutting to the east of Lodge Farm. This bed, which is about 2 ft. thick, consists of calcareous grit and sandstone with carbonaceous matter and many broken organic remains, including brachiopods and fish. Overlying the bone bed immediately north of the bridge come about 10 ft. of sandy mudstones, constituting the bottom part of the Downton Castle Sandstones. These yielded only plant fragments resembling forms referred to *Pachysporangium* and *Pachytheca*. Above these mudstones follow flaggy buff sandstones, 12 to 15 ft. thick, exposed along the east side of the tramway up to the canal. In some of these beds *Lingula lewisi* J. de C. Sowerby is abundant. Above them is a hard dark green shaly mudstone about 6 in. thick, which yielded *Pachytheca*, *Lingula lewisi* J. de C. Sowerby, *L. minima* J. de C. Sowerby, *Ledopsis barroisi* Reed ?, *Modiolopsis* cf. *nilssoni* (Hisinger) and fragments of *Eurypterus* and *Pterygotus* (Reg. Nos. T.W.11, 22–66, 233, 236–7, 239–44, 257–8, 282–86). This bed is considered by Messrs. King and Lewis[1] to represent the *Platyschisma helicites* bed of the Ludlow district.[2] It is exposed near the basin to the south of the canal and on the south side of the canal cutting east of Brewin's Bridge. Above it there are 36 ft. of massive yellow sandstones forming the top part of the Downton Castle Sandstones. The top 30 ft. of these (Fig. 3, No. 1), and the overlying beds are well exposed in the north bank of the canal giving the following section from above downwards (see Plate IIIA) :—

No. on Fig. 3		Ft.	in.
16	Coal Measures.—Basement conglomerate.		
	Red Downtonian :—		
15	Mudstones, sandy, olive, 2 ft. 1 in. and sandstones, micaceous, purple and green, 5 ft. 7 in., fragments of *Lingula cornea* ? in basal portion	7	8
	(Obscured by pier of bridge) ... ... about	3	6
14	Mudstones, rubbly, purple and green, with thin sandstones	4	0
	Temeside Shales :—		
13	Calcareous and carbonaceous bed (bone bed) 4 in. ; shales and mudstone, purple and green, 5 in. ; mudstone, pebbly (bone bed) 1½ in., with *Lingula cornea* J. de C. Sowerby, *Hemicyclaspis* cf. *murchisoni* (Egerton), *Onchus* cf. *tenuistriatus* Agassiz and other fish fragments ; mudstones, micaceous, purple, 6 in. ...	1	4½
12	Shales and mudstones, purple and green, with calcareous lenses (bone beds), *Lingula cornea*	2	10
11	Mudstones and shales, micaceous, green and purple, with thin sandstones, *Lingula cornea*	5	8
10	Sandstones, greenish...	5	2
9	Mudstones, sandy, green (*L. cornea*), with sandy mudstone breccia containing much organic matter (bone bed), 7 in. at base, *Lingula cornea* and *Onchus sp.* ...	1	7
8	Sandstones, in part siliceous, greenish, with soft partings, *Lingula cornea* and *Onchus* ?	3	3
7	Shale or marl, sandy, green, scattered quartz grains ...		9
6	Marls, purple and green, *Lingula cornea*	5	10
5	Sandstone, yellow and greenish, fragments of *Lingula* ...	1	7

---

[1] *Op. cit.*, p. 484. These authors record *Platyschisma helicites* from it.

[2] Elles, G. L., and I. L. Slater, *Quart. Journ. Geol. Soc.*, vol. lxii, 1906, p. 195

No. on Fig. 3		Ft.	in.
4	Marls, purple and green	2	1
3	Sandstones, soft, green and purple	2	2
2	Marls and mudstones, purple and green. *Lingula cornea* common about 5 ft. 6 in. and 3 ft. 4 in. above base ...	10	3
	Downton Castle Sandstones :—		
1	Sandstones, massive, cream and yellow ... ... about (Fault and Dolerite intrusion)	30	0

The fault at the west end of the above section throws down the red marls of the Red Downtonian, which crop out for about 150 yds. W.N.W. in the north bank of the canal. Similar beds are exposed in Black Brook, a stream running southward 200 to 300 yds. west of the tramway mentioned above. The Temeside Beds, with a bone bed, are exposed at about the middle of the west side of Lodge Farm Reservoir, where the bone bed yielded *Lingula cornea* and *Grammysia* cf. *triangulata* (Salter) and beds 3 ft. below it fragments of *Eurypterus*.

Lye and Wollescote.—The Netherton axis again brings up Silurian rocks at the Hayes, Lye, where a roadside section (Fig. 4) shows beds ranging from the Lower Ludlow Shales to the lower part of the Downton Castle Sandstones.[1] The Aymestry Limestone consists of massive and nodular limestones above which come flaggy limestones which have yielded *Dayia navicula* (J. de C. Sowerby) (Reg. Nos. T.W. 441-2) and may be regarded as representing the Mocktree Shales of the Ludlow district.[2] The Ludlow Bone Bed, 1 ft. or less thick, is here a rather coarse micaceous sandstone with carbonaceous matter.

A fault a short distance to the west of the section mentioned in the preceding paragraph brings up the Red Downtonian, consisting of highly micaceous purple and green sandstones, with hard calcareous bands, and purple and green mottled clays. These beds, with the overlying basal Coal Measures, were in 1922 exposed in a stream course about 150 yds. N.W. of the western end of the above-mentioned section.

Traces of the Downton Castle Sandstones were found about 300 yds. S.W. of Careless Green and of the Red Downtonian in Lusbridge (Ludgbridge) Brook, 200 yds. S.W. of the chapel of Lye Cemetery.[3]

T.H.W.

SHROPSHIRE AREA

Nordley and Morville.—The ground between Nordley and Morville northwest of Bridgnorth is occupied by the predominantly red beds that succeed the Temeside Shales. These Red Downtonian or Ledbury Beds are the oldest rocks exposed in the western or Shropshire area of the Sheet, though the outcrops of Ludlow and Temeside Beds of Linley Brook are only just beyond its northern boundary and a boring for coal 650 yds. east of the Albynes[4] entered grey Silurian rocks at a depth of 199 ft.

---

[1] King, W. W., and W. J. Lewis, *op. cit.*, p. 440. See also W. W. King, *Trans. Inst. Min. Eng.*, vol. lxi, 1921, p. 158 ; J. B. Jukes, ' The South Staffordshire Coalfield ' (*Mem. Geol. Surv.*), ed. 2, 1859, p. 107 and R. I. Murchison, ' Silurian System,' 1839, p. 482.

[2] Elles, G. L., and I. L. Slater, *op. cit.*, p. 198.

[3] See T. H. Whitehead, *op. cit.*, p. 18.

[4] ' The Country between Wolverhampton and Oakengates ' (*Mem. Geol. Surv.*), 1928, p. 222.

A 1952

A.—DOWNTONIAN SANDSTONES OVERLAIN BY COAL MEASURES SANDSTONE AND CONGLOMERATE, BREWIN'S BRIDGE, NETHERTON.

A 6903

B.—VIEW FROM ENVILLE SHEEPWALKS, LOOKING EAST TOWARDS THE CLENT HILLS.

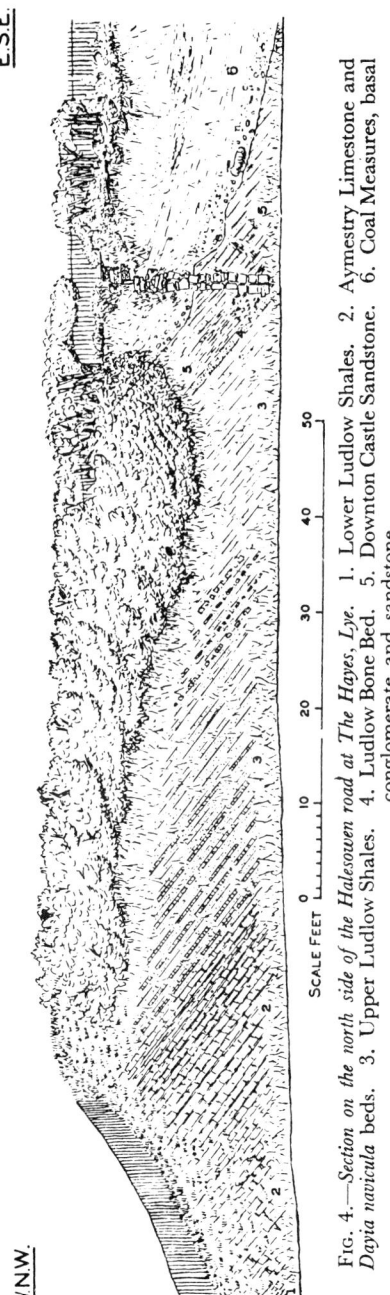

FIG. 4.—*Section on the north side of the Halesowen road at The Hayes, Lye.* 1. Lower Ludlow Shales. 2. Aymestry Limestone and *Dayia navicula* beds. 3. Upper Ludlow Shales. 4. Ludlow Bone Bed. 5. Downton Castle Sandstone. 6. Coal Measures, basal conglomerate and sandstone

The Red Downtonian beds are chiefly mudstones but several fairly persistent sandstone bands form features which are well marked in places. A band some 200 ft. above the Ludlow Bone Bed is the most conspicuous and forms the main feature of the topography of the Downtonian area near Round Hill and Nordley where it appears to coincide approximately with the 400-ft. contour.

An exposure, just north of our area (in Sheet 153), 200 yds. N.W. of the saw-mill at the Smithies, shows 4 ft. of greenish, flaggy, micaceous sandstone spotted with black. In the calcareous parts it has yielded, to Mr. W. W. King, *Pachytheca, Thyestes (Auchenaspis) egertoni* (Lank.), *T. (A.) sp.* and *Cephalaspis sp.*[1] This appears to be the equivalent of the *Auchenaspis* Sandstone of the Ledbury district.

A good road section at Haughton in what would seem to be the same bed shows some 6 ft. of current-bedded, speckled grey and black sandstone with purple bands and green marl nodules and again 500 yds. S.W. of Rhodes Farm purple and green speckled, very micaceous, platy sandstone shows a dip towards the east of about 30°.

Between Haughton and Morville and near Bridgwalton and The Leasowes much of the ground is drift covered and, where drift is absent, the exposures of solid are mostly in red and purplish marl. A small section by the side of the lane, 100 yds. S.E. of Croft, is in purple and green, micaceous, thin-bedded sandstone.

Red marls and clays of the Downton Series appear also to occupy the ground round Underton and Harpswood and crop out from below the Coal Measures west of ' The Hill.'

Between Morville and Harpswood the strong escarpment of Meadowley Hill (Survey Photographs 6865, 6867–8) is formed by the outcrop of tough bands of the *Psammosteus* Limestones and overlying sandstones (Plate IIA).

From the alluvium of Mor Brook, at the foot of the scarp and at about 300 ft. above O.D., to the Limestones, approximately at the 500-ft. contour, the strata appear to be mainly marls. South of The Lye two distinct beds of limestone can be traced and the level falls to near the 400-ft. contour at the road about 300 yds. W. of Harpswood. The thickness of the upper limestone band is from 6 to 8 ft. south-west of The Lye and of the lower band 5 ft. between The Lye and Underton. Dips measured along the escarpment are not reliable on account of slips on the steep slope.

The limestone appears to crop out again in the brook, above Criddon Bridge, one mile west of Harpswood, where it is at about 450 ft. above O.D., a somewhat higher level than at Underton. The red marl ground below the limestone of the main outcrop continues southward from Underton for about half a mile beyond Harpswood.

The Lye, Wallsbatch and Oldfield.—Above the *Psammosteus* Limestones, in the escarpment west of The Lye, sandstone bands occur in marl. One of the bands, some 20 ft. above the limestone, has yielded numerous fragments of *Pteraspis leathensis* White.[2]

A section in the brook 700 yds. upstream from Criddon Bridge, about 1½ miles south of Morville Church, has also yielded specimens of the same ostracoderm, together with *Anglaspis* ? and *Pachytheca sp.* ; this locality is just beyond the western margin of the sheet and between it and the bridge the stream has cut down to a thick limestone bed which appears to be the top bed of the *Psammosteus* Limestones.

---

[1] Robertson, T., in *ibid.*, p. 28.
[2] *Phil. Trans. Roy. Soc. Lond.*, Ser. B., vol. 225, No. 527, 1935, p. 445.

An interesting section in the stream, 120 yds. south of The Lobby, a house by the roadside about 150 yds. S.W. of Oldfield, shows :—

	Ft.	in.
Grey marly sandstone with *Pteraspis leathensis*, *Pachytheca* and carbonaceous fragments	0	4
Fine-grained grey sandy shale	1	0
Calcareous grey cornstone	0	4
Carbonaceous shale	0	1
Concretionary cornstone	1	0
Soft grey marl ... ... over	1	0

A band of dark red-brown calcareous sandstone, higher in the sequence, about 500 yds. downstream from this section is, according to Mr. King, the *Cephalaspis* Sandstone.

North of Glazeley.—Rocks believed to belong to the Ditton Series occur between the Deuxhill Fault (p. 131) and the main outcrop of the Coal Measures of Woodlands and The Hill. They are exposed mainly along the course of Borle Brook above Glazeley Bridge and dip about 12° S.E. The sequence comprises red marls ; purple, green and grey sandstones ; conglomeratic cornstones and sandstones with marl pellets. No fossils have been detected and their assignment to the Ditton Series is based on general character and structural relationship. A strong band of conglomeratic cornstone, exposed close to the main road at Glazeley, may be the Hall Barn Cornstone of Mr. King.

Overwood and Shutley.—Another area of rocks believed to belong to the Ditton Series occurs in the south-western corner of the sheet, where Mr. W. W. King has recorded *Pteraspis* from a small quarry 800 yds. S.E. of Overwood Farm. There is some indication of anticlinal structure in this tract. The general dip round Shunesley and Hall Orchard is towards the north-west, while round Lower Baveney it is towards the south-east and, with a slight pitch towards the north-east, it is possible that lower horizons, including perhaps some beds of the Downton Series, are present near Nethercott and along the valley of the R. Rea.

Eudon George and Deuxhill.—In the area west of Eudon George there are yellow, brown, and light green sandstones which appear to belong to the higher part of the Ditton Series. The dip is fairly regular towards the east but at Middleton Scriven and Coats Farm it swings towards the south-east. An old quarry about half a mile west of Eudon George shows :—red and grey rubbly cornstone with a few fragments of fish, including *Pteraspis sp.* and cf. *Phialaspis*, 3 ft. ; red and grey marl, 3 in. ; platy red-brown micaceous sandstone, 4 in. ; cornstone and grey marl, 6 in. ; on flaggy sandstone.

The dip of the rocks in Crunnel Brook south of Wadeley is uniformly towards the south-east at about 10°. A band of green calcareous sandstone 270 yds. downstream from the bridge has yielded Acanthodian spines and scales but no fossils of use in fixing the horizon.

Sidbury and Stottesdon.—The rocks around Sidbury and Stottesdon appear to belong to the higher part of the Ditton Series. At Upper Overton (Sheet 166), about a mile north-west of Sidbury, a quarry shows rubbly cornstone and marl, with fragments of fish and *Pachytheca*, resting on 4 ft. of thin-bedded brown and greenish-grey sandstone in part calcareous. A section in the lane running east from The Highlands shows 10 ft. of olive-green and brown, micaceous, platy sandstone with many carbonaceous markings dipping east at 12°. About 500 yds. west of Stanley 10 ft. of cornstone with fish fragments, including *Pteraspis sp.*, are exposed in a small quarry and dip east at 6°. This easterly dip is general in the ground around Sidbury. A quarry a third of a mile west of Hawkswood (Survey Photograph 6876) shows 10 to 12 ft. of finely-laminated, soft, yellow sandstone and yellowish sandstones with cornstone bands occur in the ground round Stottesdon, while greenish-grey platy sandstones with carbonaceous markings are exposed in the brook north-east of the village; these types suggest the Senni Beds and Brownstones of the Black Mountains and it is possible that some of the higher ground in this area should be separated from the Ditton Series and classed as Brownstones. A good section of thin-bedded, yellow, brown and grey sandstone full of carbonaceous matter is to be seen in the brook 200 yds. above the bridge west of Chorley Hall.

Yellow sandstones are frequent in the ground south of Stottesdon. At Walton Farm a quarry shows 15 ft. of fine to coarse, yellow and brown sandstone with lenticular bands of yellowish nodular cornstone (Survey Photograph 6911).

The high ground south-west of Walton is occupied by yellow sandstone of Upper Old Red Sandstone type similar to that exposed round Prescott in the adjoining district (Sheet 166), where at Farlow it has yielded *Bothriolepis* and *Holoptychius*. A band of cornstone in this yellow sandstone group, exposed in the brook half a mile south-west of Walton Farm, is succeeded by calcareous grits and conglomerates which, in the ground immediately to the west, pass up into the Carboniferous Limestone.

Between Walton and Bardley Court rocks believed to belong to the upper part of the Ditton Series crop out from beneath the Coal Measures. They include olive green, yellow and brown sandstones which, as in the road bank at Bardley, contain many carbonaceous fragments and may be on the Senni Beds horizon. Somewhat similar rocks are to be seen to the east of Walltown.

R.W.P.

WORCESTERSHIRE AREA—THE TRIMPLEY INLIER

Beds below the Psammosteus Limestones.—The lowest beds of the inlier crop out in Coldridge Wood and the meadows to the south. The structure hereabouts is obscure, and it is difficult to estimate the thickness of the strata below the ' Trimpley Fish-bed.' There can hardly be less than 150 ft., however, and it may be considerably more. These beds appear to be mainly purple and green mottled marls with some thin sandstones. The ' fish-beds ' referred to on p. 12 are exposed at the north-western edge of Birch Wood, at a point about 650 yds. E.S.E. of Witnell's End, in the bank of the stream (' Man Brook ') that forms the boundary of the wood. The section shows, from below upwards : green shaly marl, mottled purple, upwards of 2 ft. ; sandy marl, dark purple mottled with green, with, about 6 in. from the top, lenses of calcareous and micaceous sandstone, about 2 ft. ; sandy marl, with green micaceous sandstone, 3 ft. or more. At the mouth of a small tributary gully in the wood the upward succession is continued by green micaceous sandstones with a 1-ft. lens of cornstone, placed by Mr. King[1] at the base of his I.6 stage. The cornstone, in micaceous purple and green sandstones, is again well exposed about 400 yds. S.W. up the same stream, at a little waterfall.

In Birch Wood the beds between those just described and the *Psammosteus* Limestones, much obscured by vegetation in 1931, consist largely of marls, but Mr. King[2] records three sandstones and a cornstone. Some of the lowest

---

[1] *Quart. Journ. Geol. Soc.*, vol. xc, 1934, p. 532. Specimens of *Didymaspis grindrodi* Lankester, Reg. Nos. 52482-3, of *Modiolopsis complanata* (J. de C. Sowerby) var. *trimpleyensis* Reed, Reg. Nos. 52479-80, and of *M. nilssoni* (Hisinger) fide Leriche, Reg. No. 52481, presented to the Geological Survey by Mr. King and marked by him ' I.5 ' have matrices of the same lithology as a specimen of *D. grindrodi*, Reg. No. T.W.515, collected by the writer from the locality described in the text. From the cornstone and surrounding rock of this same locality Mr. Dewar obtained fish fragments comparable with *Ischnacanthus* and *Anglaspis*, Reg. Nos. Dw. 1416-27, 1428-1437. In his earlier paper (*Trans. Worc. Nat. Club*, vol. vii, 1921, p. 320) Mr. King appears to have included all the beds of the ' Man Brook ' exposure in the Trimpley Fish Zone.

[2] *Op. cit.* (1921), p. 320.

of these beds, consisting of purple and green marls and sandy marls with some thin micaceous sandstones, are exposed in 'Man Brook' between the two localities described above. Other marls, exposed farther north where the stream forms the eastern boundary of Coldridge Wood, include dark or grey-purple marls with ovoid calcareous nodules about half an inch long and smaller green nodules. It is possible, however, that those beds lie below the Man Brook 'fish-beds.'

In the central stream of the amphitheatre-like valley, at a point about 400 yds. north by east of Park Attwood, there is a waterfall over about 6 ft. of fine-grained, partly calcareous sandstones, about 150 ft. below the *Psammosteus* Limestones; Mr. King[1] records *Leperditia* and casts of doubtful shells from the top of these sandstones. A few feet above them, separated from them by marls, is a pale-green very calcareous sandstone, or sandy cornstone showing lustre-mottling and containing marl flakes.

The Psammosteus Limestones.—This group gives rise to a marked ridge in Birch Wood and in the field to the north-east. Much of the cornstone has been quarried away, but the excavation within and near the north-east edge of the wood shows compact purple and green limestone, associated with purple and green sandstones in part fairly coarse in grain. In certain beds some constituent has, to a greater or less extent, been altered to limonite, giving the more weathered portions an orange speckled appearance. Overlying the limestones are flaggy, micaceous sandstones, purple and green in colour. Near the middle of Birch Wood the outcrop is shifted by a fault, to south-west of which the position of the limestones is again marked by obscured old workings, with much cornstone debris, which continue at intervals through Man Wood, to the south-west of Birchwood (Survey Photograph 6890).

South of the Kidderminster road the *Psammosteus* Limestones outcrop forms a strong feature and is marked by old workings and cornstone debris, but there are comparatively few exposures and these show chiefly the purple and green sandstones interbedded with or overlying the limestones. Debris of grey-green sandy cornstone was seen in Long Coppice, 750 yds. N.W. of Park Attwood, and nodular cornstone in place near the south-west end of Long Coppice. Near the south-east edge of Eymore Wood, about 800 yds. N.W. of Littlegains Farm, the position of the limestones is marked by a line of workings and much cornstone debris. A short distance farther south the outcrop expands, owing to change of strike, and in the stream running westward from Bannering Cottage (650 yds. W.N.W. of Littlegains Farm) purple and green flaggy sandstones with beds of cornstone can be seen dipping to a little west of south at from 8° to 20°. A short distance farther south-west the outcrop is cut off by the Park Attwood cross-fault.

To return to the northern end of the inlier, a stream section at a point 80 yds. S.S.W. of Bodenham Farm (about 900 yds. S.W. of Castle Hill) shows from above downward: pale green sandstone; purple and green flaggy sandstone, 2 ft.; purple and green shaly marl, about 6 ft.; nodular cornstone, green-grey with purple mottling, 1 ft.; shaly marls and thin sandstones with cornstone nodules at base, about 9 ft.; purple and green micaceous sandstones, over 6 ft. These beds appear to represent part of the *Psammosteus* Limestones group. They dip east at about 36°. In the main stream, some 500 or 600 yds. N.N.W., thin sandstones and purple and green marls also dip east or south-east, and, with the beds near Bodenham Farm, are evidently separated from the main synclinal outcrop of Birch Wood by a fault.

On the south-east side of the syncline part of the *Psammosteus* Limestones group is exposed in a small quarry about 500 yds. N.W. of Horsleyhills Farm, which shows a few feet of nodular or pellety cornstone dipping north-west. Farther south the outcrop of the group can be followed only by its feature and occasional fragments. At a point 200 yds. north of the Kidderminster road

---

[1] *Op. cit.* (1934), p. 533.

the feature, here very prominent, terminates abruptly. This appears to be due to its truncation by a fault that shifts the outcrop about 340 yds. to W.N.W. to a point where old workings with fragments of cornstone can be seen on the south-west side of the Kidderminster road. From here the outcrop can readily be followed by its feature and surface debris to the point where it is cut off by the cross-fault near Park Attwood. Almost the only exposure is afforded by a gully 320 yds. N.N.W. of Park Attwood, where cream-coloured cornstone overlies sandy cornstones.

South of the Park Attwood cross-fault, in the broad outcrop extending from near Littlegains Farm to Holbeache, sandstone and rubbly cornstone are exposed in a little quarry about 300 yds. N.E. of, and cornstone and massive limestone in an old quarry by the lane 140 yds. N.E. of, Holbeache. A sliced specimen of the massive limestone (E 16830) shows that it consists of numerous very angular grains of quartz,[1] about .06 to .09 mm. in diameter, with some up to .2 mm., with a few grains of calcite and flakes of white mica and of chlorite, in a muddy, calcareous matrix in which calcite has crystallized out in large interlocking plates.

North-west and west of Mary Moors the *Psammosteus* Limestones group crops out with a westerly dip. Quarries show rubbly cornstone, with pellets of marl and sandy marl, overlain by flaggy purple and green sandstones. At Mary Moors and to the south the group has an eastward or east-south-eastward dip, and its outcrop is shown chiefly by the feature and ploughed-up cornstone debris.

Beds above the Psammosteus Limestones.—Some of these, beginning a little less than 150 ft. above the top of that group, are discontinuously exposed on the banks of the road south-east of Birch Farms. These are illustrated by Fig. 5 (p. 27), in which the sandstones and cornstones numbered 6 and 7 represent the ' *Cephalaspis* sandstone-cornstone ' (II.1 of King). The base of No. 6 is estimated to lie about 240 ft. above the *Psammosteus* Limestones as represented at the outcrop to the east (see above, p. 25). The beds between the sandstones and cornstones shown on the figure are chiefly purple marls, but other sandstones may be present and indeed traces of such were seen.

Perhaps the best section of the strata above the *Psammosteus* Limestones is provided by the stream running west-north-westward 'from Holbeache (Fig. 6, p. 28). About 140 yds. W.S.W. of the house a fault crosses the stream. To the east of the fault the beds dip north-eastward and eastward and their horizon is uncertain, but probably above that of the *Psammosteus* Limestones. West of the fault, the beds dip west or W.N.W. and are gently flexured. Near the fault is a nodular cornstone in red marls (Fig. 6, No. 1), which, we understand, Mr. King (*in lit.*) regards as part of the *Psammosteus* Limestones. Possibly the more typical beds of this group may be cut out by the fault, but it is noteworthy that the purple and green sandstones that in most places overlie the limestones seem to be absent here. The cornstone is followed, upwards, by bright red marls with green spots in which traces of another nodular cornstone (No. 2) were seen. Next comes purple and green flaggy sandstone some 10 or 12 ft. thick (No. 3) with a hard bed at the base. Assuming that the cornstone No. 1 is at the top of Mr. King's I.8 stage these sandstones would appear to lie near the top of I.9 and perhaps represent the " Eurypterid sandstones." These sandstones are separated by purple and green marls, with traces of a cornstone, from a further 10 or 12 ft. of dull purple and green sandstones (No. 4). After some marls with two purple sandstones (Nos. 5 and 6), the higher with 6 or 8 in. of rubbly cornstone at the top, the next noteworthy beds are a group of sandstones and cornstones beginning at a point about 250 yds. due west of Holbeache. Of these the sandstone (No. 7) shows surfaces that appear to be

---

[1] *Cf.* W. W. King, *Proc. Geol. Assoc.*, vol. xxxvi, 1925, p. 385, where the presence of quartz in the *Psammosteus* Limestones of Brown Clee and Westhope Hill is recorded, but is stated to be " an unusual feature in the areas so far investigated in the West Midlands."

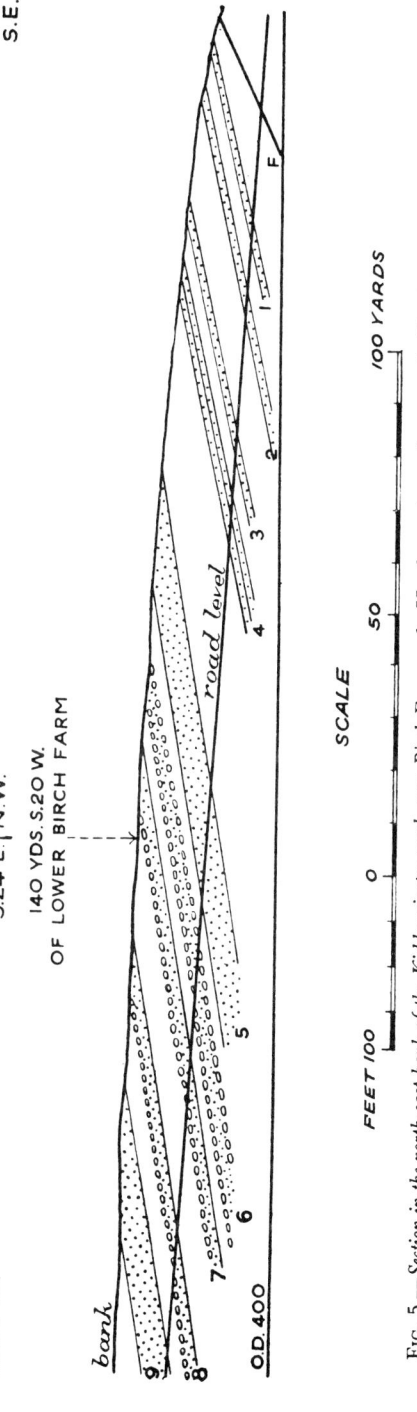

Fig. 5.—*Section in the north-east bank of the Kidderminster road near Birch Farms.* 1. Hard purple sandstone. 2. Hard sandstone, green with black specks. 3. Coarse micaceous sandstone, green, mottled purple. 4. Purple sandstone with marl parting. 5. Flaggy purple sandstone. 6. Cornstone, 6 in., on sandstone on cornstone 6 in. 7. Cornstone, 1 ft., on sandstone. 8. Cornstone, 1 ft., on dull purple sandstone. 9. Dull purple and pale green calcareous sandstone. F, fault.

28 GEOLOGY OF DUDLEY AND BRIDGNORTH: [chap.

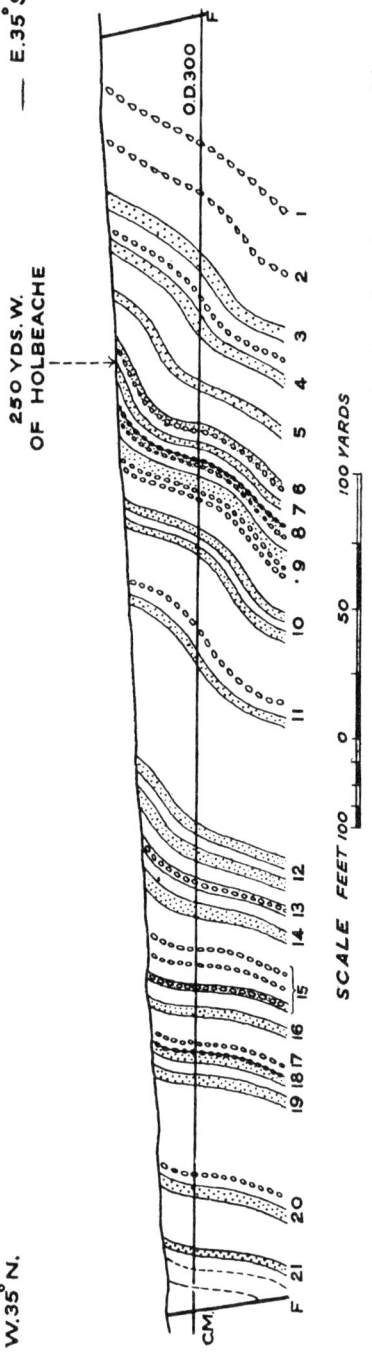

FIG. 6.—*Section along stream west of Holbeache.* C.M., Coal Measures. F, fault. For explanation of numbers see text, p. 26.

ripple-marked. It is separated by sandy marls from the purple sandstone and 3-ft. rubbly cornstone numbered 8. Green-grey marls separate No. 8 from No. 9, of which the detailed upward succession is : dull purple sandstones 12 ft. or more ; nodular cornstone, 2 ft. ; sandy marls, 2 ft. ; nodular cornstone, 8 in. to 1 ft. ; purple-mottled sandy marls, 1 ft. ; grey nodular cornstone, 3 ft. or more. These beds, together, perhaps, with those numbered 8, represent the ' *Cephalaspis* sandstone-cornstone ' (II.1 of King). The base of No. 8 is about 250 ft. above the cornstone No. 1, which is in fair agreement with the estimate of the vertical distance between the *Psammosteus* Limestones and the ' *Cephalaspis* sandstone ' obtained near Birch Farms (see p. 26 above). Other beds of sandstone and cornstone (Nos. 10 and 11) crop out farther downstream, but beyond No. 11 there is a stretch of about 50 yds. with almost no exposures. The beds there seem to be mostly marls, but comparison with other sections[1] shows that sandstones and cornstones should occur thereabouts. Two beds of hard sandstone (No. 12), the upper one giving rise to a small waterfall, crop out about 400 yds. west by north of Holbeache. Further sandstones and cornstones follow, of which those numbered 13, 14 and 15 appear to correspond in position with the Hall Barn Cornstones (p. 13). The lower cornstone at 15 is a greenish-grey cornstone breccia. Farther downstream is another group (Nos. 16, 17, 18, 19) of hard calcareous sandstones and cornstones, the latter of the grey or greenish cornstone breccia type, with calcareous nodules and pellets in a calcareous matrix. The remaining beds, up to the fault that brings in Coal Measures, are chiefly marls, with some hard calcareous sandstone and cornstone (Nos. 20 and 21).

The next stream to the south of that near Holbeache also affords exposures of beds from the ' *Cephalaspis* sandstone ' upwards. The latter crops out due south of Guildings Farm, situated 400 yds. S.W. of Holbeache. Some 40 yds. west a 1-ft. cornstone in marls and marly sandstones has yielded many fish remains to Mr. King and is called by him the "Guildings bone-bed." It seems to be represented in the Holbeache stream section by the cornstone at No. 11. Higher sandstones and cornstones can be seen, but the Guildings stream section was in 1931 much overgrown and obscured.

Some beds placed by Mr. King below the ' *Cephalaspis* sandstone ' were temporarily exposed in 1931 in the Birmingham Waterworks trench between Holbeache and Mary Moors. The dip is to south-west at a low angle. At about 140 yds. S.W. of the lane leading from Trimpley to Eymore Wood the trench exposed a bed of greenish rubbly cornstone-breccia with a 1-ft. clay parting. It is followed downwards by flaggy, purple sandstones resting on purple marl with greenish soft sandstone and sandy marls or silts. Mr. King (*in lit.*) places the cornstone and sandstone in stage I.9, and considers that the equivalent beds crop out in the Holbeache stream section at a point about corresponding to where No. 4 is shown on Fig. 6. The cornstone seen near No. 4 was not, however, of a similar type, that in the trench being more like those at 15, 17 and 19 of the Holbeache section. Specimens of *Modiolopsis* cf. *complanata* J. de C. Sowerby sp. (Reg. Nos. 52484-5) from 30 yds. S.W. of Eymore Lane, presented to the Geological Survey by Mr. King, appear to have come from the greenish silts, below the sandstones and cornstones in the trench. They are marked by Mr. King I.8.[2]

Of the beds exposed in a dingle south-east of Mary Moors, it seems most probable that it is those shown at No. 2 on Fig. 7 which represent the ' *Cephalaspis* sandstone,' but there is room for doubt. The higher beds indicated in the figure cross the dingle to south-west of the line of section, owing to the change in direction of the stream.

The harder beds of the dingle south-east of Mary Moors (Nos. 1, 2, 5 and 6

---

[1] *Cf.* e.g. Fig. 7, p. 30, of which the beds numbered 11 and 12 would appear to fall in this part of the succession.

[2] See W. W. King, *op. cit.* (1934), p. 534.

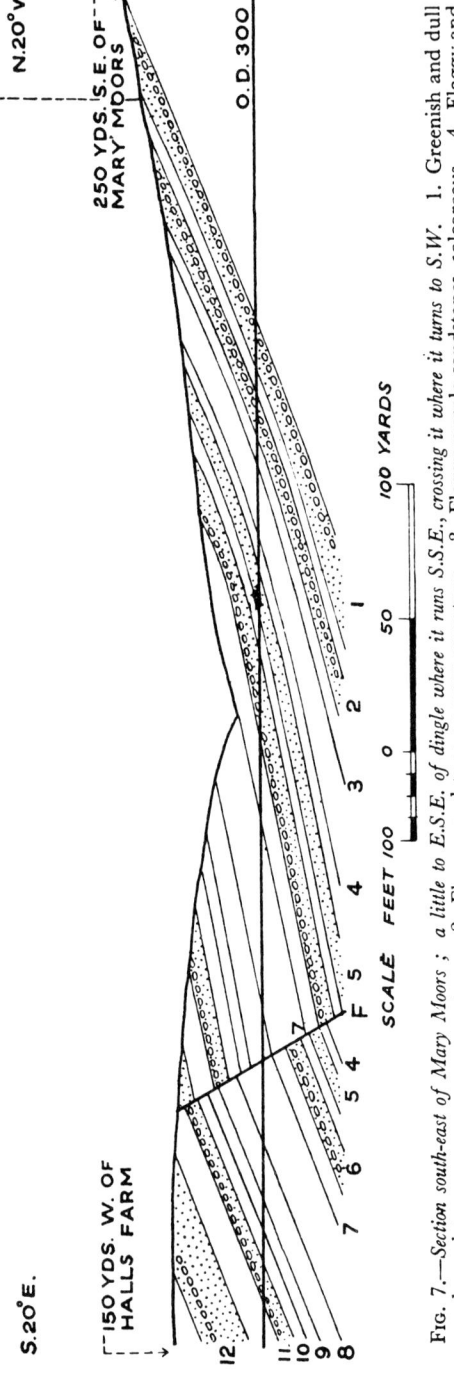

Fig. 7.—*Section south-east of Mary Moors ; a little to E.S.E. of dingle where it runs S.S.E., crossing it where it turns to S.W.* 1. Greenish and dull purple sandstone on grey-green cornstone. 2. Flaggy sandstone on grey cornstone. 3. Flaggy purple sandstones, calcareous. 4. Flaggy and thin-bedded sandstone, calcareous. 5. Purple and green calcareous and micaceous sandstones. 6. Purple and green cornstone on marl on hard calcareous sandstone. 7. Hard purple and green calcareous sandstone. 8. Sandstone. 9. Purple sandstone. 10. Hard green and dull purple sandstone. 11. Hard purple and green sandstones with cornstones. 12. Hard purple and green sandstones with cornstone. F, fault.

of Fig. 7) can be traced by almost continuous features, occasional exposures and ploughed-up fragments to the stream about a third of a mile east of Trimpley. A fault, clearly indicated by an abrupt change of dip, crosses this stream at a point about 630 yds. east of Trimpley Church, bringing in, on the east, beds above the horizon of the ' *Cephalaspis* sandstone,' with a south-westward dip.

A small quarry at Hall's Farm shows 3 or 4 ft. of flaggy micaceous sandstones on 4 ft. or more of purple and green cornstone with marly and sandy pellets. These are part of the belt of sandstones and cornstones constituting the " Hall Barn Cornstones." The Geological Survey collection includes a ventral shield of *Pteraspis crouchii*,[1] from Hall's Barn, Worcestershire, probably obtained from the quarry above mentioned. The Hall Barn cornstone group, the base of which lies about 80 or 90 ft. above No. 12 of Fig. 7, can be followed by feature to the dingle running down from south-east of Mary Moors and is exposed along a stretch of some 350 yds. beginning at a point 700 yds. S.E. of Hill Farm. The direction of the stream is here nearly parallel to the strike, and exposures show hard purple and green sandstones with partings of purple and green sandy marl and lenses of grey and greenish cornstone. Some of the sandstones are highly calcareous and show lustre-mottling.

T.H.W.

---

[1] Reg. No. 21439. It is figured by E. Ray Lankester as *Scaphaspis rectus* Lankester in ' A Monograph of the Fishes of the Old Red Sandstone of Britain, Part I, The Cephalaspidae,' *Palaeont. Soc.*, 1868, p. 23, pl. ii, Fig. 13. E. I. White (in ' The Ostracoderm *Pteraspis* Kner and the Relationships of the Agnathous Vertebrates,' *Phil. Trans. Roy. Soc.*, Sec. B., vol. 225, 1935, p. 405) states that this and other *Scaphaspis rectus* specimens are really young stages of the ventral shields of *Pteraspis crouchii* Salter MS. Lankester. C.J.S.

CHAPTER III

CARBONIFEROUS ROCKS

INTRODUCTION

Lower Carboniferous

THERE is no certain evidence of Lower Carboniferous rocks within the present district but they are present a short distance to the west in the Clee Hill area (Sheets 166 and 181).[1]

Upper Carboniferous

Lower Coal Measures do not occur in the district (see below, p. 39) and the Upper Carboniferous rocks comprise only the Middle and Upper Coal Measures. Their general succession is given on p. 3.

The Productive or ' Middle ' Coal Measures are found in the South Staffordshire Coalfield, in the east of the district, and the Forest of Wyre Coalfield, in the west. The lower parts of the Upper Coal Measures also occur mainly in these areas, but the upper parts (the Keele and Enville Beds) crop out over considerable tracts between.

The Productive Coal Measures in South Staffordshire are predominantly grey in colour and consist largely of clays and shales with impersistent beds of sandstone. Workable coal seams amount, in the aggregate, to an average thickness of 60 ft. or 70 ft. In addition, there are beds of clay-ironstone, formerly worked much more extensively than at present, and valuable seams of fireclay. In the Wyre Forest the Productive Measures (the " Sweet Coal " or Kinlet Group) contain a large proportion of red and purple clays, with beds of conglomerate and grit that resemble the " espley rocks " of the Etruria Marl Group. Grey shales and clays are interbedded with these, and, near Highley and Kinlet, such grey beds form a continuous belt in which lies the principal workable coal seam.

Following upon the Productive or ' Middle ' Coal Measures in South Staffordshire come the purple and red clays of the Etruria or Old Hill Marl Group. This is succeeded by the Halesowen Group, for the most part grey or greenish in colour, with a large proportion of sandstones.

---

[1] Dixon, E. E. L., in ' The Forest of Wyre and the Titterstone Clee Hill Coal Fields, Part II, *Trans. Roy. Soc. Edinb.*, vol. li, pt. iv, 1917, p. 1,064.

In the Wyre Forest the highest beds of the Kinlet Group may represent the lowest part of the Etruria Marl Group of South Staffordshire (p. 42), and are followed by the "Sulphur Coal" or Highley Group. This closely resembles the Halesowen Group, but differs in being productive of coal ; the thin coals of inferior quality that it contains were formerly worked within the district and are still wrought to the south, round Mamble, Bayton, etc. (Sheet 182). The Halesowen and Highley Groups pass gradually up into rocks that were formerly regarded as Permian. Of these the lowest, now known as the Keele Beds, consist of red marls and sandstones. They pass up in their turn into the Enville Beds, of which the lower part consists of calcareous sandstones and conglomerates with subordinate non-calcareous sandstones and marls. The upper part of the Enville Beds is constituted, again, of red marls and sandstones, with breccias in addition, including the well-known breccia of the Clent Hills.

The whole series outlined above appears to constitute a sequence without any time-gap of great magnitude. Unconformities do exist, however. One has been claimed[1] between the 'Middle' Coal Measures and the Etruria (Old Hill) Marl ; but the evidence appears to be in favour of a transition from one to the other except, perhaps, in certain isolated localities. It seems clear, however, that unconformity exists at the base of the Halesowen Group in South Staffordshire, and of the Highley Group in the Wyre Forest. Again, there is evidence of unconformity, at least in places, at the base of the group that includes the Clent Breccia.[2] The existence of other unconformities and non-sequences of a minor or local nature is not improbable.

The Productive Measures of South Staffordshire and the "Sweet Coal" (Kinlet) Group of the Wyre Forest were placed in the Westphalian Series by R. Kidston.[3] The Etruria Marl Group and the Halesowen and Highley Groups contain a fossil flora intermediate in character between that of the Productive Measures and that of the Keele Beds. For the beds with this transitional flora Kidston proposed the name Staffordian. The Keele Beds are now generally acknowledged to be of Upper Coal Measures age, and they were placed by Kidston in his Radstockian Series, which he considered to represent part of the highest Coal Measures (Stephanian) of the continent. E. A. N. Arber[4] maintained that the flora of the Keele Group also was transitional in character, and thought the Stephanian of the Continent might be represented by the Enville Beds. The lower part of the latter (the Calcareous Conglomerate Group)

---

[1] Kay, H., *Proc. Birm. Nat. Hist. & Phil. Soc.*, vol. xiv, 1921, p. 147.

[2] See W. S. Boulton, *Quart. Journ. Geol. Soc.*, vol. lxxx, 1924, p. 343 ; and the present Memoir, p. 93.

[3] *Quart. Journ. Geol. Soc.*, vol. lxi, 1905, p. 308 ; and *Trans. Roy. Soc. Edinb.*, vol. li, 1917, p. 1,022. The term 'Yorkian' has been suggested by Prof. W. W. Watts (*Geol. Mag.*, 1922, p. 238) as a substitute for 'Westphalian,' which Kidston used in a sense more restricted than that in which it was originally defined (A. de Lapparent, 'Traité de Geologie,' ed. 3, 1893, pp. 819, 827).

[4] *Trans. Inst. Min. Eng.*, vol. lii, 1916, p. 45.

has once more been claimed as Permian,[1] on evidence of certain fossils found in it at Hamstead (Sheet 168). On the grounds of the unconformity at its base it has been contended[2] that the upper part of the Enville Beds (the Clent Group) should be retained in the Permian System. This question is discussed in the sequel (p. 95).

T.H.W.

---

[1] Hardaker, W. H., *Quart. Journ. Geol. Soc.*, vol. lxviii, 1912, p. 680 ; and E. Dix *Geol. Mag.*, 1935, p. 557.

[2] Boulton, W. S., *op. cit.*, p. 365.

CHAPTER IV

CARBONIFEROUS ROCKS

PRODUCTIVE COAL MEASURES

'MIDDLE' COAL MEASURES
(SOUTH STAFFORDSHIRE COALFIELD)

GENERAL ACCOUNT

Sequence.—The general sequence of the 'Middle' Coal Measures in that part of the South Staffordshire Coalfield which lies within Sheet 167 is as follows (see also Fig. 11, p. 60, and Plate X):—

	Ft.
Upper Coal Measures: Etruria Marl Group (p. 58)	
'Middle' Coal Measures:—	
Measures with thin coals	45
LITTLE OR TWO-FOOT COAL	1½
Measures	25
BROOCH COAL	3½
Measures with the *Brooch Ironstone*	6
HERRING COAL	1½
Measures with the *Pins* and *Pennyearth Ironstones* and (in places) the *Ten Foot* and *Backstone Ironstones* ...	125
THICK COAL (where partings are normal)	30
Measures with the *Grains* and *Gubbin Ironstones* ...	12
HEATHEN COALS and parting	15
Measures with the *New Mine* or *White Ironstone* and the *Pennystone Ironstone*	25
*SULPHUR OR STINKING COAL	3
Measures with the *Stourbridge Old Mine Fireclay* ...	30
*NEW MINE COAL	2
Measures with the *Fireclay Balls Ironstone* and *No. 1 New Mine Fireclay*	25
*FIRECLAY COAL	2½
Measures with the *Getting Rock, Poor Robin* and *Rough Hills White Ironstones*, and *No. 2 New Mine Fireclay*	25
*BOTTOM COAL (including the Bottom Coal Holers) ...	10
Measures with the MEALY GREY or SINGING COAL, the *Gubbin and Balls, Blue Flats* and *Diamond Ironstones*, and *No. 3* and lower *New Mine Fireclays*	90

* Absent in the southern part of the area.

The thicknesses given in the above table are approximate averages. Actually many of these items vary in thickness within somewhat wide limits. The total thickness of the 'Middle' Coal Measures within the area included in Sheet 167 ranges from about 510 ft. at Tipton Moat to under 400 ft. at the southern end.

Coals, ironstones, fireclays, etc.—Of the coal seams, the Brooch and Thick are, throughout the area, the most important. The former was generally regarded as the best house-coal of the district. The Thick Coal is composed of from twelve to fourteen distinct layers varying in their quality and in the uses for which they are best suited. All of them tend to deteriorate towards the southern end of the field, where some become worthless. These layers are, in places, separated by only insignificant partings; or partings may be altogether absent, as at Dudley Woodside. Elsewhere the partings become more important. For example, near Kingswinford and Himley, and, again, near Coseley and Bilston, the top two layers are separated from the remainder of the Thick Coal as the " Flying Reed." At Highfields, near Bilston, this separation amounts to 204 ft. Again, at the northern end of the area the Thick Coal is divided into two portions by a shale parting, about 15 ft. thick, called the " Hob and Jack." Further, at the southern end of the area the Thick Coal splits up into three principal seams known respectively as the Top, Middle and Bottom coals. Near Stourbridge the Top and Middle coals are about 40 ft., and the Middle and Bottom coals about 30 ft. apart.

The Heathen Coals are generally of fair quality and even as far south as Stourbridge were worth getting in conjunction with the fireclays. North of Dudley the Lower Heathen is more usually called the Rubble Coal. In the Dudley, Gornal and Himley areas the parting between the two Heathen Coals is thin or absent; elsewhere it may amount to 15 ft. or more of measures. Near Brettel Lane and Stourbridge the measures between the two Heathen Coals include workable fireclays.

The Sulphur or Stinking Coal is in most places so impure as to be almost worthless, but was good enough to work near Gornalwood. The New Mine Coal is of better quality north of Dudley than elsewhere, and in that area was one of the principal seams below the Thick Coal, with a thickness in places of as much as 10 ft. South of Dudley it diminishes to 2 ft. or less and deteriorates rapidly, being recognizable with certainty only as far south as Bromley and Netherend (Cradley).

The Fireclay Coal is about 6 ft. thick in the Coseley and Bilston area, but is of inferior quality. Its southern limit is about the same as that of the New Mine Coal.

The Bottom Coal was an important seam in the northern area, where it is separated into two portions by measures varying from a few inches to 30 ft. in thickness. The lower portion, the Bottom Coal Holers, is of poor quality; but probably it is of this portion only that any considerable reserves remain and these are largely flooded. The Bottom Coal persists as a workable seam to the neighbourhood of Gornal and can be recognized as far south as Brierley Hill and Dudley Wood, beyond which it is either absent or represented only by " bat " (carbonaceous shale).

The Mealy Grey Coal is a seam of poor quality that has been worked near Bilston. It cannot be recognized with certainty south of Tipton.

Except in the extreme south, all the seams in the 'old' coalfield (i.e., east of the Western Boundary Fault) must now (1946) be regarded as virtually exhausted.

Of the ironstones, those below the New Mine Coal cannot be identified south of Dudley. The Pennystone exists as a workable measure as far south as Pensnett and Oldhill, and the New Mine or White Ironstone and the Gubbin still farther south. The ironstones above the Thick Coal were worked chiefly south of Dudley.

In the southern area the measures of most economic importance are the fireclays.[1] Workable fireclays are found at least as far north as Ettingshall ; but the principal seams lie in an area extending northward from near Stourbridge to Gornal, and, until lately, were not believed to persist much east of Cradley. Fireclay of good quality has been discovered, however, on the upthrow side of the Russell's Hall Fault, at the Tansley Hill shafts of Warren's Hall Colliery, about a mile south-east of Dudley (p. 38).

Clays suitable for coarse pottery are found in several places, mainly amongst the measures associated with the Bottom Coal and between the Thick and Brooch coals.

The other measures consist chiefly of grey shales and clays. In the southern area red and mottled clays occur amongst those below the Thick Coal. Sandstones are impersistent and variable in thickness. In the northern area the measures between the New Mine and Sulphur coals include, in most places, a sandstone called the New Mine or Twenty Yard Rock. A thick sandstone, sometimes called the Thick Coal Rock, is found in many places amongst the measures between the Thick and Brooch coals. A siliceous, ganister-like sandstone occurs just above the base of the Coal Measures near Gornal.

The basement beds of the 'Middle' Coal Measures consist in most places of sandstones or conglomerates ; the latter may be very coarse. In a few localities, however, the lowest bed is a clay, and in at least one it appears to be a coal seam.

T.H.W.

Fossils.—Plant remains have been found in beds associated with most of the coal seams, but the clays just above the Brooch Coal, and, still more, those of the Ten Foot Measures, above the Thick Coal, have yielded the greatest number and variety.[2]

Though non-marine lamellibranchs are not uncommon, accurately located records are somewhat scarce, as there have been few opportunities for collecting since the stratigraphical importance of these fossils was

---

[1] See T. H. Whitehead, in 'The Geology of the Southern Part of the South Staffordshire Coalfield' (*Mem. Geol. Surv.*), 1927, pp. 48-50, 55-60, *passim* ; and 193-4 ; also 'Special Reports on the Mineral Resources of Great Britain' (*ibid.*), vol. xiv (Refractory Materials : Fireclays. Resources and Geology), 1920, chap. xi ; and vol. xxviii (Refractory Materials : Fireclays. Analyses and Physical Tests), 1924.

[2] See R. Kidston, 'On the Fossil Flora of the Staffordshire Coalfields, Part III, The Fossil Flora of the Westphalian Series of the South Staffordshire Coal Field,' *Trans. Roy. Soc. Edinb.*, vol. 1, 1914, p. 73.

recognized. From the material available, however, Prof. A. E. Trueman[1] concludes that the lowest beds, possibly up to the New Mine Coal, belong to the Ovalis Zone, and the remainder of the ' Middle ' Coal Measures to the Modiolaris and Similis-Pulchra zones, the dividing line between these two being provisionally placed at the Thick Coal.

Marine bands.—Of the six marine bands now recognized in the South Staffordshire Coalfield[2] only one is recorded within the area of the present sheet. This band was noted by Jukes in the lower part of the New Mine Ironstone and the upper part of the Pennystone near Oldbury (Sheet 168). A search made underground at Baggeridge Colliery[3] in 1929 of the measures between the White Ironstone and the Stinking Coal yielded only a single specimen of *Lingula mytiloides* J. Sowerby. Its presence at Baggeridge has recently been confirmed (1943-45) and it has also been observed within the present sheet at Straits Colliery and at outcrop at Woodsetton, near Sedgley, as well as at numerous collieries in the adjoining parts of the coalfield to the north and east. At all these localities it is found that the shales immediately above the Stinking Coal carry *Lingula* together with other marine fossils in some cases, and that the thickness of the fossil band may reach several feet. This marine band is to be compared with the Seven Feet Marine Band of Warwickshire ; the Molyneux Marine Band of South Derbyshire ; the Stinking Marine Band of Leicestershire, and the Pennystone-New Mine Band of Coalbrookdale, all of which lie towards the middle of the Modiolaris Zone. Bisat[4] challenged the correlation of this marine band with the Gin Mine of North Staffordshire suggested by J. T. Stobbs,[5] correlating it instead with the Seven Feet Banbury Marine Band which lies at the base of the Modiolaris Zone.[6]

T.H.W., G.H.M.

DETAILS

For details of the ' Middle ' Coal Measures of that part of the South Staffordshire Coalfield which lies within Sheet 167 reference should be made to ' The Geology of the Southern Part of the South Staffordshire Coalfield (*Mem. Geol. Surv.*, 1927) as follows :—measures below the Thick Coal, pp. 35–44, 47–61 ; Thick Coal, pp. 62–63, 65–72 ; measures above the Thick Coal, pp. 72–74, 76–82.

Information concerning the fireclays at the Tansley Hill shafts of Warren's Hall Colliery was received from Mr. H. J. Haden and Mr. Geo. Waring after the publication of the above-mentioned memoir. From sections of the shaft and

---

[1] Trueman, A. E., ' The Lamellibranch Zones of the South Staffordshire Coalfield,' *Geol. Mag.*, 1940, p. 28.

[2] Mitchell, G. H., and C. J. Stubblefield, ' The Geology of the Northern Part of the South Staffordshire Coalfield,' *Geol. Surv.*, Wartime Pamphlet No. 43, 1945, p. 22.

[3] Thanks are due to Mr. H. D. Poole, then Managing Director, for affording facilities for this search.

[4] Bisat, W. S., ' On the Goniatite and Nautiloid Fauna of the Middle Coal Measures of England and Wales.' ' Summary of Progress for 1929 ' (*Mem. Geol. Surv.*), pt. 3, 1930, p. 75.

[5] Stobbs, J. T., ' The Marine Beds in the Coal Measures of North Staffordshire,' *Quart. Journ. Geol. Soc.*, vol. lxi, 1905, p. 495.

[6] Trueman, *op. cit.* 1940.

of a ventilation drift it appears that the highest fireclay is, in places, 11 ft. 9 in. thick. It underlies a coal that itself lies close under the White Ironstone and must therefore be the Sulphur or Stinking Coal. This fireclay thus occupies a position analogous to that of the Stourbridge Old Mine Fireclay (see Table, p. 35, and Plate X). The second and third seams of fireclay are separated by 2 ft. of bat and coal that may represent the New Mine Coal, in which case the third fireclay (which, in the drift, is 9 ft. 6 in. thick) is in the position of No. 1 New Mine Fireclay of the Stourbridge area. Two other fireclays (Nos. 4 and 5) are recorded of which the lowest, No. 5, is in several comparatively thin seams, associated with coal and bat that may represent the Fireclay Coal.

Smestow and Claverley borings.—A borehole (Appendix II, p. 193) put down, in 1912, at Smestow, near Wombourne, appears to have penetrated ' Middle ' Coal Measures from 2,710 ft. to 2,814 ft. from the surface. These measures included a seam of coal, recorded as 7 ft. thick with its base at 2,732 ft., and a bed with *Lingula* (seen by W. Gibson) at 2,743 ft. The small thickness of ' Middle ' Coal Measures suggests that part of them is faulted out (pp. 132, 196), and renders correlation with other localities very hazardous; but it may be conjectured that the 7-ft. coal represents the Thick Coal, and the *Lingula* bed the New Mine and Pennystone marine band of South Staffordshire (p. 38). The lowest bed was a very coarse grit, 9 ft. thick, which rested upon fossiliferous Silurian (Wenlock) shales with thin limestones.

In the borehole near Claverley[1] the beds from 1,797 ft. 2 in. to 2,190 ft. 6 in· were referred by W. Gibson to the Productive Measures. R. Kidston[2] considered that the plant remains indicated that above 2,032 ft. these measures were ' Westphalian ' (Yorkian[3]) in age, but that below 2,082 ft. they belonged to the Lower Coal Measures or ' Lanarkian,' the dividing line being somewhere between these depths. As Kidston himself remarked, " purely geological considerations would favour the view that the lowest portion of the bore should also be Westphalian " (i.e., ' Middle' Coal Measures). E. A. N. Arber[4] considered that the plant remains themselves supported this view, and recently Dr. R. Crookall has reviewed the palaeobotanical evidence and concludes that the whole of the measures beneath 1,805 ft. in the boring must be classed with the Yorkian Series.[5]

T.H.W.

Since their total thickness is only 393 ft. (including 23 ft. of olivine-dolerite) as compared with 473 ft. at Baggeridge Colliery, it is possible that the ' Middle ' Coal Measures are incompletely represented, owing to overlap or faulting. It is perhaps significant that the borehole passed (at 2,190 ft. 6 in.) from the Coal Measures into fossiliferous Silurian shales without revealing any basement grit or conglomerate. The comparative thinness of these measures and the absence of any coal seam thicker than one foot render detailed correlation with either the South Staffordshire or the Wyre Forest Coalfield difficult. *Lingula* was recorded in black shale at 1,811 ft. 8 in.[6] but no specimens from this depth were preserved. The following fossils were collected by Dr. J. Pringle (1905) from a depth of 1,827 ft.—Echinoid spine, *Lingula mytiloides* J. Sowerby, *Productus sp.*,

---

[1] Gibson, W., ' A Boring for Coal at Claverley, near Bridgnorth, and its Bearing on the Extension Westwards of the South Staffordshire Coalfield,' *Trans. Inst. Min. Eng.*, vol. xlv, 1913, pp. 30-48. The site is 520 yds. east of Bulwardine.

[2] In Kidston, R., T. C. Cantrill and E. E. L. Dixon, ' The Forest of Wyre and Titterstone Clee Hill Coal Fields,' *Trans. Roy. Soc. Edinb.*, vol. li, 1917, p. 999 (see p. 1,081).

[3] See p. 33, footnote 3.

[4] ' On the Fossil Floras of the Wyre Forest, with Special Reference to the Geology of the Coalfield and its Relationships to the Neighbouring Coal Measure Areas,' *Phil. Trans. Roy. Soc.*, Ser. B, vol. 204, 1914, p. 363 (see p. 409).

[5] Crookall, R., ' The Supposed " Lanarkian " in Shropshire,' *The Naturalist*, 1932, p. 37. By an evident misprint, the depth is given on p. 74 as 1,085 ft.

[6] Gibson, W., *op. cit.*, p. 33.

Palaeoneilo cf. *laevirostris* (Portlock), *Pernopecten* [*Syncyclonema*] *carboniferum* (Hind), a Conodont and a Palaeoniscid scale. Though at first sight a correlation seems to be suggested between the bands at 1,811 ft. 8 in. and 1,827 ft. with the Charles (Hamstead) and Sub-Brooch marine bands of South Staffordshire respectively, it must be emphasized that the fauna of the band at 1,827 ft. more closely resembles that of the Charles Marine Band. It is also worth noting that variegated marls occur at Claverley (Fig. 8, p. 43) and in South Staffordshire[1] at comparable levels just below these two bands.

Dr. Pringle in 1905 also found *Lingula mytiloides*, fish fragments and ? Annelid tubes in specimens from an 18-inch black shale with ironstone nodules associated with a thin coal, at 2,056 ft. 6 in. This probably represents the Pennystone marine bed of Coalbrookdale[2] and the New Mine and Pennystone marine bed of South Staffordshire (p. 38). If so, the suggestion on p. 133 that the 1-ft. coal seam at 2,044 ft. 6 in. might represent part of the Thick Coal would imply a great attenuation of the intervening measures between South Staffordshire and Claverley (cp. Fig. 8, p. 43 and Plate X). As an alternative Dr. Pocock suggests that, at Claverley, the Thick Coal horizon may be occupied by the 23 ft. of basalt or dolerite.

T.H.W., G.H.M.

KINLET GROUP

(WYRE FOREST COALFIELD)

GENERAL ACCOUNT

Introduction.—We have mentioned (p. 33) that the Coal Measures of the Wyre Forest Coalfield can be divided into an older and a younger group. The older, here named the Kinlet Group, should be regarded only as approximately equivalent to the beds formerly known as the Sweet Coal Group, the upper limit of which was never clearly defined, but it can be shown by its fossil plants to be of ' Middle ' Coal Measures (Yorkian) age.

Murchison in 1839[3] did not recognize that the measures of this area were divisible into two groups; but this was known to Daniel Jones, whose researches began in 1868 in connexion with the first Royal Commission on Coal Supplies. Cantrill in 1895[4] discussed the distribution of the Sweet and Sulphur Coal groups, and recognized the existence of an unconformity between them. In a later paper,[5] in collaboration with R. Kidston, he stated that the plant remains are adequate to distinguish the one group from the other and to show that the younger is of Staffordian age.

---

[1] Whitehead, T. H., and T. Eastwood, ' The Geology of the Southern Part of the South Staffordshire Coalfield ' (*Mem. Geol. Surv.*), 1927, p. 76 and pl. xiii (Sandwell Park No. 1).

[2] ' The Country between Wolverhampton and Oakengates ' (*Mem. Geol. Surv.*), 1928, pp. 53, 55, 69, 70.

[3] ' The Silurian System,' 1839, pp. 131-40.

[4] ' A Contribution to the Geology of the Wyre Forest Coalfield,' Kidderminster, 1895, pp. 17 and 34.

[5] ' The Forest of Wyre and the Titterstone Clee Hill Coalfields,' *Trans. Roy. Soc. Edinb.*, vol. li, 1917, p. 1,019,

'MIDDLE' COAL MEASURES : KINLET GROUP.

Sequence.—The general sequence of the Kinlet Group, based upon boring and shaft sections at Highley, Kinlet, etc., is as follows :—

[*Upper Coal Measures : Highley Group* (p. 65)]

Kinlet Group :—   Ft.

Purple and mottled marls and espley rocks with some grey beds and fireclays and, in places, a thin coal, about 200–300

Grey sandstones and mottled marls, with a *marine band* at the base at Kinlet 30

Grey and espley-like sandstones, some mottled marls and plant-bearing shales and a thin coal or fireclay at the base 60

Grey shales, sandstones and some mottled marls with a thin coal or fireclay 80

Four principal coals, the HIGHLEY-BROOCH, HALF-YARD, FOUR-FOOT and TWO-FOOT with associated grey measures and ironstones 40–70

Grey measures with some red marls, thin coals and fireclays and, at the base, a thin coal and fireclay with ironstone nodules ... ... about 100

Red and grey marls, espleys, sandstones, fireclays and thin coals up to 300

Grey ganister with rootlets and fireclays up to 40

Comparison with South Staffordshire and Coalbrookdale; marine bands.
—The sequence tabulated above shows a rough division into three parts of which the lowest includes a considerable development of red marls and espley rocks of Etruria Marl facies and the middle, mainly a grey series, includes the principal coals worked at Kinlet, Highley, etc. The highest division consists of some 200-300 ft. of red, purple and mottled marls, with coarse green grits and sandstones of a typical espley character and is lithologically similar to the Etruria or Old Hill Marl Group of South Staffordshire.

As explained below (p. 59) there is, in South Staffordshire, no marked break, either lithological or palaeobotanical, between the 'Middle' (Productive) Coal Measures and the Etruria Marl Group, and the line of division has accordingly been placed at the highest marine band known in that coalfield, or, where this has not been recorded, at a little coal which directly underlies it where the marine fossils are known, and which appears to be a tolerably persistent horizon.[1]

In the Claverley Boring, which provides a connecting link with the Highley-Kinlet area, very similar conditions obtain.

In the Highley-Kinlet area we find, in the Kinlet Boring of 1929, that the grey measures containing the Highley-Brooch Coal pass gradually

---

[1] Recent work (1943-45) in the coalfields, including zoning the Coal Measures by non-marine lamellibranchs, strongly suggests that the 'Etruria' Marl is a facies and varies in age from district to district, being in general younger in the north than in the south. Thus, the Etruria Marl of North Staffordshire is probably later in age than the 'Etruria' (Old Hill) Marl of South Staffordshire. While in this account the base of the latter is taken at the Charles Marine Band, in Cannock Chase several hundred feet of Productive Coal Measures lie above this marine band before the 'Etruria' Marl commences. C.J.S., G.H.M.

upwards into a group of red beds, some 200 ft. thick, of typical Etruria Marl facies, which comes on in force a few feet above a dark marine band, with *Lingula mytiloides* and *Productus* (" *Pustula* ") *rimberti* Waterlot (see p. 51) lithologically indistinguishable from the marine band at 1,827 ft. of Claverley (p. 39). The thin greenish silty mudstone with similar fossils found by Mr. Whitehead (p. 54) in the railway cutting at Eymore Farm, Upper Arley, seems to correspond in stratigraphical position with this Kinlet marine band.

There seems, therefore, little reason to doubt that the red beds above the marine band at Kinlet should be correlated with those above the *Lingula* band at Claverley, and also with the lower part of the Etruria Marl Group of the South Staffordshire Coalfield.

In the last-mentioned area there is no palaeontological evidence as to the age of the lowest 200 ft. or so of the Etruria (Old Hill) Marl Group (see p. 59 below); but it is natural to link them with the remainder of the group, which is shown by the plant remains to be of Staffordian age. There is equally no conclusive palaeontological evidence as to the age of the red beds above the marine band of the Kinlet area. Here, however, these beds are succeeded unconformably by the Highley Group, and it was impossible to separate them at the outcrop from the underlying measures, which, as mentioned above (p. 40), are proved by the plant remains to be of Yorkian age. In this western area, accordingly, all the Coal Measures below the Highley Group have been combined as the Kinlet Group, and classed with the ' Middle ' Coal Measures.

Of the three marine (or *Lingula*) bands detected in the Claverley Boring (p. 39) the lowest, which, as already suggested, may correspond with the New Mine-Pennystone marine band of South Staffordshire and the Pennystone marine band of Coalbrookdale, may be represented by a dark shale with ironstone nodules about 180 ft. below the Highley-Brooch Coal in the Highley Boring. A similar bed with ironstone nodules occurs in the Kinlet Boring about 130 ft. below the Highley-Brooch Coal.[1]

With regard to the coals, a comparison of the sections at Highley and Kinlet with sections of shafts in the Coalbrookdale and South Staffordshire coalfields suggests that the Brooch and associated coals of Highley may correspond with some part of the group Top to Yard coals of Coalbrookdale, and with part of the Thick Coal of the southern part of the South Staffordshire Coalfield and of the equivalent seams in the northern part of that field (Fig. 8).

Base of the Group.—The lowest Coal Measures of the western area appear to have been laid down unconformably on an evenly eroded surface of the older rocks, whereas in the South Staffordshire Coalfield there is evidence of an uneven floor, with ridges and hollows (p. 120).

---

[1] Dr. Stubblefield reports that in the J. T. Stobbs Collection now in the Geological Survey Museum there are specimens of ironstone dated 1920 from ' a crut ' at Kinlet Colliery in which he identifies crinoid columnals, a polyzoan, *Orbiculoidea* cf. *nitida*, *Productus* (" *Pustula* ") *piscariae* Waterlot, and *Aviculopecten* cf. *gentilis* (J. de C. Sowerby). The productid is a species characteristic of the Pennystone Ironstone of Coalbrookdale and South Staffordshire (Portway Hall, Dudley).

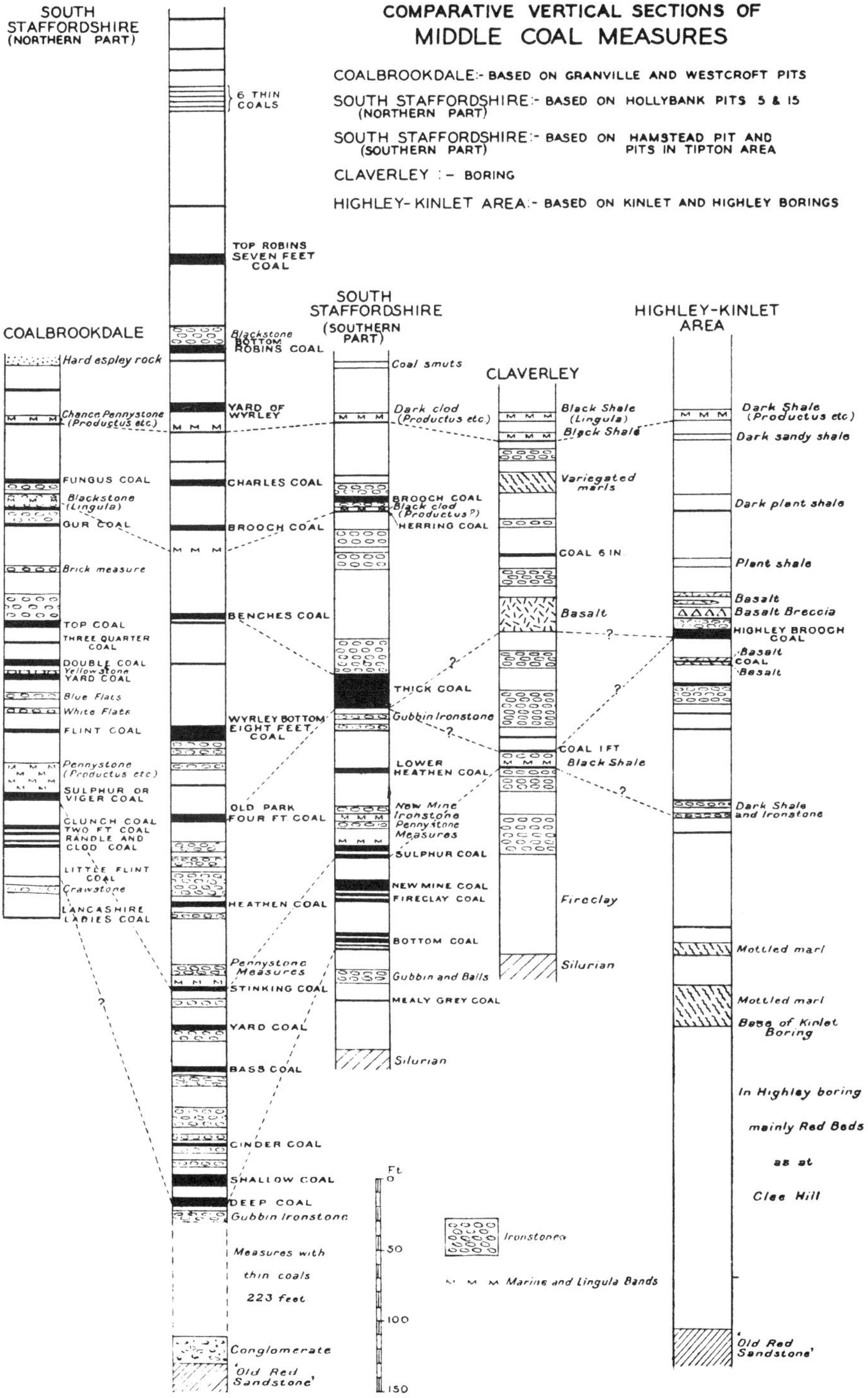

'MIDDLE' COAL MEASURES: KINLET GROUP.

FIG. 8.—*Comparative Vertical Sections of Middle Coal Measures.*

The basement beds of the Kinlet Group are marked by the presence of a hard grey or white sandstone of ganister type containing rootlets. No similar rock has been found at any other horizon in the Coal Measures of the Wyre Forest area.

The outcrop of this basal sandstone can be traced by its debris from the point, three-quarters of a mile east of Deuxhill, where it is unconformably overlapped by the Highley Group, southwards by Billingsley and Bagginswood to Baveneywood and beyond the margin of the district. Small outlying patches of Coal Measures, with the same ganister-like rock, rest on the Old Red Sandstone near Middleton Scriven and west of Chorley.

On the Old Series map (Sheet 55 N.E.) a small outcrop of this rock at Bagginswood is shown as Millstone Grit. The small outlier at Cleobury Mortimer (New Series, Sheet 182) was believed by T. C. Cantrill[1] to belong to the Upper Coal Measures ; but it also contains the ganister type and so would appear to be another patch of the lowest Kinlet beds. It may be mentioned, too, that at Catherton Common, Clee Hill (Sheets 181 and 166), at Brown Clee (Sheet 166) and in the outlier of Shirlett Hill (Sheet 152) the same type of rock is present in the lowest beds of the Coal Measures. In the Highley Boring the Coal Measures rest on Old Red Sandstone, and in the Claverley Boring on Silurian rocks, but in both cases the lowest beds of the Coal Measures are fireclays and grey sandstones with plant remains. Again, in the Dowles Valley region (Sheet 182), where the Coal Measures appear to rest on Old Red Sandstone, the lowest bed is a fine sandstone with rootlets.

It would thus appear that over a large area Coal Measures sedimentation was ushered in by the deposition of the same type of sandstone. Taken in conjunction with other points of correspondence in the sequence, this fact suggests that the surface of deposition was continuous and approximately level.

Upper limit and variation in thickness.—The deposition of the Kinlet Group appears to have proceeded without any marked interruption, there being no clear evidence of any serious break ; though variation in thickness shows that some movements of a differential character were taking place (p. 45). There is a marked unconformity, however, between the Kinlet Group and the overlying Highley Group (Fig. 9, p. 46) and north of the latitude of Deuxhill and Kinlet the Kinlet Group is overlapped unconformably by the Highley Group, which oversteps onto the Old Red Sandstone (p. 68, and Plate IV).

In the Eardington Deep Pit, about 2 miles south of Bridgnorth, the Kinlet Group is absent, beds of the Highley Group resting directly on the Old Red Sandstone.

The upper part of the Kinlet Group, above the Highley-Brooch Coal, amounts to approximately 170 ft. at Billingsley Colliery, at the Borle Mill Boring 340 ft., at the Kinlet Boring 300 ft., at Kinlet Shaft 440 ft.,

---

[1] ' A Contribution to the Geology of the Wyre Forest Coalfield,' 1895, p. 19.

at the Highley Boring 450 ft. and at Highley Colliery 460 ft. (see Fig. 1, p. 5, for localities).

The total thickness of the group in the Highley Boring is about 950 ft., whereas the Coal Measures occupying a corresponding position, *i.e.*, below the Halesowen Group (p. 65), in the Claverley Boring are only 590 ft. thick. Since probably only part of the difference in thickness is due to denudation at the unconformable junction with the overlying group, we may surmise that at Highley the floor of the Coal Measures was sinking at a greater rate than at Claverley. In fact, such evidence as is available suggests that during the deposition of the Kinlet Group a syncline was forming with its axis directed north-eastwards and situated probably somewhat to the south-east of Kinlet.

Coals.—The Kinlet Group contains several thick seams of good and sweet-burning (non-sulphureous) coal and some ironstones formerly worked at Billingsley and Chorley. The sweet coals crop out and have been worked at Billingsley, Harcott, Bagginswood and Baveneywood. Those of Baveneywood are said to have shown signs of deterioration in number and thickness as compared with those at Harcott and Billingsley. A similar deterioration appears to take place southwards from the Highley-Kinlet area. The measures that should include the Highley coals must crop out between the southern border of the present district, near Pound Green and Buttonbridge, and the Dowles Valley (Sheet 182), but there is no trace of workable coals in this area. Cantrill[1] was, therefore, no doubt correct in his view that the 'Middle' Coal Measures become barren towards the south ; though he may have been mistaken in supposing that borings in the Dowles Valley passed through beds equivalent to those which contain the Highley seams. The thickness of measures in the Highley Boring below the workable coals is about 480 ft. and at the Town Mill Boring in the Dowles Valley a similar thickness of Coal Measures was proved to overlie the Old Red Sandstone. It is possible, therefore, that in the latter case all the measures belong to a lower horizon than the Highley coals and this interpretation is adopted in the section across the coalfield shown in Fig. 9, p. 46.

The measures in the Pound Green, Buttonbridge and Dowles Valley area show a development of Etruria Marl facies, with espley rocks, comparable with that found in the lowest part of the Middle Coal Measures of Clee Hill and with the red beds in the lower part of the Highley and Kinlet borings.

Correlation with the Clee Hills Coalfield.—A close comparison can be made between the four principal coals at Billingsley, Harcott and Clee Hill, and, in the case of the three upper coals, with Brown Clee Hill.[2] The ironstones associated with the coals are also closely comparable (see Fig. 10).

The four coals at Titterstone Clee are known as the Great Coal, Three-quarter Coal, Smith Coal and Four-foot Coal. The corresponding coals

---
[1] Cantrill, T. C., *op. cit.*, p. 18, and *Trans. Roy. Soc. Edinb.*, vol. li, 1917, pp. 1,017-8.
[2] Jones, D., *Geol. Mag.*, 1871, p. 364.

46   GEOLOGY OF DUDLEY AND BRIDGNORTH :   [chap.

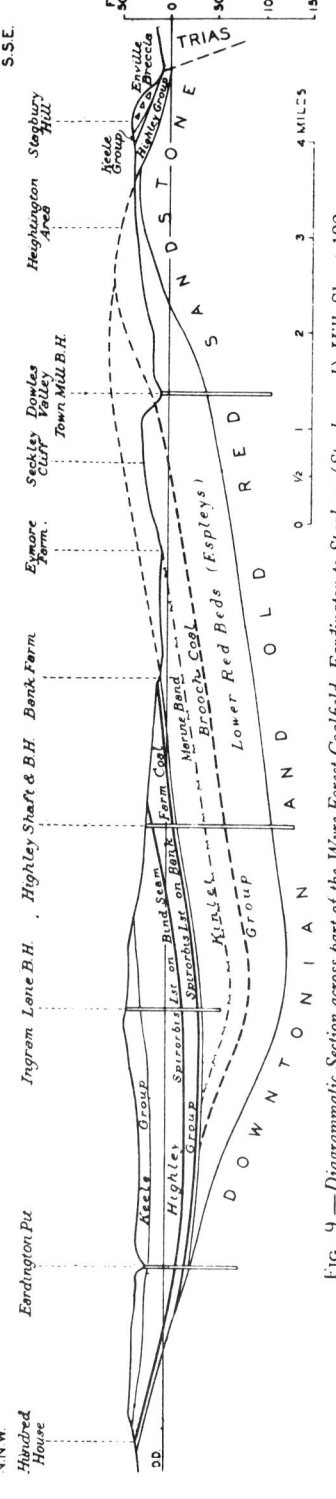

Fig. 9.—*Diagrammatic Section across part of the Wyre Forest Coalfield, Eardington to Stagbury (Stagborough) Hill, Sheet 182.*

'MIDDLE' COAL MEASURES : KINLET GROUP.

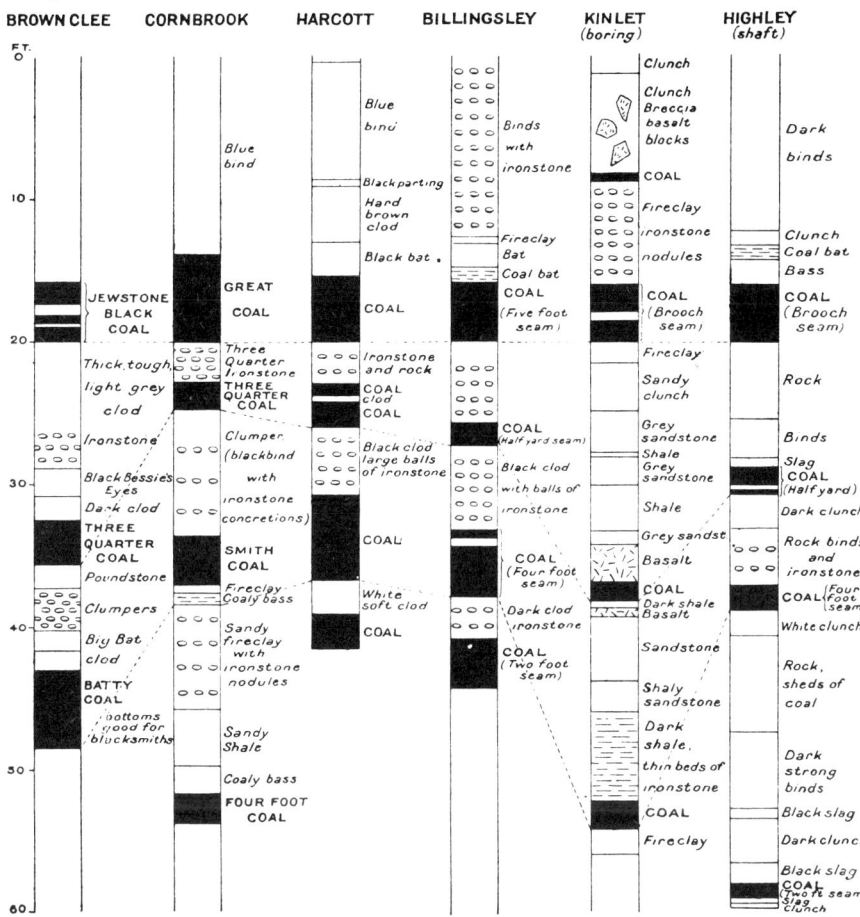

Fig. 10.—*Comparative Vertical Sections of the Principal Coals ; Clee Hills and Wyre Forest Coalfields.*

at Billingsley would appear to be the Five-foot, Half-yard, Four-foot and Two-foot seams. At Highley we have the same names as at Billingsley but the Five-foot seam is generally called the Brooch and is here referred to as the Highley-Brooch to distinguish it from the Brooch Coal of South Staffordshire. At Brown Clee the three upper coals are known as the Jewstone Black, Three-quarter and Batty Coals. Daniel Jones also considered that the Bottom Coal of Brown Clee represents the Four-foot Coal of Titterstone Clee Hill ; although it is separated from the overlying Batty Coal by 64 ft. of measures.

Red beds.—The peculiar facies of red beds (marls with espley rocks) that occurs both in the lowermost and in the uppermost division of the Kinlet Group, and is strongly developed in the Etruria Marl of South and North Staffordshire and its equivalents elsewhere, is regarded by Dr. T. Robertson[1] as closely connected with igneous activity.

---

[1] 'The origin of the Etruria Marl,' *Quart. Journ. Geol. Soc.*, vol. lxxxvii, 1931, p. 13.

It is known that intrusions and extrusions of basic igneous rock took place in the Midlands during Lower Carboniferous times[1] and again at intervals during the deposition of the Coal Measures,[2] although the extrusive character of some of these rocks has recently been questioned by Dr. C. E. Marshall,[3] and it may be inferred that the existing basalt masses of Little Wenlock, and of the Clee Hills, Kinlet, Rowley Regis, etc., are relics of originally much more extensive sheets.

Dr. Robertson supposes that the red beds in question were produced by the contemporaneous denudation of this igneous material.

This theory is not, however, generally accepted and some further research would seem to be necessary before it can be regarded as definitely established.

R.W.P., T.H.W.

DETAILS

West of the Severn the main outcrop of the beds of the Kinlet Group extends from near Wadeley, east of Deuxhill, southwards by Billingsley and Kinlet to the southern margin of the sheet near Baveneywood and Seckley.

Outliers near Middleton Scriven.—To the north-west of the main outcrop two small outliers of Coal Measures rest on the Old Red Sandstone, one half a mile north of Middleton Scriven Church and the other three-quarters of a mile south-west of the church, near the head of Sidbury Dingle. The first shows yellow and grey clay with many blocks of white ganister containing rootlets ; the second shows yellow and reddish clay in an old pit which has been worked for brick-making and here again fragments of the ganister are abundant. The presence of this ganister (see p. 44) is taken as evidence that the measures belong to the base of the Kinlet Group, preserved by pre-Upper Coal Measures faulting or folding.

Billingsley and Chorley.—In the main outcrop the lowest beds of the Kinlet Group appear from beneath the unconformable cover of Upper Coal Measures (Highley Group) east of Wadeley, their presence being indicated by many fragments of ganister at the surface. In Crunells Brook, a tributary of Borle Brook, a thin coal close to the base of the Kinlet Group crops out 400 yds. N.N.E. of Billingsley Hall Farm. The high ground between Billingsley Church and the smithy is capped by the lowest beds of the group which consist of yellow and grey clays with ganister. The old brick and tile works by the side of the road to Chorley and a quarter of a mile south-east of Billingsley Church shows the following section of the lowest beds in the north-east corner of the pit :—

	Ft.	In.
Yellow and grey sandy shale	4	0
Red-brown ferruginous sandstone	0	6
Deep-purple marl	4	0
Greenish-yellow sandstone, impersistent ...	2	0
Purple marl and dark-grey shaly marl... ...	6	0

---

[1] Pocock, R. W., ' The Basalt of Little Wenlock,' Summary of Progress for 1925 (*Mem. Geol. Surv.*), 1926, Appendix I, p. 140.

[2] Pocock, R. W., ' The age of the Midland Basalts,' *Quart. Journ. Geol. Soc.*, vol. lxxxvii, 1931, p. 1.

[3] ' Field Relations of certain of the Basic Igneous Rocks associated with the Carboniferous strata of the Midland Counties,' *Quart. Journ. Geol. Soc.*, vol. xcviii, 1942, p. 1.

The sandstones are gritty and of espley type. At the floor of the pit there is a foot of ganister in yellow clay. The outcrop of the ganister can be traced by fragments near the base of the Coal Measures to the north-east, north and west of Chorley. Two old shafts 200 yds. north of Brooksmeeting Bridge at the north end of Chorley Covert would appear to be near the site of a boring made in 1866 on the Hawkswood Estate which passed through the lowest beds of the Coal Measures into the Old Red Sandstone.

Downstream from Brooksmeeting Bridge traces of coal crops occur at intervals and the principal coals of the Kinlet Group cross the brook about 200 yds. W.S.W. of Southall Bank, the farm 500 yds. S.W. of the Cape of Good Hope inn. The coals are associated with purple and red mottled marls and ironstones and can be traced to High Green through Chorley Covert, where they have been extensively worked in the past, together with the ironstones. A record dated 1799 states that the ironstone mines in the Coppice of Common Heath (now Chorley Covert) were leased to the proprietors of Old Willey Furnace near Broseley about the year 1707. They appear to have been worked all through the coppice nearly to High Green. The ore was carried on horses to Willey Furnace and later to Charlcot Furnace (Wrickton Forge ?) near Cleobury North. The same record says that there was an old furnace or ' bloomery,' in very ancient times, at Ned's Garden at the bottom of Common Heath by the Southall Bank Brook and large quantities of slag were carried thence to Charlcot for re-smelting.

Coals are said to have been worked in the Ebleys, about 300 yds. N.E. of High Green. The seams recorded are a thin coal, the Rider (3 to 5 in.), from 22 to 24 ft. above the Yard Coal, here usually 3 ft. to 3 ft. 8 in. thick ; the Four-foot Coal from 3 to 6 ft. below the Yard Coal and a coal (1 to 2 ft. thick) 3 ft. or less below the Four-foot Coal.

One or two beds of ironstone occurred above the Yard Coal. Within recent years further attempts have been made to mine the coals in the Chorley area but with little success. Two shafts were sunk 300 and 350 yds. E.S.E. of the Smithy at High Green ; No. 2 Shaft reached a depth of 324 ft. They were abandoned in 1939.

The coals were also worked, about 1768–1790, to the south of High Green between the road and the brook, and coal seams associated with ironstone are exposed in the brook 400 yds. S. by W. of High Green, just W. of the Brock Hall Fault.

Between High Green and Knowle Hill opencast mining of the ' Brooch ' Coal was begun in 1943-4 but, owing to the frequency of old workings, the project was abandoned.

The coals which crop out in Chorley Covert were mined at Billingsley Colliery, 250 yds. E. of the Cape of Good Hope inn. The shaft there passed through the following measures below the depth of 458 ft. 9 in. :—

	Ft.	In.
COAL " Five-Foot Seam "	6	9
Fireclay with balls of ironstone	5	3
COAL " Half-Yard Seam "	1	8
Black clod with balls of ironstone (black)	6	3
COAL " Four-Foot Seam "	4	8
Dark clod with pins of ironstone	3	0
COAL " Two-Foot Seam "	3	5

These coals may be correlated with the Highley-Brooch and associated seams of Highley and Kinlet. Over much of the area worked the dip was about 7° E. but on the south-east side of the workings the dip steepened rapidly towards the Brock Hall Fault, the surface position of which appears to be about 130 yds. S.E. of the shaft, in which the dip is recorded as 31°.

The throw of the Brock Hall Fault is probably about 200 yds. S.E. The fault divides about half a mile south-west of the colliery and between the two branches the coals have been worked at the Harcourt or Harcott Pit by the brookside near Lower Harcourt a quarter of a mile south of High Green. The principal coals are associated with ironstones as at Billingsley (see fig. 10, p. 47). The record of the shaft below 274 ft. is as follows :—

	Ft.	In.
Brown clunch	30	0
COAL	1	3
Fireclay, binds, clod and bat	17	6
COAL [Five-Foot Seam]	4	6
Ironstone and rock	3	0
COAL (sulphurous)	0	9
Clod	0	6
COAL [Half-Yard Seam]	1	8
Black clod and large balls of ironstone	5	0
COAL [Four-Foot Seam]	6	0
White soft clod	2	6
COAL [Two-Foot Seam]	2	6

About 100 yds. S.E. of the shaft there are indications of an outcrop of basalt.

Kinlet.—Between the Brock Hall Fault and the Kinlet Hall Fault, the Kinlet Beds crop out in Old Coppice and Bush Wood and towards Bagginswood and Bardley Court. They are interbedded with basalt, the Kinlet basalt mass, which appears to lie a little above the horizon of the principal coals of the Harcott Pit section.

The higher beds of the group comprising the red and mottled marls and greenish espley-type sandstones are exposed along the brook in Old Coppice. A thin coal crops out just above the basalt at several points near Mass House and again west of Kinlet Hall and near the Dog Kennels. Outlying patches of measures lie on the surface of the main outcrop of basalt and the thin coal occurs in the one at Knowle Hill. Coals below the basalt horizon have been worked north of Knowle Hill and just under the hill on its west side. A coal worked at Bagginswood appears to be near the base of the Coal Measures. There are traces of a coal crop close to Mires Farm, 400 yds. north of Bardley Court, in the slope below the base of the basalt, and along the brook at the northern side of Birchen Park.

Blocks of ganister are present at the surface near Bagginswood and, near the base of the Coal Measures, between Bagginswood and Bardley Court.

Between the Kinlet Hall Fault and the Tiphouse Fault a large area of Kinlet Beds extends from the base of the Highley Group in the north through Kinlet Park to Baveneywood in the south-west and Southall House in the south-east.

The Borle Mill boring, 400 yds. N.E. of Nortonsend Farm, passed through about 180 ft. of Highley Beds into the red marls and sandstones and underlying grey beds of the Kinlet Group. The total depth of the boring was 603 ft. 8 in. The record gives the following details below the depth of 518 ft. 10 in. :—

	Ft.	In.
COAL	0	6
Strong dark binds	7	2
COAL [Highley-Brooch]	4	6
Fireclay, sandstone, ironstone balls, etc.	22	2
BURNT COAL	0	7
Sandstone, ironstone and shales	13	7
COAL	0	5

The higher beds of the Kinlet Group, mottled marls and espley-type sandstones, are exposed in the brook west of Nortonsend Farm and again north-east of Winnal Farm in a tributary of Borle Brook.

In 1929 a boring was put down alongside this tributary at a point about 250 yds. S.W. of Tiphouse Cottages (Appendix II, p. 197). The boring commenced in beds just below the base of the Highley Group and passed first through about 200 ft. of mottled marls and espley grits, the upper red beds of the Kinlet Group, and then entered more typical Middle Coal Measures. A dark shale was met with between 236 ft. 5 in. and 242 ft. 2 in. the lower part containing ironstone nodules. From 237 ft. A. Templeman collected fish remains including *Megalichthys sp.*, *Pleuroplax rankinei* (Hancock and Atthey) and *P. attheyi* (W. J. Barkas), identified by Prof. D. M. S. Watson ; also *Lingula mytiloides* J. Sowerby, and *Orbiculoidea* cf. *nitida* (Phillips). From 239 to 240 ft. he collected *Chonetes sp.*, *L. mytiloides*, *L. sp. nov.*, *O.* cf. *nitida*, *Productus carbonarius* de Koninck, *P.* ("*Pustula*") *rimberti* Waterlot, of. *Rhipidomella michelini* (Léveillé), indeterminate lamellibranch, a bellerophontid and *Serpulites sp.* This is the marine band which we correlate with the marine band of the Claverley Boring at 1,827 ft. and with the Charles Marine band in the upper part of the Middle Coal Measures of South Staffordshire (but see pp. 39 and 40).

A 2-in. coal was found at 309 ft. 2 in. and a group of coals and fireclays, apparently the horizon of the Brooch or Five-Foot Coal of Highley, between 383 ft. 10 in. and 396 ft. 6 in. Other coals below may be considered as representing the Half-Yard, Four-Foot and Two-Foot coals of Highley.

The coals are associated with beds of basalt and basalt-breccia. The details of this peculiar part of the section are given in Fig. 17 (p. 145) and a description of the igneous rocks will be found on p. 147.

From a depth of 450 ft. to the base of the boring at 669 ft. 4 in. there was a return to beds of red and mottled marl with espley-type grits, conditions similar to those found at Clee Hill and in the lower part of the Highley Boring.

The top 200 ft. of red beds proved in the boring are exposed in the banks of the brook upstream as far as a point 700 yds. due east of the smithy, and an old quarry on the slope east of the brook 500 yds. N.E. of Winnal Farm shows a 10-ft. greenish espley-type sandstone and rather coarse false-bedded conglomerate.

About 300 yds. E.N.E. of Winnal Farm a coal of about 6 in. thick in the bank of the brook and associated with grey clays and sandstones is referred to the workable coal group of Kinlet and Highley collieries. These grey measures can be traced up the brook to a point east of Catsley, where they may run out against the southward continuation of the Tiphouse Fault, or may swing north and west of Catsley, where there are traces of old surface workings, and thence towards Bradley and Birchen Park.

At the north end of Kinlet Park the brook through High Wood follows the strike of the grey measures and a coal 1 ft. thick in grey marl is exposed three-quarters of a mile north by east of the Hall.

There is probably a fault, parallel to the Kinlet Hall Fault, along the valley south-east of the church.

The grey measures with coal seams and an outcrop of basalt are exposed in the stream that flows through the fish ponds south of the Hall. An old quarry on the south side of the stream about 300 yds. above the fish ponds shows : vesicular basalt 5 ft. ; on a thin coal ; fireclay 1 ft., on basalt.

At Baveneywood there are a number of old pits and surface workings along the outcrop of the coals but very little is now to be seen. An old record gives the following section :—

	Ft.	In.
From the surface to the first coal	90	0
Sweet Coal (Very good) ... 2 ft 6 in. to	2	9
Ground	90	0
Sweet Coal	4	0

Between the Tiphouse Fault and the Highley-Kinlet Fault the Highley-Brooch Coal has been worked from Kinlet Colliery shaft, about a quarter of a mile east of Tiphouse, over an area extending a mile and a half north, and half a mile south of the shaft. The maximum width of the workings between these faults is a little over half a mile. The record of Kinlet Shaft below 869 ft. 9 in. is as follows :—

	Ft.	In.
COAL [Highley-Brooch]	3	11
Light Clunch	1	1
COAL	1	0
Tough clod	0	5
COAL and clod	0	5
Rock binds	4	2
COAL	1	8
Rock with streaks of coal	6	8

The total depth of the shaft is 889 ft. 1 in.

The throw of the Tiphouse Fault near Netherton is about 50 ft. to the west and perhaps as much as 120 ft. half a mile to the north, where it commences to swing round into a north-north-easterly direction.

Towards the south the Kinlet Beds crop out about 500 yds. south of Birch Farm, where there are traces of old workings of a thin coal seam. To the east of Birch Farm mottled marls and espley-type sandstones are exposed in the brook that runs through Lodge Coppice, a third of a mile south of Severn Lodge, to join the stream that forms the County boundary near Bank Farm (see section Fig. 9). These beds would appear to belong to the upper red beds of the Kinlet Group.

R.W.P.

Buttonbridge and Pound Green.—South of Lodge Coppice the outcrop of the Kinlet Group extends southwards to the southern border of the district and beyond, upfaulted between the Station Fault (p. 129) on the east and the Tiphouse Fault (p. 130) on the west. Most of this tract is wooded, but stream sections afford some exposures.

Coarse green sandstone and grit, with conglomerate, can be seen in a ford, about 4 furlongs south of Severn Lodge, across the brook that forms the county boundary. In the path about 50 yds. east of the ford greenish grey shale and fine-grained sandstones, higher in the sequence than the grit and conglomerate, yielded plant remains (Reg. Nos. T.W.540–44, 563) including *Neuropteris sp.*, *Annularia stellata* (Schlotheim), ? *Poacordaites gentilis* P. Bertrand and *Samaropsis sp.* Dr. Crookall, who named the plants, considers that the beds could be Yorkian in age, though *Annularia stellata* is commoner in the Upper Coal Measures. The beds must lie in the upper part of the Kinlet Group.

Farther up the county boundary brook (i.e., south of the ford) greenish grits and sandstones with green and purple-mottled sandy mudstones and some grey shales are exposed at intervals. In grey shales, about 600 yds. south of the ford, plant remains were found but only a *Neuropteris* was identifiable. Still farther south, opposite the entry of a tributary that runs between Postens Plain and Blackgroves Copse, green grits with lenses of coarse conglomerate are exposed. These must lie at a slightly lower horizon then the conglomeratic beds at the ford, as the general dip is north-north-eastward at a low angle.

The brook between Fastings Coppice and Coldwell Copse affords some exposure of beds above those at the ford. Some of the arenaceous beds are espley-like, gritty sandstones which show a concretionary structure, while others, harder and more compact, approach the Halesowen type of sandstone. Between the sandstones are occasional exposures of grey clay or shale. Sandy shale with coal streaks may be seen in the south bank of the brook about 880 yds. west of its mouth.

The stream between Postens Plain and Blackgroves Copse shows similar beds, and here again such argillaceous strata as are exposed are mainly grey shales and mudstones. About 100 yds. west of the junction with the county boundary brook grey shales yielded *Sphenopteris dilatata* Lindley and Hutton along with indeterminate plant remains. A few yards upstream the shale includes coal streaks up to half an inch in thickness.

The grits and sandstones can be followed to some extent between the streams by means of features and occasional exposures; but the difficulty of tracing outcrops through the thick oak-coppice renders the identification of particular beds from place to place uncertain. Moreover, individual beds are certainly impersistent. Some grits and sandstones may be seen in quarries about 650 and 700 yds. N.W. of Bannurt Tree Farm, and in New Pool Quarry, on the south-west side of the Bewdley road, about half a mile south-east of Buttonbridge. This quarry shows about 10 ft. of false-bedded greenish sandstones, some with coaly streaks with occasional partings of sandy shale. From it R. Kidston recorded[1] *Calamites sp.*, *Lepidodendron aculeatum* and *Cordaites sp.*

T.H.W.

Highley.—East of the Highley-Kinlet Fault, which separates the workings of the two collieries by a downthrow west averaging about 200 ft., the Highley-Brooch Coal has been worked over a considerable area ranging from half a mile east of Little Woodlands in the north to the neighbourhood of Severn Lodge in the south.

The record of Highley Colliery Shaft, 300 yds. west of Highley Station, gives the depth to the top of the Highley-Brooch Coal as 869 ft., below which we have :—

	Ft.	In.
COAL, Five-Foot or Highley-Brooch Seam ...	6	3
Rock binds and slag	9	2
COAL, Half-Yard Seam	1	9
Dark clunch, rock binds and ironstone ...	4	2
COAL, Four-Foot Seam	1	8
Clunch, rock and slag	19	4
COAL, Two-Foot Seam	1	5
Clunch	0	3

The total depth of the shaft is given as 915 ft. 4 in. The strata dip east at 1 in 27.

The Highley-Brooch Seam is said to deteriorate in quality towards the south but with no great variation in thickness.

A boring put down in 1924–5 at Schoolhouse Lane 400 yds. W.S.W. of the Colliery passed through 350 ft. of the Highley Group into the upper red beds of the Kinlet Group. The gob of the Highley-Brooch Coal was entered at a depth of 810 ft. 6 in. A considerable thickness of the lower red beds of the group, red marls and espley-type sandstones, was passed through below a depth of about 1,070 ft. and the base of the Coal Measures was reached at about 1,306 ft. The boring was continued to 1,663 ft. in red marls and red and grey sandstones belonging to the Old Red Sandstone which yielded, at depths of 1,489, 1,530 and 1,610 ft., fragments of fish, of which Prof. D. M. S. Watson reports : " These fragments present no resemblance to any Carboniferous fish and appear to be most readily regarded as Ostracoderms."

R.W.P.

---

[1] *Trans. Roy. Soc. Edinb.*, vol. li, 1917, p. 1,027 ; where the beds are assigned to the Staffordian Series (see below, p. 70, footnote 4).

Seckley Wood, etc.—To the east of the Station Fault and of the Arley Park Fault (p. 129) and west of the Severn the Kinlet Beds crop out in Cliff Wood and Seckley Wood. Amongst the highest beds seen are thick greenish grits with lenses of breccia and of conglomerate, which form strong features in the woods and are well exposed in a gully that runs down to the Severn nearly opposite the " moat " near Eymore Farm. Some of the grits can be traced northwards to the railway cutting south-south-east of Arley Station, where they are interbedded with purplish, green and yellowish shales, sandy shales and mudstones. Near the mouth of the gully traces of a thin coal seam were observed.[1]

South of the gully the Kinlet Beds form a cliff or steep bluff on the south-west side of the Severn, reaching a height of over 200 ft. above the river (Survey Photograph 6905). This cliff is capped by some of the coarse grits interbedded with purple marl. Slips obscure the sequence in several places in the upper part of the bluff. The lower part is formed of thinner and, for the most part, less coarse grits, with grey shales and dull purple and greenish marls and mudstones. Plant remains, including *Neuropteris sp.* and *Asterophyllites equisetiformis* (Schlotheim), were found in grey sandy shales just south of the gully above referred to, and about 20 ft. above the tow-path. This must be near the locality from which Cantrill[2] obtained plant remains. Some of the beds forming the lower part of the bluff can be seen better in a gully at its south end, which shows purple and greenish marls and sandy mudstones with some beds of grey clay and of greenish, soft sandstone.

The coarse grits can be traced to the line of the Elan Aqueduct across the south end of Seckley Wood, where some of them appear to have split. Along the line of the trench, which in 1932 had recently been filled in, traces of these grits could be seen with purple and ochreous clays. Some purple beds were seen in 1931 in a part of the trench, from 100 yds. to 600 yds. west of the Severn. Amongst the coloured beds were some grey clays and dark carbonaceous shales, and there was a coal streak about 420 yds. west of the river ; this would be on about the same horizon as the coal seen at the north end of Seckley Cliff.

Eymore Wood and Shatterford.—In the south-western part of Eymore Wood, to the north and north-east of Eymore Farm, a group of greenish, espley-like grits crops out. Though coarse and pebbly in parts, these grits are less so than those of the upper (western) part of Seckley Wood, with which, however, they probably in general correspond. Stream sections show purplish, brown and grey clays between the grits in places. The superimposed grits form a cliff above the east bank of the Severn to the north of the railway bridge over the river. In the railway cutting W.N.W. of the bridge leading to Eymore Farm, the following section can be made out :—

		Ft.
Sandy clays, largely overgrown ... ... about		15
Purple clays ... ... seen for		6
Sandstone, flaggy		2
Purple rubbly mudstones, with beds of sandstone ... about		4
Sandstone, flaggy		6
Clayey sandstone, fine-grained, purple and green mottled, concretionary, flaggy		2
Silty shale with plant remains, including a thin (1-in. ?) bed of greenish-grey sandy mudstone with *Lingula mytiloides* J. Sowerby, *Productus* (" *Pustula* ") *rimberti* Waterlot and *Anthracoceras* cf. *aegiranum* H. Schmidt[3] ... ... about		1

[1] See T. C. Cantrill in R. Kidston, T. C. Cantrill and E. E. L. Dixon, ' The Forest of Wyre and the Titterstone Clee Hill Coal Fields,' *Trans. Roy. Soc. Edinb.*, vol. li, 1917, p. 1,017.

[2] *Op. cit.*, pp. 1,017 and 1,023.

[3] Reg. No. T.W.573. The plant remains in the shale include *Mariopteris sp.*, *Neuropteris sp.*, *Lepidophyllum sp.* and *Sphenophyllum sp.* Reg. Nos. T.W.545-556.

	Ft.
Grey clays with black streaks	2
Purple and brown mottled clays	3
Thin-bedded sandstone, purple and green rubbly sandy mudstone and green shaly sandstone ... ... about	5
Sandstone, green, espley-like, with coarse pebbly lenses, thick-bedded to flaggy ... ... seen for	$5\frac{1}{2}$

The marine bed of the above section was found in the north bank of the cutting at a point 100 yds. N.N.W. of Eymore Farm. Its position is below the main group of coarse espley grits which crop out in the wood to the north. The marine bed should, therefore, presumably crop out amongst the beds near the foot of Seckley Cliff (see p. 54 above) on the opposite side of the Severn, but has not been found.

The group of grits in the south-western part of Eymore Wood dips at comparatively low angles to north-north-west; but north of Folly Point the dip suddenly increases and the strike changes to north-north-east, under the influence of the Trimpley axis. The increase of dip causes the grit outcrops apparently to coalesce, so that along the flank of the anticline they can only be traced as a few thick belts.

North-east of the road from Trimpley to Eymore Farm two stream-courses in Eymore Wood afford some exposures of grits and grey or purple-mottled clays. The dip is very high, in places even inverted. An exposure of the Shatterford basalt in the more north-easterly stream, at a point about 200 yds. downstream from the south-east boundary of Eymore Wood, is the most south-westerly outcrop of this rock so far detected. It lies about 670 ft. below the base of the highest belt of grits, which includes all those which crop out in the south-western part of Eymore Wood (p. 54) together, probably, with lower beds.

The beds a short distance above the basalt consist of sandstones, some of which are coarse, pebbly and espley-like, with some pale grey and purple clay. The basalt rests on hardened shale, below which discontinuous exposures of grey clay and shale and greenish sandstone may be seen to the junction with the Downtonian rocks, which, along the border of Eymore Wood, is considered to be a fault. The coarse grits near the top of the Kinlet Group are exposed in the stream south of Gunhill Wood, the northern prolongation of Eymore Wood, just south of Shatterford, where the dip again is very high or inverted and some of the higher beds can be traced by features to Shatterford.

In the Deep Pit,[1] situated 220 yds. south by west of Bellman's Cross Inn, Shatterford, there appears to be a thickness (after allowing for dip) of about 750 or 760 ft. of beds between the base of the Highley Group and the basalt, in which a boring from the bottom of the shaft ended. In the upper half of these beds 'red marl,' 'mottled ground,' and 'red and blue ground' predominate amongst the argillaceous measures, with beds of 'rock,' some of which is described as conglomerate or as gritty. In the lower half the argillaceous measures are chiefly grey binds and shale, but red and mottled beds occur sporadically nearly to the bottom. A 2-in. coal is recorded at a depth of 1,097 ft. 4 in. from the surface, an 8-in. seam at 1,159 ft. 4 in., coal and bat 6 in. thick at 1,201 ft 4 in., coal 13 in. thick at 1,222 ft. 4 in., and coal and bat 1 ft. at 1,251 ft. 4 in. These seams lie in the lower, predominantly grey portion of the measures, and appear to be the only possible representatives at Shatterford of the Brooch Coal group of Highley. Assuming that the 8-in. coal represents the Highley-Brooch Coal itself it would appear that at Shatterford there are about 140 ft. of beds at the top of the Kinlet Group not present in the Highley or the Kinlet shafts. These beds consist of red and mottled ground interbedded with an approximately equal

---

[1] For section see G. E. Roberts, *The Geologist*, 1861, pp. 422-6, and E. Lees, *Proc. Dudley and Midland Geol. & Sci. Soc.*, vol. iv, 1893, p. 13, and frontispiece, facing p. 1. See also, D. Jones, *Trans. Fed. Inst. Min. Eng.*, vol. vii, 1893-4, p. 299.

aggregate thickness of ' rock,' some of it conglomeratic. It is probably these rocks that constitute the highest belt of grits seen at the outcrop in Eymore Wood. The total thickness of the Kinlet Group in the neighbourhood of Shatterford and Eymore Wood must be over 1,400 ft.

Amongst the beds above the horizon of the Shatterford basalt are fireclays that were formerly worked for the local pottery industry, their outcrops being marked by trenches parallel to the strike. An exposure by the roadside about 300 yds. S.S.E. of Bellman's Cross Inn shows, from above downwards : dark shale, 18 ft or more ; coarse sandstone with gritty and pebbly lenses and coaly streaks, 6 ft. 6 in. ; rubbly sandy shale and sandstone, 6 ft. ; sandstone, in part gritty and pebbly, with shaly partings, 17 ft. The beds are vertical or inverted. They appear to lie immediately above the basalt, though the actual junction is not here seen.

Witnells End, Arley Wood, etc.—A stream course about 300 yds. south of Witnells End affords discontinuous exposures of beds below the basalt. The latter rests on rubbly sandy shales, below which there is considerable disturbance of the beds (see p. 140). The remainder of the Coal Measures, to the junction with the Downtonian rocks, consist of grey clays with thin coal seams and sandstones.

The grits in the upper part of the Kinlet Group form well-marked features from Witnells End to the dingle about 3 furlongs N.N.E. where some of them can be seen interbedded with purple sandy clays, in part somewhat micaceous, and some grey beds.

In the basalt quarries 500 yds. N.N.E. of Witnells End the following section was noted. Grey clay, 2 ft. ; black carbonaceous shale, 2 ft. ; sandstones, 1 ft. 4½ in. ; dark grey shales, in part sandy, 2 ft. 4 in. ; black, soft shale, 1 ft. ; on basalt. Plant fragments occur in several of these beds.

In the stream between Coldridge and Arley Woods about 150 ft. of Coal Measures below the basalt are discontinuously exposed. Immediately below the basalt are dark grey sandy mudstones followed downwards by grey and chocolate sandy clays, from which the following plant remains were collected by Mr. Dewar (Reg. Nos. Dw.1438-49A, 1450-54) :

Eupecopteris volkmanni (*Sauveur*)	Neuropteris sp.
Lonchopteris rugosa *Brongniart*	Calamites sp.
Mariopteris nervosa (*Brong.*)	Cordaicarpus cordai (*Geinitz*)

For the next 100 yds. the section shows chiefly sandstones, some of which are espley-like, with pink and green grains. About 50 yds. still farther down, or some 80 or 90 ft. below the basalt, an 8-in. coal was seen, in grey clays ; and similar clays, with indications of buff sandstone below them, appear near the junction with the Downtonian rocks, which here, as elsewhere along the west side of the Trimpley anticline, is believed to be a fault.

In Arley Wood there are few exposures. Traces of coarse espley-like grit were found to the west of the basalt outcrop and grey clays, in part purple-mottled, with beds of yellowish sandstone, near the base of the Kinlet Group, are exposed in a stream about half a mile west-north-west of Castle Hill. A ridge of relatively high ground south-east of Starts Green is due to the outcrops of grits in the upper part of the Kinlet Group. Grey and purplish clays and sandy shales with beds of sandstone near the base of the Kinlet Group are exposed in the stream about 600 yds. S.W. of Castle Hill. There, as in the south-eastern part of Arley Wood, the junction with the Downtonian rocks seems to be unfaulted ; the beds near the base of the Coal Measures dip at about 30° to the east, and unless the dip in the upper part of the Kinlet Group is much higher (unfortunately no exposures were found upon which it could be measured), there is room here for a total thickness of about 850 ft. only. Some 400 to 600 yds. N.N.E. of Lower Birch Farm, much debris of buff and brown sandstones, unlike those in the Lower Old Red Sandstone, seems to indicate a faulted outlier of Coal Measures.

The Compton Sinking.—In the Compton shaft and boring,[1] 900 yds. W. 5° N. of Pidgeonhouse Farm, Compton, the beds below 453 ft. 11 in. may be regarded as belonging to the Kinlet Group. They begin with a bed of white rock below which come 39 ft.[2] of ' red marl or mottled ground.' This is followed by hard grey rock, and this, in its turn, by clay and fireclay with some marl or mottled ground and seams of coal. From a depth of 574 ft. 5 in. to the bottom of the sinking the record shows : coal, 4 in. ; sulphur coal, 1 ft. 6 in. ; little coal, 10 in. ; smut or coal, 2 in. ; little coal, 10 in. ; little coal, 3 in. This is an aggregate thickness of 3 ft. 11 in. of coal, which is unknown in the Kinlet Group except in the Brooch Coal group of Highley. Assuming that the seams at Compton represent the Highley-Brooch Coal, there are only about 120 ft. of measures between them and the base of the Highley Group, as compared with 460 ft. at Highley. Though there may be some overstep by the Highley Group at Compton it seems more probable that this small thickness is principally due to the sinking having passed through a fault which cuts out part of the Kinlet Group (see p. 80).

Trimpley.—To the east and south-east of Trimpley the Coal Measures crop out in a narrow strip, faulted against the Bunter and Enville Beds in the east. On the west their junction with the Old Red Sandstone is in part also a fault ; but in a stream south-west of Eastham's Farm (700 yds. south of Lower Barns Farm), they rest unconformably upon the pre-Carboniferous rocks. Here the Coal Measures consist of grey clays, in places mottled with purple, and olive sandstones. A coal seam was observed by T. C. Cantrill[3] in this stream section, but was not found here in 1931, though about 270 yds. to the north grey clay with traces of coal was seen in a ditch near an old, shallow excavation.[4]

Cantrill appears to have regarded the Coal Measures near Eastham's Farm as belonging to the Sulphur Coal Group, since he thought the coal that occurs amongst them was probably the ' Main Sulphur ' seam[5] ; but in view of the fact that about a mile to the south-west, in North Wood[6] (Sheet 182), the lowest Coal Measures have yielded plants that show them to be of Yorkian age, it seems more probable that those near Eastham's Farm belong to the Kinlet Group.

T.H.W.

---

[1] Section communicated by the late T. C. Cantrill from Daniel Jones's MSS. The boring began at a depth of 385 ft. 2 in.

[2] The thickness is given as recorded. There are no data available as to the dip of the measures in the sinking.

[3] The record is not published, but was found on Cantrill's field-maps.

[4] This is perhaps the site of the small pit north of Eastham's Farm referred to by Cantrill in ' A Contribution to the Geology of the Wyre Forest Coalfield,' 1895, p. 21.

[5] Cantrill, T. C., *loc. cit.*

[6] See T. C. Cantrill, *op. cit.*, pp. 37, 38, and R. Kidston, T. C. Cantrill and E. E. L. Dixon, *Trans. Roy. Soc. Edinb.*, vol. li, 1917, pp. 1,017, 1,022. In 1931 Mr. Dewar collected plants in Gills Rough, north-west of North Wood, and these have been named by Dr. Crookall, who considers they indicate a Yorkian age.

CHAPTER V

CARBONIFEROUS ROCKS

UPPER COAL MEASURES

Etruria (Old Hill) Marl Group
(South Staffordshire Coalfield)

GENERAL ACCOUNT

Introduction.—The beds that succeed the mainly grey Productive Coal Measures form a predominantly red and purple group called by Jukes[1] the Red Coal-measure Clays and by Lapworth[2] the Espley or Brick Clay Group. They have also been called the Old Hill[3] or Oldbury Marl. Their correspondence in position and lithological character to the Etruria Marl Group of North Staffordshire is sufficiently close to warrant the use of the same name ; but the term Old Hill Marl may conveniently be retained for local purposes.

Lower limit.—The line of division between the Productive Measures and the Etruria or Old Hill Marl Group is ill defined. There seems to be no evidence of unconformity[4] between them, except possibly of a strictly local nature, and the red and mottled clays of Etruria Marl type appear to come in quite gradually and not necessarily everywhere at the same horizon.[5] The absence in South Staffordshire of anything that can be correlated with the Blackband Group, which in North Staffordshire lies between the ' Middle ' Coal Measures and the Etruria Marl, might seem to constitute *prima facie* evidence of an unconformity in the former locality. This absence might, however, equally well be due to the great attenuation of the Coal Measures, accompanied possibly by changes in the facies of portions of them, that takes place as they are followed from north to south.

The plant remains do not give any assistance in drawing the line of division between the Productive Coal Measures and the Etruria (Old Hill) Marls, nor do they, so far as our knowledge of them at present goes, afford any evidence of an abrupt change of flora such as might result from a stratigraphical break. The plants of the Productive Measures

---

[1] ' The Geology of the South Staffordshire Coalfield ' (*Mem. Geol. Surv.*), ed. 2, 1859, pp. 20, 30.

[2] ' The Birmingham Country, its Geology and Physiography,' in *Brit. Assoc. Handbook*, 1913, p. 565.

[3] Kay, H., *Quart. Journ. Geol. Soc.*, vol. lxix, 1913, pp. 433, 436.

[4] See, however, H. Kay, *Proc. Birm. Nat. Hist. & Phil. Soc.*, vol. xiv, 1921, p. 147.

[5] See p. 41, footnote.

show them to be of ' Middle ' Coal Measures, or Yorkian,[1] age (Westphalian of Kidston); those of the Old Hill Marl, considered as a whole, indicate the Transition Coal Measures or Staffordian Group of Kidston's classification.[2] But no plant assemblage has been described from any horizon between that of the clays overlying the Brooch Coal (p. 37) and one some 280 ft. higher, well up in the Old Hill Marl and some 150 or 200 ft. above its base as defined in the following paragraph.

There being, so far as the evidence goes, no sharp lithological or palaeontological break between the ' Middle ' Coal Measures and the Etruria (Old Hill) Marl of South Staffordshire, it becomes necessary to select a conventional line of division. In order to exclude from the ' Middle ' Coal Measures as much as possible of the red and mottled beds more characteristic of the Etruria Marl, the best place for division would be the top of the Two-foot Coal. This has, however, the objection that it would leave the marine horizon of Hamstead and Sandwell Park (the Charles Marine Band, see p. 40) in the Etruria Marl Group. In view of the fact that marine horizons are almost unknown in the Transition or Upper Coal Measures elsewhere,[3] this seems undesirable. The most convenient course, therefore, appears to be to draw the base of the group at the top of the marine bed. At the places mentioned the latter rests upon a little coal that appears to be identifiable over the part of the coalfield with which this memoir is concerned, in which the marine bed itself has not been found.

Near Tipton the coal in question seems to be that called in certain sections the Upper Sulphur Coal (see Fig. 11, section of Wednesbury Oak, No. 8 Pit). It lies, however, from 30 to 40 ft. above the Two-foot Coal, and is evidently not the Upper Sulphur Coal that Jukes adopted as the line of division between the Productive Measures and the Red Coalmeasure Clays.[4] The Upper Sulphur Coal of Jukes lies about 150 ft. above the Two-foot Coal, well up in the beds of Etruria Marl facies, and does not form a convenient place for division.

Lithological characters.—The Old Hill Marl Group consists mainly of red, purple, ochreous or greenish clays; but grey clays, more like those of the Productive Measures, occur in places. The red and purple clays provide the raw material for the well-known Staffordshire blue bricks. Amongst the clays there are lenticular beds of greenish grit and conglomerate that weather to a yellowish-buff colour. The coarser varieties

---

[1] See above, p. 33, footnote 3.

[2] Kidston, R., ' On the Various Divisions of British Carboniferous Rocks as determined by their Fossil Flora,' *Proc. Roy. Phys. Soc. Edinb.*, vol. xii, 1894, p. 183; and ' On the Divisions and Correlation of the Upper Portion of the Coal Measures, with special reference to their Development in the Midland Counties of England,' *Quart. Journ. Geol. Soc.*, vol. lxi, 1905, p. 308. See also E. A. N. Arber, ' On the Fossil Floras of the Coal Measures of South Staffordshire,' *Phil. Trans. Roy. Soc.*, Series B, vol. 208, 1916, p. 127, and T. H. Whitehead in ' The Geology of the Southern Part of the South Staffordshire Coalfield ' (*Mem. Geol. Surv.*), 1927, pp. 84, 85.

[3] One has been claimed to occur in the Whitehaven Sandstone of Cumberland. See T. Eastwood, in ' Summary of Progress for 1930 ' (*Mem. Geol. Surv.*), 1931, p. 55, and ' Northern England ' (*Geol. Surv. Reg. Handbook*), 1935, p. 52.

[4] *Op. cit.*, pp. 20, 31.

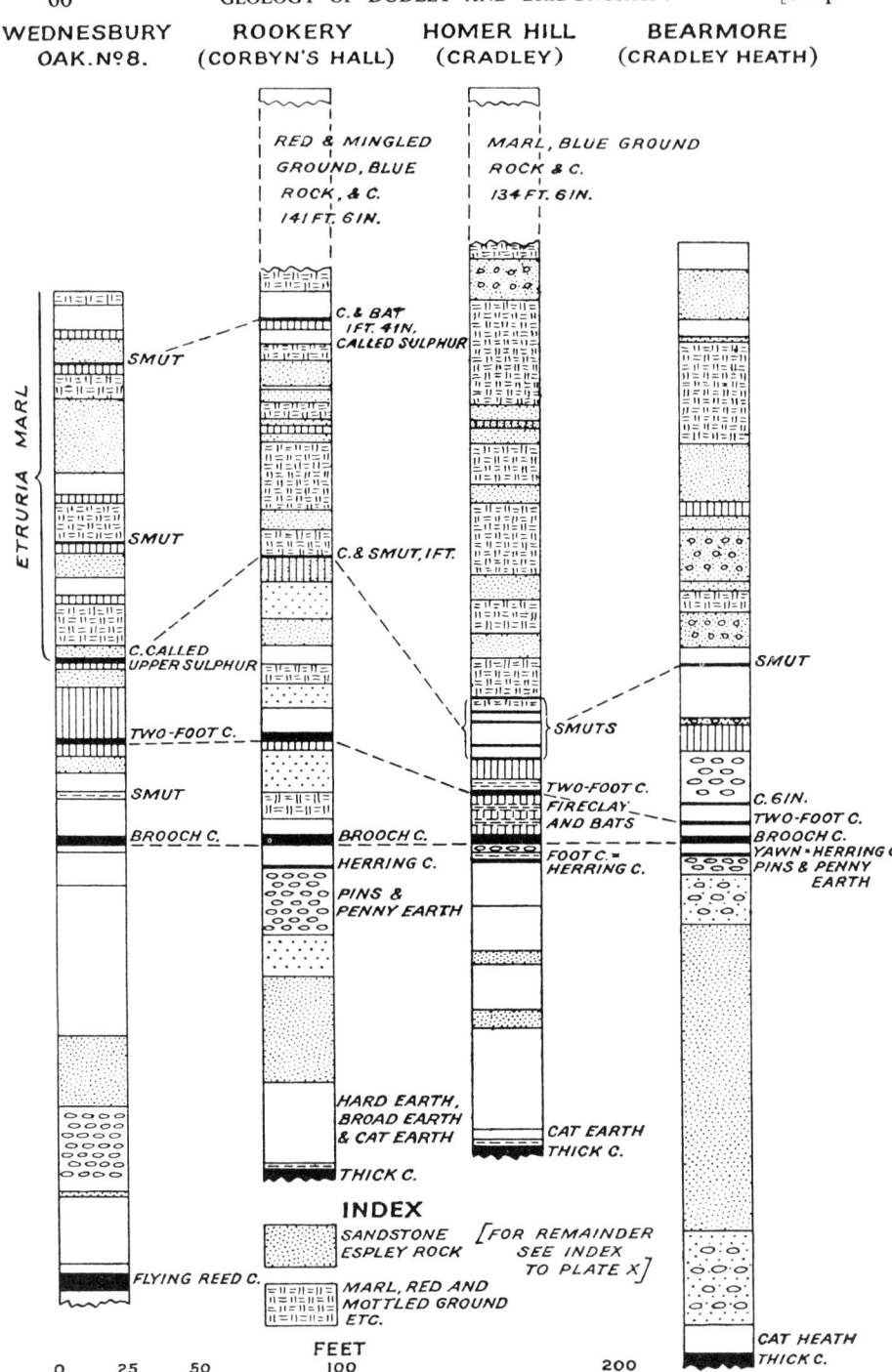

Fig. 11.—*Vertical Sections of Measures above the Thick Coal and of Etruria Marl in the part of the South Staffordshire Coalfield within Sheet* 167.

UPPER COAL MEASURES: ETRURIA MARL GROUP.

of these are frequently described by shaft-sinkers as "espleys" or rough rock. The fragments in the conglomerates are chiefly quartzite of Lickey (Cambrian) type; but Llandovery sandstone and quartzite, vein quartz, and decomposed igneous rocks also occur.[1] The finer grits contain a large proportion of material that seems to be of pyroclastic or igneous derivation. The clays of the group contain about 9 to 10 per cent. of iron oxides, but usually less than 2 per cent. of lime, so that, as pointed out by Prof. W. S. Boulton,[2] the term marl usually applied to them is not strictly appropriate.

Coal seams.—Thin coal seams occur in the group, and some of them appear to be fairly persistent. That called by Jukes the Upper Sulphur Coal can be found in most of the shaft sections from Brockmoor to Pensnett, lying from 130 to 150 ft. above the Two-foot Coal, and it appears to be recognisable near Tipton; though there, as remarked above, a lower seam is actually called the Upper Sulphur. At Baggeridge Colliery a coal from 3 to $4\frac{1}{2}$ ft. thick is recorded in No. 2 shaft immediately below, and in No. 1 Shaft a short distance below, what appears to be the base of the Halesowen Group. It has not been found possible to identify this coal anywhere else. Its position is reminiscent of that of the "Bank Farm Coal" of the Arley and Highley area (p. 66), but, as it is not recorded in the Claverley borehole, such a correlation would be unsafe.

Thickness and upper limit.—The thickness of the Etruria Marl Group is remarkably variable. It is greatest (700-800 ft.) in the area extending from Old Hill to Halesowen (Sheet 168), and lying between Blackheath (Sheet 168) and Cradley. To the west a gradual attenuation sets in as the Netherton axis is approached. In the shafts of Baggeridge Colliery the Etruria Marl Group, showing, to judge from description, all its usual characters, has a thickness of about 200 ft. Other parts of the coalfield show similar variations; and it is probable that these variations in thickness are due to crustal movements during deposition. These movements, particularly near pre-existing anticlinal axes, may even have resulted in some degree of denudation, and overstep on the part of the succeeding Halesowen Group.

Distribution.—The Etruria or Old Hill Marl Group forms a cover to the Productive Measures over more than half of that portion of the coalfield lying within Sheet 167. The most extensive areas of outcrop are (*a*) near Old Hill and Cradley and (*b*) near Brockmoor, Pensnett and Shut End.

DETAILS

For details of the Etruria Marl Group in South Staffordshire the following pages of the memoir on 'The Geology of the Southern Part of the South Staffordshire Coalfield' (*Mem. Geol. Surv.*, 1927) should be consulted:— Tipton, etc., p. 88; Old Hill, etc., p. 90; Tansley Hill, p. 91; Quarry Bank,

---

[1] See also W. O. Williamson, 'Some Grits and Associated Rocks in the Etruria Marls of North Staffordshire,' *Geol. Mag.*, 1946, p. 20.

[2] 'The Clays of South Staffordshire and its Borders,' *Trans. Ceramic Soc.*, vol. xvi (1916–17), p. 237.

Cradley and Wollescote, pp. 91-2 ; Brierley Hill, pp. 93-4 ; Brockmoor, Pensnett and Kingswinford, pp. 94-5 ; Shut End and Himley, p. 95.

Smestow and Claverley.—In the borehole at Smestow (Appendix II, p. 193) the Etruria Marl Group appears to be present with its usual character, but with a thickness of only about 50 ft. Much core was lost from this part of the boring, and it is probable that part of the group is cut out by faulting.

In the Claverley Boring the Etruria Marl Group (called by W. Gibson[1] the ' Brick-Clay or Espley Group ') extends from 1,604 ft. 2 in. to 1,797 ft. 2 in. from the surface. It presents the usual characters, including green grits and breccias of espley type. Mottled clays, interbedded with grey and black shales, extend down to 1,863 ft., but as these include the *Lingula* bed at 1,811 ft. 8 in. they were excluded by Gibson from the Brick Clay Group (p. 39).

HALESOWEN GROUP
(SOUTH STAFFORDSHIRE COALFIELD)

GENERAL ACCOUNT

Introduction.—The Halesowen Beds, like the Newcastle-under-Lyme Group of North Staffordshire, to which they may be considered generally equivalent,[2] are essentially a grey or greenish group, largely arenaceous, that intervenes between the predominantly red beds of the Etruria (Old Hill) Marl below and of the Keele Group above.

Lithological characters and sequence.—In the type area, which may be regarded as that extending from Halesowen (Sheet 168) to the Western Boundary Fault near Pedmore, the lower part of the group is fairly constant in character and consists of alternations of sandstones and clays, each member somewhat variable in thickness, the sandstones being as a rule thicker than the clays. In the clays that succeed the lowest or Basal Sandstone (Survey Photographs 2208-9) a coal seam from 1 ft. to 18 in. in thickness has been found in several places. This seam will be referred to as the Halesowen Coal. In the clays above the Second Sandstone another coal seam, 1 ft. thick, or perhaps more in places, has been found and traced over much of the western half of the type area. To distinguish it from the lower seam this coal will be referred to as the Wassel Grove Coal.

Somewhat above the middle of the Halesowen Group comes a band of *Spirorbis* limestone, first found by W. Gibson,[3] and selected by Charles Lapworth for subdividing the group. To distinguish this limestone from that believed to occur below the Halesowen Coal it is called the Illey Brook Limestone, after the type locality of Illey Brook.[4] It lies in the clay-belt that succeeds the Second Sandstone, and has been traced

---

[1] *Trans. Inst. Min. Eng.*, vol. xiv, 1913, p. 31 (see p. 37).

[2] Gibson, W., ' On the Character of the Upper Coal-Measures of North Staffordshire, Denbighshire, South Staffordshire, and Nottinghamshire ; and their Relation to the Productive Series,' *Quart. Journ. Geol. Soc.*, vol. lvii, 1901, p. 251

[3] *Op. cit.*, pp. 261–2.

[4] Sheet 168. See T. Eastwood in ' The Geology of the Country around Birmingham '· (*Mem. Geol. Surv.*), 1925, p. 44 ; and in ' The Geology of the Southern Part of the South Staffordshire Coalfield ' (*Mem. Geol. Surv.*), 1927, p. 110.

over the greater part of the eastern half of the type area. It has not been found in the western half of that area, but its horizon is believed to lie in the clays amongst which the Wassel Grove Coal occurs. From Hayley Green westward, sandstones in the upper part of the Halesowen Group form a thick and persistent belt above these clays. This belt of sandstones may conveniently be referred to as the Third Sandstone (Survey Photographs 2206-7). It is to be seen in the village of Hayley Green, where it is stained red, in the quarries near Bogs Farm, and in the upper part of the dingles north of Hagley Wood. It forms the conspicuous feature of Hodge Hill (Survey Photograph 2203) and the high ground upon which Wollescote Hall stands, and is finally see n Ham Dingle, near the Western Boundary Fault.

East of Hayley Green the beds above the horizon of the Illey Brook Limestone are more variable, and in particular, the Third Sandstone is replaced by impersistent beds and lenses separated by grey or greenish-grey clays. The highest beds of the group are clays that become mottled with red and purple near the top. Thin coals near the top of these clays have been seen in Uffmoor Wood.

The following table summarizes the sequence of the Halesowen Group in the western half of the type area (*i.e.* within Sheet 167).

Halesowen Group

	Ft.
Clays, grey, purple and red with thin coals of Uffmoor Wood	? 100
Third Sandstone	50-100
Clays, grey, with WASSEL GROVE COAL (1 ft.) ? including horizon of *Illey Brook Limestone*	30
Second Sandstone	20-100
Halesowen Coal Group : clays, grey, with HALESOWEN COAL (1 ft. to $1\frac{1}{2}$ ft.)	10-50
Basal Sandstone	100

With the exception of a possible outlier near the Lays, Brockmoor,[1] the Halesowen Beds do not appear at the surface along the western border of the South Staffordshire Coalfield. The group can, however, be recognized in the sections of the shafts of Baggeridge Colliery, in which it appears possible to identify some of the subdivisions observed in the type area.

Relation to underlying beds.—As mentioned on p. 61, the variation in thickness exhibited by the Etruria Marl Group is such that it seems probable that the Halesowen Group is definitely unconformable,[2] except, possibly, in the immediate neighbourhood of Halesowen. Here, where

---

[1] See T. H. Whitehead in ' The Geology of the Southern Part of the South Staffordshire Coalfield ' (*Mem. Geol. Surv.*), 1927, p. 118.

[2] The existence of such an unconformity was recognized by Mr. H. Kay (' On the Halesowen Sandstone Series of the South Staffordshire Coalfield,' *Quart. Journ. Geol. Soc.*, vol. lxix, 1913, p. 449), who, however, believed it to be confined to the area west of Cradley and the Lutley valley.

the lowest beds of the Halesowen Group show a considerable resemblance to the conglomeratic espleys of the Etruria Marl, and where the latter group has its maximum ascertained thickness, there may perhaps be an actual passage from one group to the other. Elsewhere such a passage seems nowhere to exist, and the line of division between the two groups is well marked. The effect of this unconformity appears most clearly in the Wassel Grove Pits, where the beds that can be assigned to the Etruria Marl are at most 60 or 70 ft. thick. It seems probable that a short distance south of these pits the Halesowen Group must overstep the Etruria Marl altogether and come to rest upon the Middle Coal Measures. The pre-Halesowen movements are indicated also by the remarkable rise in the Thick Coal that was found in following that seam in a south-westward direction from Witley Colliery,[1] a structure that has no counterpart in the Halesowen Beds at the surface. That the existence of this unconformity may have practical implications is shown by the fact that Jukes,[2] who appears to have been unaware of it, predicted that the Thick Coal would be found at a depth of not less than 800 or 900 ft. at Wassel Grove, whereas the base of the measures that appear to represent that seam, or part of it, was actually found at a depth of only 495 ft.

Upper limit and thickness.—The sections in Uffmoor Wood show that the Halesowen Beds pass gradually and conformably up into the Keele Beds. At outcrop the line of division has been drawn at the base of the lowest sandstone of definitely Keele type. This is the sandstone that crops out near the middle of Uffmoor Wood. Upon this basis the thickness of the Halesowen Group at outcrop becomes a little over 400 ft. whereas in the Baggeridge shafts about 370 ft. may be attributed to the Group.

Fossils.—Although plant remains are fairly common in certain parts of the Halesowen Group, there are comparatively few published records. Indeed, the only one from the part of the South Staffordshire Coalfield with which this memoir is concerned appears to be that of the fossil wood (Survey Photograph 2029) discovered by Mr. H. Kay[3] at Witley Colliery. R. Kidston[4] has published a list of plants obtained from this group in the Claverley Boring.

Economic products.—The Halesowen Beds are of little economic importance. The sandstones were formerly quarried for building stone ; the clays have evidently been dug and the *Spirorbis* limestones have been burnt for lime ; but these practices also have died out. Good springs are in many places thrown out at the base of the sandstones, and to some extent are used as a local source of water supply.

---

[1] Mathews, W., ' The Halesowen District of the South Staffordshire Coalfield,' *Proc. Birm. Phil. Soc.*, vol. v, 1886–87, p. 313 (see p. 322). See also H. Kay, *op. cit.*, p. 450.

[2] ' The Geology of the South Staffordshire Coalfield ' (*Mem. Geol. Surv.*), ed. 2, 1859, p. 30.

[3] *Op. cit.*, p. 453, and appendix, by E. A. N. Arber, ' On the Structure of *Dadoxylon kayi*, sp. nov., from the Halesowen Sandstone at Witley (Worcestershire).'

[4] See p. 65, footnote 1.

DETAILS

For details of the Halesowen Group reference should be made to 'The Geology of the Southern Part of the South Staffordshire Coalfield' (*Mem. Geol. Surv.*, 1927) as follows :—southern border of the South Staffordshire Coalfield : Basal Sandstone, pp. 106-8 ; Halesowen Coal Group, pp. 108-9 ; Second Sandstone, pp. 109-10 ; Clays with the Wassel Grove Coal, pp. 111-13, 115-16 ; Third Sandstone, pp. 113, 115-17 ; beds above the Third Sandstone, pp. 117-18 ; western side of the South Staffordshire Coalfield, pp. 118-19.

Smestow and Claverley.—In the Smestow Boring (Appendix II, p. 193) the Halesowen Group appears to be about 400 ft. thick and seems to have the usual characters, though the subdivisions of the type area are not clearly recognizable from the record. It includes four thin coals.

In the Claverley Boring the beds from 1,240 ft. 2 in. to 1,604 ft. 2 in. from the surface were placed by W. Gibson[1] in the 'Halesowen Sandstone Group,' giving a thickness of 364 ft. They consist mainly of grey and blue shales, with beds of grey sandstone of which the thickest lie at the base and seem to correspond with the Basal Sandstone of the type area. No *Spirorbis* limestones were found, and only two coal seams, each 3 in. thick, were noted. These both lie in the upper half of the group and one of them may conceivably represent the Brock Hall Coal of the Forest of Wyre Coalfield (p. 66). On the whole, however, the Halesowen beds of the Claverley section resemble more closely those of Baggeridge than they do the corresponding beds of the Highley area.

T.H.W.

HIGHLEY GROUP

(FOREST OF WYRE COALFIELD)

GENERAL ACCOUNT

Introduction.—The strata included in the Highley Group are approximately those to which the name " Sulphur Coal Group " has hitherto been applied ; but, as remarked by T. C. Cantrill,[2] the base of the " Sulphur Coal Group " has never been defined or located. Moreover, some confusion appears to have arisen as to the details of its stratigraphy. There can be little doubt that the Highley Group is equivalent generally to the Halesowen Group of South Staffordshire, which it resembles to a considerable degree ; but it is more doubtful whether the stratigraphical limits of the two groups respectively exactly correspond, and it therefore seems advisable, at least for the time being, to adopt a local term for the Wyre Forest beds. The Highley Group corresponds also, in a general manner, with the Coalport Group of the Coalbrookdale Coalfield.[3]

Lithological characters and sequence.—The Highley Group consists of grey, purple and mottled clays, with thick beds of sandstone like those of the Halesowen Group. The thickest and most persistent sandstones are

---

[1] ' A Boring for Coal at Claverley, near Bridgnorth,' *Trans. Inst. Min. Eng.*, vol. xlv, 1913, p. 37. For a list of plants (Reg. Nos. J.P. 4293-4382) from the Halesowen Group in this boring see R. Kidston, *Trans. Roy. Soc. Edinb.*, vol. li, 1927, p. 1,079.

[2] *Trans. Roy. Soc. Edinb.*, vol. li, 1917, p. 1,019.

[3] ' The Country between Wolverhampton and Oakengates ' (*Mem. Geol. Surv.*), 1928, p. 82.

near the base and can be recognized over most of the area to the west of the Severn, at Upper Arley and along the west and north sides of the Trimpley Anticline. Moreover, these sandstones near the base of the Highley Group are linked by the basal sandstone of the Halesowen Group in the Claverley borehole (p. 65) to that of the Halesowen Group in South Staffordshire (p. 63). Here, however, a difficulty arises ; for, as will be seen, there are, in the Highley Group, certain beds below these sandstones that seem to have no counterpart in South Staffordshire.

Thin seams of " sulphur." coal are numerous in the Highley Group and the three principal ones seem to have been worked to some extent,[1] though they are variable in thickness, and liable to split. Of these three seams the highest is that known as the Brock Hall (or Brockholes) Coal, up to 18 in. thick, about 300 ft. above the base of the group. The next, about 100 ft. lower, just above the " basal sandstones," is represented by a seam, 1 ft. 3 in. thick, called " Bind Seam " in the Highley Colliery shaft (Plate IV) and by a seam 1 ft. 7 in. thick in the Kinlet shaft. It can be recognized in most of the other sections in the district, including Shatterford and Compton. This seam, or one near it in position, has been locally called the " Main Sulphur Coal " and that name was applied to it in an abridged section of the Kinlet Pit published by E. A. N. Arber.[2] Further, by comparing the Kinlet and Highley sections with others in the Broseley district, using the Eardington Deep Pit (p. 74) and the Albynes borehole[3] as connecting links, it becomes probable that the seam in question corresponds with the " Main Sulphur Coal " of the Broseley and Coalport area.[4]

Below the sandstones near the base of the Highley Group comes, however, a group of clays, of varying thickness, in which lie one or more seams of " sulphur " coal. One of these seams crops out, and has been worked, at Bank Farm, about half a mile south-east of Severn Lodge Farm, and to avoid confusion it will herein be referred to as the Bank Farm Coal. This coal is present in the Eardington Deep Pit ; but both it and the thick sandstones above are absent in the Harpsford (or Harpswood) Pit,[5] probably owing to overlap (see below, p. 75). In the Bayton and Mamble area,[6] Sheet 182, the " Main Coal " lies below thick sandstones that seem to correspond with those above the Bank Farm Coal in the Highley area, suggesting a probable correlation.

The evidence from *Spirorbis* limestones in the Highley Group supports the above conclusions. Hitherto it has been supposed that there was only

---

[1] *Cf.* R. I. Murchison, ' Silurian System,' 1839, pp. 131-2.

[2] *Phil. Trans. Roy. Soc.*, vol. 204, 1914, p. 373.

[3] Situated 650 yds. east of the Albynes, near Astley Abbots. See ' The Country between Wolverhampton and Oakengates ' (*Mem. Geol. Surv.*), 1928, Appendix I, p. 222.

[4] ' Country between Wolverhampton and Oakengates ' (*Mem. Geol. Surv.*), 1928, p. 99.

[5] Situated 1⅜ m. north-west of Glazeley Church. See T. C. Cantrill, ' A contribution to the Geology of the Wyre Forest Coalfield,' 1895, p. 19.

[6] Cantrill, T. C., *op. cit.*, p. 21.

one such limestone within the present district.[1] It now seems clear that there are at least two. Arber[2] interpreted as a *Spirorbis* limestone a bed of " hard rock," 11 in. thick, 64 ft. 7 in. above the " Main Sulphur " coal in the Kinlet Shaft (see above, p. 66) ; in this he was probably right, for a shaly, impure limestone containing *Spirorbis* was found, in a corresponding position, in the borehole put down in 1932 near Ingram Lane (p. 74), about half a mile west-south-west of Hampton. A " hard light rock " is recorded in a similar position in the Highley Shaft (Plate IV), but no such rock can be recognized in the record of Eardington Deep Pit.

The Ingram Lane limestone is probably the same as the one recorded in the Albynes boring and at Harpsford which is clearly the same as that long known at Broseley, above the " Main Sulphur Coal " of that area.

At Bank Farm, on the other hand, a *Spirorbis* limestone crops out between the thick sandstone belt and the Bank Farm Coal (p. 66). This limestone occurs again at Skeets Farm (300 yds. west of Arley Station) and can be traced thence to Woodhouse Farm. It has also been found near Winwoods (Sheet 182), about a mile south of Buttonbridge, where, again, it lies below a thick sandstone of Halesowen type and above a coal seam (p. 77). This Bank Farm limestone has hitherto been assumed to be the same bed as that of Harpsford and Tasley, and thus correlated with that of Broseley ;[3] but it would seem clear that they are distinct beds. On the other hand it is probable that the Bank Farm limestone does correspond with the *Spirorbis* limestone of the Bayton and Mamble area (Sheet 182), which lies above the local " Main Coal " (Plate IV) and below the Thick Sandstone of that area.

To sum up, the name " Main Sulphur Coal " has apparently been applied to two distinct coal seams in that part of the Wyre Forest Coalfield which lies within the present district, and each of these coal seams is associated with a *Spirorbis* limestone. The higher coal seam and limestone, which lie above the thick sandstone belt already several times mentioned, may be correlated with the " Main Sulphur Coal " and limestone of Broseley. The lower coal seam and limestone, below the thick sandstone belt, may be correlated with the " Main Coal " and limestone of the Bayton area. It follows that the " Main Sulphur Coal " and limestone of Broseley are not the same as the " Main Coal " and limestone of Bayton.

These conclusions, and their bearing upon the stratigraphy of the Highley Group, are summarized in the subjoined table, and are further illustrated by the comparative sections in Plate IV.

---

[1] See D. Jones, ' The Spirorbis Limestone of the Forest of Wyre Coalfield,' *Trans. Manchester Geol. Soc.*, vol. x, 1871, p. 37 ; and ' The Structure of the Forest of Wyre Coalfield,' *Trans. Fed. Inst. Min. Eng.*, vol. vii, 1894, p. 287 ; also T. C. Cantrill, *op. cit.*, pp. 17, 22.

[2] *Loc. cit.*

[3] See D. Jones, *op. cit.*, and T. C. Cantrill, *op. cit.*

Highley Group
Generalized Sequence

	Ft.
Red, mottled and blue clays, with thin sandstones and one or more thin seams of sulphur coal ...	100 or more
Blue clays and sandstones, with one or more thin sulphur coals	100
SULPHUR COAL = BROCK HALL COAL, in places in more than one seam, with partings	up to 1½
Blue, red and mottled clays, with one or more thin sulphur coals and a *Spirorbis* limestone	100
SULPHUR COAL, ' BIND SEAM ' (= ' Main Sulphur Coal ' of Broseley and Coalbrookdale), in places in more than one seam, with partings...	up to 3
Grey clays, with thin sulphur coals	up to 40
Thick sandstones and pebbly sandstones, with thinner beds of grey, mottled or red clay... ...	up to 150
Grey clays, with red and mottled clays in places, a *Spirorbis* limestone and thin sulphur coals, including the BANK FARM COAL (= ' Main Coal ' of Bayton and Mamble) ... thickness variable	20—110
Average thickness of group	400

Associated unconformities.—It has long been held that an unconformity exists between the " Sulphur Coal Group " and the " Sweet Coal Group " of the Forest of Wyre Coalfield, though, as the base of the former had never been clearly delimited, it was not possible to demonstrate this unconformity or its effects with any certainty. With the Highley Group defined as above the effect of an unconformity at its base can be clearly seen. The thickness of beds between the base of that group and the Brooch Coal at Highley and Kinlet shafts is about 450 ft. At Billingsley Colliery the thickness is reduced to 170 ft. It may be noted that, notwithstanding this reduction in thickness, the clays with the Bank Farm Coal are present at Billingsley ; which indicates that they belong to the Highley Group and not to the underlying Kinlet Group.

The overstep of the Highley Group takes place towards the west, and northward towards the Linley Anticline.[1] In the Eardington Deep Pit the whole of the Kinlet Group is absent, the Highley Group resting directly upon the " Old Red Sandstone." Nevertheless the thick sandstones near the base of the Highley Group remain, and also the clays below, with a coal, 2 ft. 3 in. thick, that no doubt represents the Bank Farm Coal. The overstep by the Highley Group is accompanied by overlap within that group, for in the Harpsford Pit, about one and three-quarter miles west of the Eardington Deep Pit, the lower part of the Highley Group, including the thick sandstones and the clays with the Bank Farm Coal, is absent. These clays are absent in the Albynes borehole (p. 20) also, and the top part only of the thick sandstones seems to be present, resting directly upon Silurian (Downtonian) rocks.

---

[1] Robertson, T., in ' The Country between Wolverhampton and Oakengates ' (*Mem. Geol. Surv.*), 1928, p. 161.

UPPER COAL MEASURES : HIGHLEY GROUP.

If comparison be carried farther north, to the Broseley and Madeley areas (Sheet 153), it is found that the lower part of the Highley Group, including the thick sandstones and the clays with the Bank Farm Coal, occupies the position of the brick and tile clays and the red clays with espley-like grits that form the lowest part of the Coalport Group.[1] This lends support to the suggestion[2] that the red brick clays, with espleys, of the Coalbrookdale Coalfield, which lie above the " Symon Fault " unconformity, represent, stratigraphically, part of the Halesowen Group, rather than of the Etruria Marl Group.

Overstep of the Highley Group also occurs towards the south where, in the Bayton and Mamble area (Sheet 182), the Kinlet Group is absent and the Highley Group, with the " Thick Sandstone " and underlying *Spirorbis* limestone and " Main Coal," rests on the Old Red Sandstone.

In addition to unconformity at the base of the Highley Group there may be some within the group. The variable thickness of the clays with the Bank Farm Coal, and the apparent absence of the Bank Farm *Spirorbis* Limestone in the Highley and Kinlet pits and elsewhere north of its outcrop at Bank Farm, require explanation. They are perhaps due to slight unconformity, with penecontemporaneous erosion, at a plane represented by the base of the thick sandstone belt. Another possible explanation is that the Highley Group was deposited upon an uneven surface, and that the lowest beds were laid down only in the hollows, deposition failing on the intervening ridges. Against this, however, is the circumstance that the Bank Farm Coal appears to be considerably more persistent than the associated *Spirorbis* limestone, although it lies below the latter.

Upper limit.—Stream sections near Arley and south-west of Nomans Green show that the Highley Group, like the Halesowen Group, passes up gradually into the Keele Group, there being a transition series of grey, greenish, lavender and purple marls and sandstones. Near Bridgnorth, however, it is possible that some overstep on the part of the Keele Group may take place, for the thickness of the upper part of the Highley Group, i.e., above the Brock Hall Coal. seems to be rather less than it is to the south and south-east. Due allowance must be made, however, for the difficulty of fixing a consistent base for the Keele Group amongst the transitional beds ; and it is probable that the narrowness of the outcrop of the Highley Group northward from Billingsley is due less to any overstep by the Keele Group than to overlap of the lower beds of the Highley Group (see also p. 73, below).

Fossils.—Of the fossils in the Highley Group, animal remains[3] have, at present, been found only in the *Spirorbis* limestones. The annelid *Spirorbis pusillus* (Martin) is generally abundant in both beds. The fauna of the lower limestone at Bank Farm differs from that at Winwoods in containing many fish-fragments, including scales of a species of *Rhabdo-*

---

[1] *Ibid.*, p. 82.
[2] *Ibid.*, p. 31, footnote 6.
[3] Of the fossils collected by the Geological Survey the fish remains have been named by Prof. D. M. S. Watson and the others by Dr. C. J. Stubblefield.

derma and of Platysomids, but comparatively few ostracods. In the limestone at Winwoods ostracods are abundant and include *Carbonita fabulina* (Jones & Kirkby), *C. rankiniana* Jones & Kirkby sp. (both also found at Bank Farm) and *C. secans* (Jones & Kirkby). At Winwoods the non-marine mollusc *Anthracomya sp.* was also found. The difference in fauna of the lower limestone at Bank Farm and Winwoods respectively may probably be correlated with the difference in lithology, the former being a pale putty-coloured limestone of considerable purity, the latter a dark-grey, less pure limestone. The limestone of the Deuxhill outlier, which is the upper bed (p. 75), contains a fauna similar to that at Winwoods, with abundant ostracods, including *Carbonita fabulina*, and fragments of *Anthracomya* and of *Anthraconauta* cf. *phillipsii* (Williamson).[1] It need, perhaps, occasion no surprise that the two limestones, believed to be distinct beds, should contain the same fauna ; for they are separated by only some 250 ft. of strata of a type the deposition of which would hardly occupy sufficient time for a perceptible faunal change to take place.

Lists of plant remains from the " Sulphur Coal Group " have been published by E. A. N. Arber[2] and by R. Kidston.[3] Of the localities from which these fossils came, none of Arber's and only two[4] of Kidston's lie within the present district. The species[5] recorded by Kidston are :—

Sphenophyllum emarginatum Brongniart, from Borle Brook, 850 yds. N.15°E. of Wadeley Farm, three quarters of a mile N.N.E. of Billingsley Church ; *Neuropteris rarinervis* Bunbury and *Sphenophyllum emarginatum* Brongn., from Borle Brook, 1,000 yds. N.10°E. of Wadeley Farm.

The horizon at both localities is stated to be not far above the " Main Sulphur Coal," which in this case probably indicates the " Bind Seam " (p. 66).

From these, and other species obtained outside the present district, chiefly from beds associated with the " Main Coal " in the Bayton, Mamble and Pensax areas, Kidston[6] concluded that the beds of the " Sulphur Coal Group " must be referred to the Newcastle Group, and that " . . . the suggestion made by Cantrill, Prof. Lapworth and Prof. Watts,[7] that the Sulphur Coals belonged to the Newcastle Group or Halesowen Sandstones . . . is quite correct." He expressed himself as

---

[1] The same fauna, including *A.* cf. *phillipsii*, was found in the limestone of the Broseley district at an exposure in Linley Brook 220 yds. west-south-west of Linley station. See T. Robertson in ' The Country between Wolverhampton and Oakengates ' (*Mem. Geol. Surv.*), 1928, p. 104.

[2] ' On the Fossil Floras of the Wyre Forest . . . ,' *Phil. Trans. Roy. Soc.*, Series B, vol. 204, 1914, p. 363 (see pp. 387–8).

[3] ' The Forest of Wyre and the Titterstone Clee Hill Coal Fields,' *Trans. Roy. Soc. Edinb.*, vol. li, 1917, p. 999 (see p. 1,020).

[4] Specimens from New Pool Quarry, W. side of main road, ½ mile S.E. of Button Bridge (Kidston, *op. cit.*, p. 1,021), are excluded as detailed mapping shows that the beds in that locality lie in the Kinlet (' Sweet Coal ') Group.

[5] The specimens are in the collection of the Geological Survey, Reg. Nos. R.E. 243–7, R.E. 223–42.

[6] *Op. cit.*, p. 1,035.

[7] ' Shropshire ' in ' Geology in the Field,' *Jubilee Vol. of Geol. Assoc.*, 1910, p. 739.

unable to accept Arber's view that " the Sulphur Coal Series of the Wyre Forest belong to the same horizon as the Red Clay Group (Etruria Marl) of South Staffordshire, and may be the actual representative of that very series."

During the recent survey of the district plant remains were collected by Mr. Dewar from the Borle Brook at localities not far from those mentioned above (p. 70). Dr. Crookall, who has named these plants (p. 74), considers that they indicate a horizon " high in the Staffordian," which is in harmony with Kidston's conclusion.

Economic products.—Apart from the coals, the only beds of economic importance in the Highley Group are the *Spirorbis* limestones, which have been worked and burnt for lime, and the sandstones, some of which have been rather extensively quarried for building stone.

T.H.W., R.W.P.

DETAILS

Severn Lodge, Tiphouse and Highley.—The distribution of the outcrops in this area is determined by three principal faults, the Tiphouse, the Highley-Kinlet, and the Station faults. The first has a westerly throw of about 50 yds. and on its western side the thick sandstone near the base of the Highley Group forms the banks of the stream 600 yds. west of Tiphouse Farm and dips at about 18° N.E. Below the sandstone coal about a foot thick crops out and a few feet below this a brecciated calcareous bed, possibly the equivalent of the Bank Farm *Spirorbis* Limestone, can be seen. The outcrop of the sandstone runs southwards between Winnal Farm and Birch Farm, and north-westwards to form the steep bank overlooking the drive to Kinlet Hall, just west of Nortonsend Farm, and is underlain by a 9-in. coal at the junction of the stream from Kinlet Park with that which flows through the northern side of Old Coppice.

The Borle Mill boring, 400 yds. N.E. of Nortonsend Farm, passed through the sandstone, here 51 ft. thick, and just below it a 4-in. coal was recorded.

To the east of the Tiphouse Fault, which crosses the brook 350 yds. W.S.W. of Tiphouse Farm, the thick sandstone is exposed again in the steep bank south of the farm with a dip of between 10° and 20° N.N.E. Below the sandstone calcareous tufa in the stream bank 300 yds. S.W. of the farm may indicate the outcrop of the Bank Farm *Spirorbis* Limestone. A coal about 18 in. thick crops out at several points along the brook-course and there is probably a second coal at a slightly lower level. These coals appear to be on the horizon of the Bank Farm Coal. They have been worked in the southern slope of the valley north of Birch Farm, in the rough ground known as Coal Pit Slang north-east of the farm and also to the north of Earnwood 550 yds. S.E. of Birch Farm. Cantrill[2] noted large ironstone cakes on the spoil banks near Earnwood and regards the coal worked there as probably the Brock Hall Coal, but its position below the thick sandstone indicates that it is on the horizon of the Bank Farm Coal nearly 300 ft. lower, as shown by the record of Kinlet Shaft east of Tiphouse Farm.

Eastwards from Earnwood the coal horizon cannot be followed with any certainty but there is little doubt that it is represented by exposures in the brook south of Severn Lodge. In the shaft of Kinlet Colliery about 130 ft. of sandstone, in part pebbly, are recorded above two coals, but no *Spirorbis* limestone is mentioned.

---

[1] Arber, E. A. N., *op. cit.*, p. 429.
[2] ' A Contribution to the Geology of the Wyre Forest Coalfield,' 1895, p. 23.

West of the Tiphouse Fault a higher bed of sandstone dipping about 10° N.E. can be seen along the stream 800 to 900 yds. N.E. of Nortonsend Farm. This would appear to be the sandstone recorded in the Highley and Kinlet shafts above the Brock Hall Seam. It is exposed in the new road to Highley, where it is cut off by the Tiphouse Fault.

A third and higher sandstone can be traced from the mill north-westwards to New England near where a large quarry shows some 40 ft. of massive grey and yellow sandstone dipping about 20° N.E. This is apparently the same bed as that which crops out at Highley Station.

On the east side of the Tiphouse Fault, an outcrop of coal in the bank of the new road to Highley some 500 yds. N. 10° W. of Tiphouse Farm is somewhat obscured by slips, but the seam appears to be a foot or more in thickness and its position in relation to the sandstones above and below suggests that it is the Bind Seam.

The Borle Brook cuts across the outcrop of the sandstone which, in the Kinlet and Highley shafts, lies above the Brock Hall Seam. A quarry 550 yds. N.W. of Kinlet Colliery, close to the new road, shows :—

	Ft.	in.
Shaly and massive grey-brown sandstone with plant films	15	0
Ferruginous marly sandstone	1	0
Coal [Brock Hall]	1	0
Grey and yellow fireclay ... ... seen for	1	0

The road cutting between 300 and 400 yds. N.W. of the colliery shows a similar section (Survey Photograph 6880).

This sandstone outcrop trends south-eastwards towards Severn Lodge, just south-west of which place it is cut off by the Highley-Kinlet Fault, while northwards it follows the eastern bank of Borle Brook.

A higher bed of sandstone is exposed in road banks round Netherton.

To the east of the Highley-Kinlet Fault the Brock Hall Seam, with the sandstone overlying it, is again exposed in the new road 800 yds. east by north of Kinlet Colliery and the sandstone occupies the rising ground south of Borle Brook, in the angle between it and the River Severn. It is cut off to the east by the Station Fault. In the Highley Colliery shaft the Brock Hall Seam, 1 ft. 10 in. thick, occurs at a depth of about 100 ft. and at 240 ft. sandstone about 150 ft. thick overlying the Bank Farm Coals was entered.

East of Highley Colliery a higher sandstone crops out. Some 30 ft. of this rather massive yellowish-grey sandstone have been quarried at Highley Station and in several quarries (Survey Photograph 6881) south-east of the colliery as far as the northern end of Stanley railway-cutting, where the outcrop ends against the Station Fault. A similar sandstone is exposed at several points, west of the railway, for about three-quarters of a mile north of the Vicarage.

A block of Highley beds is let in between the Kinlet Hall Fault and the Brock Hall Fault. The thick sandstone near the base is not well exposed but appears to form the steep northern side of the valley south-west of Rays Farm. It was passed through in a borehole put down about 250 yds. N. 25° W. of the farm. The thickness proved was 75 ft. between the depths of 235 and 310 ft. and it was underlain by a group of three coals with fireclays. The lowest coal, 1 ft. 4 in. thick, is correlated with the Bank Farm Coal, while a fireclay ' lime rib ' between the two upper coals may represent the Bank Farm *Spirorbis* Limestone.

In the same boring a limestone was struck at a depth of 20 ft. and a group of coals at between 120 ft. and 150 ft. The position of these coals above the thick sandstone strongly suggests that they represent the Bind Seam (Main

UPPER COAL MEASURES : HIGHLEY GROUP.

Sulphur of Cantrill), while the limestone is supposed to be that between this coal and the Brock Hall Seam.

The Brock Hall Seam, with sandstone above it, crops out in the road bank 750 yds. south of the Cape of Good Hope Inn and was worked by adit and shaft.

At Bind Farm and westwards to the plantation the soil is a red marly clay and this ground appears on the map as Keele beds ; it seems more reasonable, however, to regard it as the outcrop of red and mottled marls in the Highley Group such as were encountered in the boring at the Reservoir, Ingram Lane, near Sutton.

The Brock Hall Fault, at a point 500 yds. south of the Cape of Good Hope Inn, divides into two branches which come together again near Bagginswood and a small area of Highley Beds is included in the loop. The thick sandstone near the base of the group is crossed by the brook about 600 yds. S.E. of High Green and there are old workings, said to have been in a sulphur coal, alongside the little brook, known as Tiddle Brook, near Scot's Farm, 700 yds. W.N.W. of Rays Farm ; here also, according to an old record, a thick bed of limestone crops out by the brook and is probably that above the Bind Seam met with in the boring, mentioned above, near Rays Farm.

Billingsley.—An old record of the Billingsley Engine Pit, the site of which is probably in the rough ground south-west of the Cape of Good Hope Inn, gives the depth to the Sulphur Coal as 144 ft. and mentions soil, clay, binds, grey rocks (the one above the coal bringing water) and clunch containing poor ironstone. The details of the seam, which is apparently the Bank Farm Coal, are :—Top sulphur coal, 8 in. ; working clod, 3 ft. ; sulphur coal, 6 in. ; clod, 9 in. ; sulphur coal, 2 ft. 7 in. ; clod, 10 in. ; strong hard coal, 4 in.

At the Billingsley Colliery the record of No. 2 Shaft shows a 2-ft. coal, the Brock Hall Seam, at a depth of about 40 ft., a group of coals at about 100 ft., which may be correlated with the Bind Seam, and at about 270 ft. a sulphur coal, 3 ft. 6 in. thick, known as the " Stanley Seam," which we correlate with the Bank Farm Coal. The Stanley Seam, together with the Rider Coal, some 12 ft. above it, lies below 84 ft. of rock, apparently the thick sandstone of the Highley region. The dip of the beds in the shaft is stated as 31°, so that the sandstone would have a true thickness of about 71 ft.

There are many old workings in the rough ground west of the main road, but it is not possible to trace the actual outcrops of the Brock Hall and Bind seams.

Northwards towards Billingsley Hall Farm the lower beds of the Highley Group, including the Bank Farm Coal and the thick sandstone, appear to be overlapped by the higher beds of the group, which also overstep the greater part of the Kinlet Group with the result that, in the brook north of the farm, beds of the Highley Group almost rest on the Old Red Sandstone, and still farther north, at a point six furlongs east of Deuxhill, the Kinlet Group is completely overstepped by them.

Borle Brook and Chelmarsh Ridge.—From the Brock Hall Fault for two and a half miles northwards to Glazeley Bridge the Borle Brook follows the strike of the Highley Group and almost the outcrop of the Bind Seam (see p. 66), which can be detected at intervals in both banks. On the steep eastern side of the valley, deeply notched by small tributary streams which drain the western flank of Chelmarsh Ridge (Survey Photographs 6877-8) and capped by the Keele beds, there crops out a thick grey sandstone, just above the Brock Hall Coal, which can be seen at many points. Between the Brock Hall Coal and the Bind Seam in the bottom of the valley the crop of the *Spirorbis* limestone is obscured by the broken nature of the ground and the debris from old crop workings in the Brock Hall Seam. Murchison, however, appears to have seen the limestone, for he says[1]: "In the bed of the brook and dipping beneath this coal is a band about a foot and a half thick, made up of concretions of impure limestone . . . The lowest coal [Bind Seam] has been worked at a depth of about forty yards beneath the limestone, and is two feet six inches thick."

At the head of the dingle half a mile north-west of Brock Hall there are indications of another band of *Spirorbis* limestone above the sandstone.

A boring put down, on the Chelmarsh Ridge, in 1932, at the Reservoir, Ingram Lane, Sutton, passed through about 130 ft. of Keele beds into Highley Group measures and found a coal, probably the Brock Hall, about 15 in. thick, at a depth of 498 ft. Two thin beds of *Spirorbis* limestone were encountered at 557 ft. and 578 ft. and a coal 3 ft. 6 in. thick at 643 ft., which is probably the Bind Seam. Below this seam came a thick group of sandstones with fireclays, totalling some 150 ft., comparable with the thick sandstone above the Bank Farm Coal. No coal was found at the Bank Farm horizon, though fireclays are present, and the sandstone group rests on some 30 ft. of mottled marl, the highest bed of the underlying Kinlet Group.

Shales associated with the Bind Seam and well exposed in the bank of Borle Brook 160 yds. south of Covert Lane Bridge and 600 yds N.E. of Hook Farm, furnished :—*Asterotheca oreopteridia* (Schlotheim), *Eupecopteris sp.*, *Neuropteris* cf. *rarinervis* Bunbury, *N. scheuchzeri* Hoffmann, *Annularia stellata* Schlotheim, and *Sphenophyllum emarginatum* (Brongniart). A section upstream, at the junction with Crunells Brook, yielded the same forms with the addition of *Linopteris obliqua* Bunbury, *Mariopteris sp.* and *Neuropteris ovata* Hofmann.

North of Glazeley Bridge, in the coppice half a mile west of Woodlands, a 6-in. coal is exposed, and at the Foxholes, 400 yds. east of The Hill, a thin coal overlies sandy grey shale and flaggy sandstone. Traces of coal are also to be seen in Mor Brook north-east of The Hill.

Eardington Pit and Boring.—The Eardington Deep Pit, sunk in 1843 in the valley of Mor Brook, 450 yds. upstream from Eardington Mill, passed through 456 ft. of Coal Measures and 330 ft. of red and blue clay with rock ; probably Old Red Sandstone. A boring continued a further 180 ft. in similar beds.

The Coal Measures belong to the Highley Group, with the exception of perhaps the top 100 ft., which were mainly red, brown and blue binds and clunches. A bright coal at 250 ft., probably the Brock Hall Seam, was slightly worked, although only 1 ft. 6 in. thick. A group of thin coals between 320 ft. and 347 ft. is regarded as representing the Bind Seam, while a sandstone about 80 ft. thick may be correlated with that above the Bank Farm Coal. A coal 2 ft. 3 in. thick occurs below the sandstone at a depth of 446 ft. from the surface ; it rests on 10 ft. of fireclay and dark binds, which appear to be the lowest Coal Measures in the shaft. Between the shaft and the nearest outcrop of Highley Beds at Eudon Burnell the lowest beds of the Highley Group have been cut out by overlap.

Deuxhill Outlier.—A fault running northward from Deuxhill to near Harpswood brings in to the west an outlier of the Highley Group about two and a half miles long and less than half a mile wide. The best section of the measures is furnished by the shaft of Harpsford Colliery near the northern end of the outlier ; the exact position of this shaft is not known.

The only coal known to have been worked lies only 6 ft. above the base of the measures, which rest on the Old Red Sandstone, and has been got all along the outcrop at Hollycott, Little Scotland, Eudon George, Tedstill and

---

[1] 'Silurian System,' 1839, p. 132.

UPPER COAL MEASURES: HIGHLEY GROUP.

Deuxhill. It has the following section, which is similar to that of the Bind Seam of Eardington with which it is correlated:—

	Ft.	In.
Coal (best quality)		6
Clod		4
Coal		6
Parting		½
Coal (second best)	1	6
Tough clod		6
Coal (bottoms, very sulphureous)		6
Parting		½
Coal (Poundstone coal)		3

The coal rests on 3 ft. of fireclay, which was often heaved up by the pressure of water percolating from the outcrop.

The *Spirorbis* limestone, some 63 ft. above the coal, is 3 ft. thick in the Harpsford Shaft. Its outcrop can be traced mainly by surface debris, but in the Borle Brook, which crosses the outlier, it is seen *in situ*. Following the stream eastward from Down Mill Bridge, which is on the Old Red Sandstone, we find at 160 yds. from the bridge the bastard cornstone which is recorded at the base of the Coal Measures in the Harpsford Shaft. At about 220 yds. from the bridge the Bind Seam is seen resting on fireclay. At 320 yds. the limestone is exposed, associated with grey shale and sandstone. A higher coal, 3 in. thick, crops out 440 yds. eastward of the bridge and at 680 yds. a still higher coal is indicated by fragments in the bank. The eastern boundary fault of the outlier is evident about 150 yds. S.E. of the ford; and again in the tributary stream from Eudon George, 150 yds. S.W. of its junction with Borle Brook. One of the higher coals can be seen about 300 yds. upstream from Borle Brook.

At Deuxhill there is some evidence of the outcrop of the Bind Seam and traces of old workings, but there is not here a sufficient thickness of strata above the coal to include the *Spirorbis* limestone.

The absence of the thick sandstone horizon and of the Bank Farm Coal below the Bind Seam indicates overlap of the higher measures westwards from Eardington and overstep onto the Old Red Sandstone.

In the main outcrop northward from the Mor Brook valley there is no sign of the *Spirorbis* limestone or of the underlying coal until we cross the Bridgnorth-Morville road, and this may be due to absence of still more of the lower beds of the Highley Group owing to overlap. It is significant that in this region the base of the Keele Beds comes very close to the Old Red Sandstone outcrop.

Tasley.—From the Racecourse Farm near Tasley, northwards to the Albynes (Hobbins of the Old Series Map), there is good evidence of the outcrop of the *Spirorbis* limestone with, below it, occasional indications of a coal crop. There are also signs of a coal a short distance above the limestone. Murchison[1] gives a section of the beds exposed in an open work near Tasley as follows:—

	Ft.	in.
Green and yellow shale, decomposing to stiff clay, passing into a thick ferrugino-calcareous layer	6	0
Limestone, compact and cream-coloured, with *Microconchus carbonarius* and *Cypris*. (The rock of Pontesbury and Le Botwood)	3	0
Blue and grey shale	3	0

---

[1] 'Silurian System,' 1839, p. 100.

	Ft.	In.
Sandstone of greyish colour with carbonaceous matter and fragments of plants	4	0
Blue bind or carbonaceous shale forming the top of coal	2	0
Smut or impure coal		6
Clod or argillaceous shale		2
Coal	1	4
Stiff mottled red and green shale, depth unproved		

In the Albynes boring (p. 20) the limestone, 3 ft. 2 in. thick, was met with at a depth of 124 ft., and a coal only 4 in. thick about 40 ft. below the limestone. Below the coal there are some 30 ft. of blue shale and grey sandstone resting on Silurian rocks.

R.W.P.

Upper Arley (west of R. Severn).—In the north-west bank of the brook that forms the county boundary south-east of Severn Lodge Farm greenish and buff sandstones crop out which are part of the thick sandstone belt near the base of the Highley Group. Below them, in the brook at a point 450 yds. S.S.E. of Severn Lodge Farm, the Bank Farm *Spirorbis* Limestone (p. 67) crops out. It is here a putty-coloured limestone in which *Spirorbis* is abundant and in which fish remains also occur. It is probably about 2 ft. thick, but the full thickness is not seen owing to slipping of clay in the banks. Below the limestone a few feet of grey clay, mottled with purple, can be seen.

An outlier of the sandstones just north of Bank Farm[1] has been trenched by the railway, and the cutting and an old quarry show greenish-grey, medium-grained, false-bedded sandstones. An old adit to the coal beneath the sandstone opens on the west side of the railway. Many traces of coal workings occur between the railway and Bank Farm, and can be followed to the county boundary brook, where further spoil-banks and heaps of burnt clay can be seen about 460 yds. nearly due south of Severn Lodge Farm. The coal seam is not actually exposed in the stream, but its position is evidently below (i.e., west of) that of the limestone. Traces of the latter occur below the outcrop of the overlying sandstone, between the stream and Bank Farm, where, according to T. C. Cantrill,[2] it has been burnt for lime. The clays in which the limestone and coal lie were formerly dug for brick-making. All the beds above described are cut off north-east of Bank Farm by the Station Fault (p. 129) but reappear near Arley Station, and a seam, probably the Bank Farm Coal, crops out in the railway cutting a few yards south-east of the road bridge. The outcrop is now largely overgrown, but Cantrill[3] estimated the thickness as 2 ft. 7 in. He states that coal was formerly worked at Skeets, i.e., Skeets Cottages, 300 yds. west of Arley Station. In a field 350 yds. S.W. of Arley Station coal was ploughed up in 1932, and lumps of *Spirorbis* limestone were found on the surface. A strong feature immediately to the west is evidently due to the bottom part of the overlying sandstone. The outcrops, shifted slightly in two places by the Arley Park Fault (p. 129), can be traced by old workings and occasional fragments of limestone to Woodhouse Farm, near which both coal and limestone were cut through by the Elan pipe-trench. The position at which a coal seam, said to be 1 ft. thick, had been found in the trench for a third pipe-line was indicated by a workman in 1932 and much coal debris remained on the surface at that point.

Many pieces of the limestone, here dark grey in colour, lie on the spoil bank 400 yds. S.E. of Woodhouse Farm, with *Spirorbis* hardly less abundant than at Bank Farm.

---

[1] Bank Farm, not named on the one-inch map, is 670 yds. S.W. of Severn Lodge Farm.

[2] ' A Contribution to the Geology of the Wyre Forest Coalfield,' 1895, p. 22.

[3] *Op. cit.*, p. 20.

UPPER COAL MEASURES : HIGHLEY GROUP.

The outcrop of the coal can be traced by old workings to the Bewdley road about half a mile E.S.E. of Button Oak (Sheet 182), where it appears to be cut off by the Station and other faults.

The sandstone overlying the clays that contain the Bank Farm Coal crops out at Bannut Tree Farm, where it is exposed in a roadside quarry, and to the east of Pound Green Common. Higher beds of sandstone crop out and have been quarried west of Skeets. The outcrops of all these beds are cut off to the west by the Station Fault.

Buttonbridge.—Beds of the Highley Group crop out in a down-faulted wedge between the Tiphouse Fault (p. 130) and another fault that meets it a short distance north of Buttonbridge. Greenish-grey, flaggy sandstones, on grey shales, are exposed in a stream about 400 yds. south of Buttonbridge, and are folded in a shallow syncline. Farther south thick sandstones crop out on the southern border of the present district. They are well exposed in the brook just south of the border and in a lane 300 to 450 yds. east of Winwoods (Sheet 182). They are grey-green sandstones, weathering buff, mainly in rather thick beds, resembling those of the Halesowen Group. About 20 ft. below them a dark grey *Spirorbis* limestone crops out in the stream 600 yds. N.E. of Winwoods. In the corner of the wood south-east of where the limestone was found, traces of coal were seen. There is little doubt that the limestone and coal correspond to those at Bank Farm.

T.H.W.

Bradley and Meaton.—South of Bradley beds of the Highley Group form an outlier a mile and a half long from north to south and three-quarters of a mile in maximum width. Only the northern half of the outlier lies within the present district and the beds are poorly exposed. They overlie purple-red marls and sandstones of espley type with conglomerates and their lowest division consists of clays with thin sandstones and a seam of coal, said to be 3 ft. 6 in. thick where cut by the Elan Aqueduct at the south-eastern end of the outlier. The outcrop of this coal, believed to be the Bank Farm Coal, although no trace of the *Spirorbis* limestone has been found above it, can be traced at intervals all along the eastern margin of the outlier but would appear to be overlapped towards the west, as it is not seen on the western side of the outlier.

These beds are succeeded by a fairly thick yellow sandstone of Halesowen type, which also does not appear to be represented on the western side. This sandstone, presumably that above the coal at Bank Farm, is followed by clays and thin sandstones with a seam of coal at the top which can be traced by crop workings all along the east side of the outlier and also on the north-west.

There thus appears to be evidence in this area of the unconformity at the base of the Highley Group and of an overlap of the higher beds towards the west, as in the case of the Deuxhill outlier (p. 75).

R.W.P.

Upper Arley (east of R. Severn) and Shatterford.—The position of the Bank Farm Coal near Arley village is uncertain but its outcrop probably crosses the river near the ferry. Cantrill[1] was informed that coal had been raked from the river opposite Severnfield Cottages, which are 400 yds. E.S.E. of the ferry, and he evidently considered this to come from one of the three thin seams that he found in the brook above Worralls.[2] Traces of two seams were seen there in 1932. The lower, some 10 in. thick, about 250 yds. above the mouth of the brook, lies in grey clays resting upon greenish gritty sandstones which appear to extend northward along the east bank of the Severn to a point opposite Severnfield Cottages. Any coal raked from the Severn hereabouts would

---

[1] *Op. cit.*, p. 20.

[2] Worralls is near the mouth of the brook that runs along the north-west border of Eymore Wood, and forms a parish boundary.

presumably lie below them and thus not correspond to the seams in the Worralls brook. It is, however, possible that a small fault may run down the gully, just north of where these sandstones are last seen, and bring the coal down to river level.

South of Worralls brook the outcrop is obscure, but it probably runs west and south of Huntsfield Cottage (700 yds. N.N.E. of Eymore Farm) where, as recorded by Cantrill, two thin seams were found in a well; one 8 in. thick, at a depth of 60 ft., the other 30 ft. lower and said to be thicker. Traces of coal occur in the brook 150 yds. S.S.E. of Huntsfield Cottage, and from there the coal can be followed by numerous old workings. About 300 yds. S.E. of Huntsfield Farm (850 yds. N.N.E. of Eymore Farm) the outcrop is shifted west-north-westward by a fault, beyond which it can be followed by old workings north-eastward to a tributary of the Worralls brook where, at a point 350 yds. E.N.E. of Huntsfield Farm, a 10-in. coal was found in grey clays. From here the outcrop can be followed, by old workings, through Eymore Wood to Shatterford.

The subjoined section[1] shows the lower part of the Highley Group in the Shatterford Deep Pit :—

	Recorded thickness	Corrected thickness	Depth from surface
	Ft. In.	Ft. In.	Ft. In.
26. Brown and white rock	75 6 ⎫	57 0	
27. Conglomerate ...	6 0 ⎭		374 8
28. White rock binds	18 0	12 6	
29. Limestone balls	1 0	8	393 8
30–33. Dirt, mottled ground and binds...	53 0	37 0	
34. Cakes of ironstone	3	2	
35. COAL	1 8	1 2	448 7
36–38. Fireclay and dark binds ...	54 0	38 0	
39. Thin rock, brown	1 6	1 0	
40, 41. Dark binds and clay	19 0	13 0	525 1
42. ROOF COAL	9 ⎫		
43. Holing clay	1 6 ⎬	3 6	
44. COAL	2 8 ⎭		530 0
Fireclay	5 0	3 6	535 0

No. 44 in the above section is the Bench Coal, which D. Jones[2] regarded as representing the 'Main Sulphur.' Cantrill[3] considered that Nos. 42–44 together represented the Main Sulphur Coal. Further, he thought the thin brown rock (No. 39) might be the *Spirorbis* limestone, and identified the coal above it (No. 35) as the Brock Hall Coal. This can hardly be accepted, for there can be little doubt that Nos. 26 and 27 are part of the thick sandstone belt that, in the Highley area, lies near the base of the Highley Group and well below the Brock Hall Coal (Plate IV). No. 35 may possibly represent the Rider Coal of the Bayton and Mamble area; whilst Nos. 42–44 may probably be identified with the 'Main Coal' of that area and with the Bank Farm Coal. It is noteworthy that at Shatterford the thickness of the beds below the thick

---

[1] From G. E. Roberts, 'The Geologist,' vol. iv, 1861, pp. 422–23. The numbering of the items is Roberts'. The figures in the second column from the right are thicknesses corrected for an average dip of 45° (see D. Jones, *loc. infra cit.*, and a section by J. M. Fellows in *Proc. Dudley and Midland Geol. and Sci. Soc.*, vol. iv, 1893, facing p. 1).

[2] *Trans. Fed. Inst. Min. Eng.*, vol. vii, 1893, p. 299. By a presumed oversight the depth to the Bench Coal is there given as 288 ft.

[3] *Op. cit.*, p. 21.

UPPER COAL MEASURES : HIGHLEY GROUP.

sandstones[1] corresponds more nearly with that of the apparently equivalent beds in the Bayton and Mamble area than with that of those in the Highley area, where they are considerably thinner.

Nodules of purplish-brown limestone, which did not yield *Spirorbis*, were found in the Worralls brook about 120 yds. above the outcrop of the 10-in. coal mentioned on p. 77 ; but, apart from this possible indication, the Bank Farm *Spirorbis* Limestone has never been found between Arley and Shatterford.[2] It seems probable that it is represented by the ' limestone balls ' forming item No. 29 of the section on p. 78, and recently (1932) a grey-brown *Spirorbis* limestone, 6 in. thick, was found by Mr. W. W. King[3] in a pipe trench along the Bridgnorth road, at a point about 100 yds. S.E. of Bellman's Cross Inn.

The Brock Hall Coal is perhaps represented in Shatterford Pit by a seam about 1 ft. thick at a depth of 26 ft., whilst another seam, of which the thickness is given as 1 ft. 8 in., at a depth of 83 ft., may correspond to the seams that, at Highley and Kinlet, appear to represent the ' Main Sulphur Coal ' of the Broseley area (see p. 66).

Beds above the horizons of the Bank Farm Coal and limestone are exposed in the Worralls brook along the north-western border of Eymore Wood. For a short distance above the coal (p. 77) these include some purple and mottled clays, but higher up, between beds of greenish, somewhat micaceous sandstone, grey shales predominate. Several sandstones can be traced towards Arley village and park, and they are exposed in the streams joining the Severn at the ferry. Three of these sandstones, which form distinct beds near Eymore Wood, appear to coalesce in the dingle north of Arley Mill, about 700 yds. E.N.E. of Arley Castle. About 70 yds. higher up this dingle a red Keele-like sandstone crops out, and some 80 yds. still higher up a greenish sandstone of ' Halesowen ' type appears. It would seem that the red sandstone must be regarded as belonging to the Highley Group, for it is difficult to imagine what complication of faulting could bring Keele beds into this position.[4]

In a branch stream in Batemans Dingle, 500 yds. east of Arley Mill, a 6-in. coal was found in the upper part of the Highley Group, perhaps about 100 ft. below the top (see p. 68). Near Castle Hill, 280 yds. north of Arley Mill, ' coal pits ' are indicated on the Old Series geological map (Sheet 55 N.E.). These must have been sunk in the clays associated with the little coal seen in Batemans Dingle, but they probably worked a seam somewhat lower which, as Cantrill suggested, was probably the Brock Hall Coal.

The dingle from Nash Elm to Arley Mill shows the transitional beds from the Highley to the Keele Group. The highest greenish sandstone of ' Halesowen ' type crosses the dingle about 500 yds. E.N.E. of the mill, and between this and the red sandstone taken as the base of the Keele Group (p. 86) the beds consist of lavender and dull red marls with thin sandstones of like colours. Similar transitional beds are less well exposed in a stream, south of Bromley Farm, that forms the boundary of Arley Park ; they also crop out near the road south-west of Bellman's Cross.

The sandstones above the horizon of the Bank Farm Coal make well-marked features near Shatterford. Owing to the high dip, from 60° to 90°, the outcrops are narrow and it is not possible to sub-divide them freely. The map, in consequence, shows fewer sandstone outcrops near Shatterford than in the area of lower dip near Arley.

Witnells End, Hightrees, Compton, etc.—North-east of Shatterford few traces of the Bank Farm Coal occur ; but it was probably the seam worked

---

[1] I.e., items 28 to the fireclay below No. 44 in the section on p. 78, with a total (corrected) thickness of 110 ft. 6 in.

[2] See T. C. Cantrill, *op. cit.*, p. 22.

[3] *In. lit.* Some specimens were kindly supplied by Mr. King.

[4] *Cf.* T. C. Cantrill, *op. cit.*, p. 27.

from some shafts and a level in the west side of Arley Wood, about 400 yds. E.S.E. of Brittle's Farm. The coal here, according to Cantrill, is said to be 2 ft. 8 in. thick.

In the stream south of Witnells End some loose fragments of *Spirorbis* limestone may have come from the same bed as that found at Shatterford (p. 79) or from the band in the Keele Beds a short distance to the west (p. 86); but, apart from this, no traces of the *Spirorbis* limestone have been found northeast of Shatterford.

A higher seam of coal was observed in the dingles between south and southeast of Brittle's Farm. In an exposure 500 yds. N.N.E. of Witnells End the seam is 8 in. thick, and is an iridescent, ' peacock ' coal in grey clays. Whether this is the Brock Hall Coal or another is quite uncertain. Here, as at Shatterford, the dip is high and minute subdivision is not possible; moreover, parts of the sequence may, in places, be cut out by strike faults, as is almost certainly the case near Witnells End, where some of the higher beds of the Highley Group seem to be absent at outcrop.

North of Arley Wood the lower beds of the Highley Group are cut out by a cross-fault that, as mentioned on p. 138, terminates the outcrop of the Shatterford basalt south of Hightrees Farm. Sandstones in the higher part of the group, near Hightrees and Park farms, make strong features changing direction from north-east to south-east, owing, no doubt, to the north-eastward pitch of the Trimpley Anticline.

The stream in Rough Park Wood, about 600 yds. north of Park Farm, shows the lavender and purple marls and thin sandstones of the transitional beds at the top of the Highley Group. At a point in this stream, 640 yds. N.N.E. of Park Farm, a coal seam 13 in. thick crops out, in grey clay underlying a greenish sandstone ; the dip is over 60° N.W. Between 20 and 30 yds. below (south-east of) the coal outcrop a fault may cross the dingle, for the dip changes suddenly to south-east. From this point for about 200 yds. the stream runs in a ravine cut in olive, flaggy sandstones, which are believed to underlie the coal.

In the Compton sinking (p. 57), a short distance to the east, a coal seam 3 ft. 6 in. thick (including an 18-in. parting of fireclay) was found at a depth of 66 ft. 10 in. This may be the seam observed in Rough Park Wood, and some thick sandstones below it may be those of the ravine, while the Brock Hall Coal may be represented by a 10-in. seam, at a depth of 184 ft. 4 in. Another 10-in. coal, at a depth of 286 ft. 7 in., may represent the seam at Highley and Kinlet that, as suggested on p. 66, corresponds to the ' Main Sulphur Coal ' of Broseley. ' Bleeding rock ' (sandstone), 69 ft. thick, beginning at a depth of 309 ft. 2 in., seems to occupy the position of the thick sandstones near the base of the Highley Group, whilst two coals, 1 ft. 2 in. thick, and 6 in. thick, separated by 1 ft. 6 in. of fireclay, may correspond with the Bank Farm Coal. Above these coals is a bed, 18 ft. thick, described as ' white rock containing balls of stone, very hard.' These ' balls of stone ' may possibly represent the Bank Farm *Spirorbis* Limestone. All these interpretations, however, are given with reserve. No data as to the dip of the strata in the shaft and borehole are available, and it is probable that one or more faults were passed through. Moreover, the site is in an area of which the detailed structure is obscure, the exposures being inadequate to unravel the complications.

A buff sandstone, exposed near Compton Court Farm (about 1,000 yds. W.S.W. of Compton), dips north-westward, and its outcrop is cut off, just east of the Compton sinking, by a fault that probably intersects that shaft. Near the Enville Fault (p. 128), west of Compton and Pigeon House Farm, the ground is largely occupied by purple and mottled clays that probably belong to the transitional beds at the top of the Highley Group.

In a stream about 400 yds. S.W. of the Lydiates several exposures of coal occur, all probably of the same seam, the strata being here bent into a number of sharp folds. This coal appears to lie well up in the Highley Group and may be the seam found in Rough Park Wood. Coal, possibly the same seam, is said to have been found in the Wilderness, about 400 yds. E.S.E. of Park Farm. A coal seam about 1 ft. thick crops out in Round Hill Covert, 400 yds. S.S.E. of Greyfields Court ; it is probably the same seam as that in the stream near the Lydiates.

Coal is reported to occur in Starts Green Coverts, about 400 yds. E.S.E. of Starts Green. This would be considerably lower in the sequence than those mentioned in the preceding paragraph, and is, perhaps, on about the horizon of the Bank Farm Coal. Traces of what may be the same seam were observed in a gully about 350 yds. S.W. of Castle Hill.

South of that near Castle Hill there appears to be no outcrop of the Highley Group on the east side of the Trimpley Anticline unless the beds near Trimpley, described on p. 57, should be placed in that group.

KEELE GROUP

GENERAL ACCOUNT

Introduction.—The Keele Group succeeds the Halesowen Group, and its approximate equivalent the Highley Group, conformably, and by gradual passage, everywhere in the district, except, possibly, in the area west of the Severn, near Bridgnorth (p. 69). The determination of the base of the Keele Group is, in consequence, a matter of some difficulty and at the outcrop it has, in most places, been drawn at the base of the lowest considerable bed of red sandstone having the characters typical of the group. This usually leaves some red and purple beds, chiefly marls, in the group below.

Lithological characters.—The colour of the Keele Group measures is predominantly red, but varies from crimson to dull brownish-red, chocolate, purple and lavender. Green beds are occasionally found, and small green spots or " fish-eyes " are common both in the sandstones and in the marls. All the beds are more or less calcareous. The marls vary from tough clay to fissile shale. The sandstones are characteristically flaggy, but in places are in thick or even massive beds ; they are everywhere current-bedded, usually strongly so. The sandstones frequently contain pellets of marl which, in lenticular masses, may be very numerous and give rise to a " pellet rock " or marl-breccia. " Cornstones " arise from the inclusion of small lumps of calcareous matter (" race "), where these are numerous enough to predominate over the matrix.

Spirorbis limestones.—*Spirorbis* limestones occur in the Keele Group, and are commonly of a dark blue-grey colour in contra-distinction to the cream or putty-coloured limestones more usual in the Halesowen Group. This, however, is not a reliable criterion, for both types of limestone are found in both the Keele and the underlying groups, and even in the same bed (p. 76 above and p. 87 below). At the southern border of the South Staffordshire Coalfield two beds of *Spirorbis* limestone were mapped by the Geological Survey.[1] One of these lies about 360 ft. above the base of the group as mapped, the other being about 50 to 100 ft. lower. Later the existence of a third, and lower, bed was established by Prof. W. S. Boulton,[2] who estimates its position as 217 ft. above the base of the Keele Group. Of these three beds, only the highest one has been traced into the present district.

Two beds of *Spirorbis* limestone[3] are known in shafts and boreholes on the east side of the South Staffordshire Coalfield,[4] outside the area dealt with in this memoir. Of these the lower can be correlated with considerable probability with the lowest bed of the outcrop found by Prof. Boulton. The correlation of the higher with either of the other two beds at the outcrop is, however, very doubtful. In the sinkings at Baggeridge Colliery certainly one and probably two beds of *Spirorbis* limestone were met with and these may correspond with the two higher beds of the outcrop at the southern end of the coalfield. There are thus at least three, and possibly four, beds of *Spirorbis* limestone in the Keele Group of the South Staffordshire Coalfield. Three limestones were identified amongst the Keele strata in the borehole at Claverley (p. 85) ;[5] of these the middle one evidently corresponds with the higher bed at Baggeridge, whilst the lowest bed may be equivalent to the lower one there. The highest limestone at Claverley is only a few feet below the top of the Keele Group.

In the Arley and Alveley area the existence of *Spirorbis* limestone in what is now called the Keele Group was first discovered by T. C. Cantrill.[6] There now prove to be at least two persistent beds, one about 100 ft. and the other about 400 ft. above the base of the Keele Group. Further, there is probably a third bed between these two ; i.e., about 200 ft. above the base of the group (p. 87). In addition to these, traces of limestones that appear to lie at still other horizons have been noticed in several places. Indeed, there are few marl belts of any considerable thickness that do not yield, somewhere, evidence of a limestone of " *Spirorbis* " type. It would thus appear that such limestones are at once more numerous and individually less persistent than has been generally supposed,

---

[1] Eastwood, T., in ' The Geology of the Country around Birmingham ' (*Mem. Geol. Surv.*), 1925, p. 50. See also T. C. Cantrill, ' *Spirorbis* Limestones in the " Permian " of the South Staffordshire and Warwickshire Coalfields,' *Geol. Mag.*, 1909, p. 447. The existence of a bed of limestone about 50 ft. above the base of the Keele Group, as mentioned by Cantrill (*op. cit.*, p. 449), has not been confirmed.

[2] ' The Red Rocks between the Carboniferous and the Trias, and the *Spirorbis* Limestones of the Keele Beds, in the Birmingham District.' *Geol. Mag.*, 1928, p. 313 (see p. 318).

[3] According to some observers (e.g., W. S. Boulton, *op. cit.*, p. 322) three beds. The supposed third bed, above the other two, is, however, not actually recorded in any shaft sinking or boring, and its apparent occurrence in a railway cutting near Handsworth Station (see W. W. King, *Proc. Birm. Nat. Hist. & Phil. Soc.*, vol. xv, 1923, p. 45) may be due to repetition by strike-faulting.

[4] See T. H. Whitehead in ' The Geology of the Southern Part of the South Staffordshire Coalfield ' (*Mem. Geol. Surv.*), 1927, p. 121.

[5] Gibson, W., *Trans. Inst. Min. Eng.*, vol. xlv, 1913, pp. 23, 36.

[6] *Trans. Fed. Inst. Min. Eng.*, vol. vii, 1894, p. 577 ; and ' A Contribution to the Geology of the Wyre Forest Coalfield,' Kidderminster, 1895, p. 25.

and that they should, accordingly, be used with extreme caution for purposes of correlation, except within severely restricted areas.

Fossils.—Apart from the annelid *Spirorbis pusillus* (Martin) which, though sometimes absent, is in places abundant in the limestones, fossils are rare in the Keele Group. Teeth of a fish referred by the late Sir A. Smith Woodward to the genus *Diplodus* were found by Mr. W. W. King[1] near Hagley. The same observer found, also near Hagley, specimens of a gastropod described by Dr. L. R. Cox[2] under the name *Anthracopupa britannica* Cox. Some of these gastropods came from a stream course at a point 900 yds. east of Hagley church.[3] The beds here lie in an outcrop between the probable south-eastward continuation of the Wychbury Fault, beyond where it is met by the Hodge Hill Fault, and the base of the Clent Breccia in Hagley Park. In the Survey memoir on the Southern Part of the South Staffordshire Coalfield and on the published six-inch map (Worc. 9 N.E.) these beds were placed in the Enville Group; but, except for the presence in some of definite pebbles, which is more typical of the Enville Group, they have, on the whole, the characteristics of the Keele Group. In view of the evidence for an unconformity at the base of the Clent Breccia (pp. 93 and 97) it is possible to regard the Calcareous Conglomerate Group of the Enville beds as being here completely overstepped, and it is, thus, perhaps more reasonable to relegate the beds under discussion to the Keele Group, probably rather high in that group, and they have been so coloured on the one-inch map, Sheet 167.

In certain of the quarries near Alveley Dr. F. Raw has found footprints and tracks, as well as plant remains. Of those found in a quarry about 750 yds. west of May House, he has kindly contributed the following account :—" In the Butts quarry the chief interest centres in a single bed of sandstone yielding abundant tracks of vertebrates and arthropods. As frequently obtains, the footprint tracks are casts, and are associated with sun cracks. The main foot-printed surface was the red clay beneath it, and the underside of the bed is thickly covered with tracks, every piece of more than an inch or two showing ' footprints.' The creatures had walked over the underlying clay, as it lay exposed, at all stages of its drying.

" The vertebrate tracks present considerable variety in character and size. Six different types are recognised, but all are of primitive character. The right and left lines of prints are wide apart. All have five digits in the pes impressed, and in the manus four or five. They are perhaps mainly, if not wholly, amphibian. None is comparable with the more specialized of the Hamstead species,[4] a fact in harmony with their lower

---

[1] *In. lit.* and *Quart. Journ. Geol. Soc.*, vol. lxxxii, 1926, p. 404, footnote 2.

[2] ' *Anthracopupa britannica*, sp. nov., a Land Gastropod from the Red Beds of the Uppermost Coal Measures of Northern Worcestershire,' *Quart. Journ. Geol. Soc.*, vol. lxxxii, 1926, p. 401.

[3] King, W. W., *in lit.* See also L. R. Cox, *op. cit.*, p. 402. The other specimens were found on a spoil dump in Hagley Wood.

[4] Hardaker, W. H., ' On the Discovery of a Fossil-bearing Horizon in the "Permian" Rocks of Hamstead Quarries, near Birmingham,' *Quart. Journ. Geol. Soc.*, vol. lxviii, 1912, p. 639.

geological horizon. From the tracks it is possible to estimate the size of the animals, and assuming they were long-tailed I estimate they would range from a length of 5 feet, i.e., one-third the length of a full-grown crocodile, down to that of a small newt. Several tracks of a web-footed and hence aquatic animal occur.

" Within the footprint bed are the abundant tracks of two very different arthropods, one the tracks of a hopping creature using seven pairs of limbs, probably a Syncarid Malacostracan, the other of a minute creature represented by very numerous short tracks only 1 mm. wide, probably an Entomostracan."

Our knowledge of the fossil plants of the Keele Group is at present derived almost solely from those collected from shaft sinkings and boreholes, mostly in the South Staffordshire Coalfield, but in the Claverley borehole this group yielded some plants of which a list has been published by R. Kidston.[1]

Thickness.—The average thickness of the Keele Group in the present district is about 800 ft. ; but in some localities it is less than this and in others it appears to rise to 1,000 ft. The difficulty of fixing a consistent base for the group detracts from the significance of apparent variations in thickness. Nevertheless, a decrease in all directions from Alveley is not affected by this consideration, and is evidently due to the thinning out of the massive sandstones that include the Alveley grindstone beds and building stones (p. 88).

Distribution.—In the present district the Keele beds crop out in two main areas, eastern and western, which may be subdivided. The eastern area includes the outcrop along the south-western border of the South Staffordshire Coalfield, near Uffmoor Wood and Hagley Hill, with, in addition, small outcrops against the Western Boundary Fault of that coalfield near Pedmore and Amblecote, and a larger one near the Straits, Himley Park and Baggeridge Wood.

The western area is divided into three by the N.W.—S.E. Romsley Fault (p. 129), which throws down the overlying Enville Beds on the west, and by other faults that bring up the underlying beds in the Severn valley near Highley. East of the Romsley Fault the Keele Group crops out in a wide belt from near Broad Lanes and Wooton to the flanks of the Trimpley Anticline near Shatterford and Compton ; west of that fault the outcrop extends from near Wooton to Upper Arley and west of the Severn from near Astley Abbots to Highley.

DETAILS

For details of the eastern area reference should be made to the following pages of the memoir on ' The Southern Part of the South Staffordshire Coalfield ' (*Mem. Geol. Surv.*, 1927) :—Uffmoor Wood, Hagley Hill, etc., pp.129–31 ; Pedmore and Amblecote, pp. 131–2 ; The Straits, Himley Park and Baggeridge, pp. 132–3.

Smestow and Claverley.—In the Smestow Boring (Appendix II, p. 193) the Keele Group appears to have a thickness of 750 ft. or more. A black

---

[1] *Trans. Roy. Soc. Edinb.*, vol. li, 1917, p. 1,079. Reg. Nos. J.P. 4222–4542.

Spirorbis limestone, about 340 ft. above the base, and a grey limestone with entomostraca near the bottom were noted by W. Gibson.

In the Claverley Boring the strata from 472 ft. to 1,240 ft. below the surface are placed by W. Gibson[1] in the Keele Group, giving a thickness of 768 ft. The beds consist predominantly of red and mottled marls with subordinate red, lavender and grey sandstones ; about 56 ft. of beds near the middle of the group are mainly sandstones. These may represent part of the grindstone beds and building stones of Alveley (p. 88) ; but, in general, there is a closer resemblance to the Keele beds of Baggeridge than to those of Alveley, although the site of the borehole is nearer the latter. The highest 16 ft. of the Keele Group in the Claverley borehole consist of sandy red marl containing rounded pieces of grey *Spirorbis* limestone. At 775 ft. from the surface, or 303 ft. 10 in. below the top of the group, a thin bed of dark blue *Spirorbis* limestone occurs, corresponding with that recorded in No. 1 shaft of Baggeridge Colliery. Further, the mottled marls between 950 ft. and 972 ft. at Claverley contain rounded blocks of grey *Spirorbis* limestone, which may be equivalent to the limestone found in the Straits Farm borehole of South Staffordshire.

Wooton and Shatterford.—In the outcrop extending east of the Romsley Fault from the neighbourhood of Broad Lanes to near Shatterford the Keele beds consist largely of marls and give rise to an area of damp, clayey pasture and woodland. The proportion of sandstone increases southward, in part due to the presence of sandstones in the lower part of the group which are cut out at their northern end by the Romsley Fault, and in part by the increase southward in number and thickness of sandstones in the upper part. A prominent belt of sandstones crops out in the angle between the Romsley and Pattingham faults near Wooton, where there is some disturbance by minor faults. It consists of two beds, well exposed in Wooton Dingle and in a disused quarry at Kingsnordley. The sandstones are red, purple or lavender in colour, for the most part calcareous, with lenses of calcareous breccia made up of marl pellets, nodules of race, fragments of compact limestone, and, in some cases, a few quartz pebbles, in a sandy matrix cemented by crystalline calcite. The sandstones are mainly thick-bedded and, especially in the quarry, have a certain resemblance to the Alveley grindstone beds ; it is probable that they are on approximately the same horizon. Between Kingsnordley and Astley the lower of the two sandstones is cut out by the Romsley Fault ; but from Astley southwards both beds can be followed, becoming more widely separated owing to decreasing dip and thickness. The two beds are well seen in Perryhouse Dingle (which runs south-westward by Fillets) and again in the upper part of Bowhills Dingle, where they consist of red, lavender and greenish sandstones, generally in thinner beds than near Wooton.

Below the lower of these two sandstones, in Perryhouse Dingle, 400 yds. S.W. of Fillets, dark grey *Spirorbis* limestone 4 in. or 5 in. thick crops out, in red marl. A similar *Spirorbis* limestone, in the same position relatively to the sandstone, appears in a branch of Bowhills Dingle 550 yds. N.W. of Hartsgreen. These two outcrops are presumably of one limestone, perhaps corresponding to the middle bed of the area south of Alveley (p. 87 below). In the unnamed dingle between Perryhouse and Bowhills dingles, a bed, 1 ft. or more in thickness, of rubbly limestone of '*Spirorbis*' type was found 780 yds. S.E. of Fillets, while in the upper part of Bowhills Dingle there are large blocks of dark grey *Spirorbis* limestone. These two occurrences may indicate another bed lying in the marls above the two sandstones mentioned in the preceding paragraph, and possibly equivalent to the highest of the three limestones near Alveley (p. 88), though it lies nearer to the top of the Keele Group.

To the east of Coton and Astley other sandstones appear above those already mentioned, and two are fairly distinct and persistent. In a stream, at a point 940 yds. E.N.E. of Fillets, a short distance below the higher of these two sand-

---

[1] *Op. cit.*, p. 356. For the site of this boring see p. 39, footnote 1.

stones, blocks of cream-coloured concretionary limestone were found imbedded in purple marl. Again, in a stream north-west of Rough Park Wood (see p. 80), fragments of dark grey *Spirorbis* limestone were found just below the outcrop of the same sandstone. These two occurrences do not correspond to any bed hitherto found in the Arley and Alveley area.

Two lenses of ' pellet rock,' of which one is exposed in an old gravel pit 620 yds. S.W. of Tuckhill, are of interest in showing, by the presence of pebbles of red and yellow chert, an approach to the characters of the Enville beds.

North-east of Hartsgreen the. outcrop turns sharply north-eastwards, and the dip increases suddenly to about 50°. The breadth of outcrop, however, clearly leaves insufficient room even for the thickness of beds exposed to the north-west, and still less for the full thickness of the Keele Group, as seen west of the Romsley Fault. This might at first sight suggest the possibility of unconformity at the base of the Keele beds along the north-west flank of the Trimpley Anticline, accompanied by conformable overlap within the group. This explanation, however, is negatived by the presence of the highest beds of the underlying Highley Group (p. 80) ; and it seems clear that the sandstones that make the prominent and continuous feature from just west of Compton to a point about half a mile east of Hartsgreen and are exposed in the stream in Rough Park Wood must include a bed approximately equivalent to the basal sandstone of the Keele Group elsewhere. There is the further possibility of some unconformity at the base of the Enville Group ; but this explanation would be insufficient by itself, since the beds missing presumably belong to the lower two-thirds of the Keele Group and it is, in fact, not possible to demonstrate that any are missing from the top of the group. The absence of part of the Keele Group here must, therefore, probably be ascribed to strike-faulting.

The basal sandstone of the Keele Group can be followed from west of Hightrees Farm nearly to Bellman's Cross, though its outcrop is shifted in places by faults. It is exposed in gullies and dingles near Brittle's Farm, and consists of flaggy purple and greenish sandstone with some lenses of ' pellet rock ' in places.

In a field 500 yds. N.W. of Witnells End are two shallow excavations evidently due to the working of a *Spirorbis* limestone, of which many nodules lie on the surface.[1] The limestone can be traced by debris nearly to the Shatterford and Romsley road where it, together with the basal sandstone below, appears to be cut off by a fault. The limestone; however, seems to crop out south of the fault, west-north-west of Bellman's Cross, its presence being indicated by shallow pits and fragments. There is little doubt that this limestone is the same as the bed exposed in Hallclose Coppice and elsewhere near Arley (p. 87).

The basal sandstone appears to form an outlier at Hillhouse Farm, 400 yds. S.W. of Bellman's Cross ; but the sandstone that makes a prominent hill, and is exposed in a quarry 130 yds. north of Hillhouse Farm is probably a higher sandstone brought down by a strike fault. The rock in the quarry is brownish-red in colour and in rather thick beds. The lower part consists largely of highly calcareous ' pellet rock ' and cornstone.

Arley and Alveley.—To the west of the Romsley Fault the basal sandstone of the Keele Group is probably responsible for a well-marked feature north-east of Popehouse Farm ; but it appears to be shifted southward to the farm itself by a N.N.W.—S.S.E. fault. From Popehouse Farm it can be traced to the dingle in Nash Elms Wood, which it crosses just over 1,000 yds. S.W. Hillfields House. The outcrop can be followed for about 450 yds. W.N.W. from the dingle but the feature that it forms then dies out rather abruptly. Another feature sets in near the same place and is due to a sandstone that can be followed to Nash End and beyond, but it seems clear that this sandstone is not the

---

[1] See T. C. Cantrill, *Quart. Journ. Geol. Soc.*, vol. li, 1895, p. 536.

basement bed of the Keele Group, for there is room for only 700 ft. of strata, at most, between it and the base of the Enville Beds.[1] The lowest 200 or 300 ft. of the Keele Group are, in fact, missing at the outcrop north of Arley Park and Pickard's Farm owing, it is believed, to a fault that runs west-north-westward between the Romsley and Arley Park faults. North of where this cross-fault meets it the Arley Park Fault (p. 129) has a much diminished throw.

The basal sandstone of the Keele Group reappears on the west side of the Arley Park Fault 400 yds. S.W. of Bromley Farm, and is traceable to the Severn west of Hexton's Farm. Here the outcrop crosses the river ; but it returns to the east bank opposite to Stanley and can be followed through Little London and Hallclose coppices (about 900 yds. S.W. and 700 yds. W.S.W. respectively of Hallclose), where it is much broken up by slips owing to the unstable conditions produced by wet marl-outcrops on the steep slope. Southwest of Hadleys the basal sandstone is believed to be cut off by a continuation of the Station Fault (p. 129), but it reappears, on the west side of that fault, near Moor House.

The basal sandstone is succeeded, upwards, by a marl-belt in which lies the lowest bed of *Spirorbis* limestone (p. 82). Several large blocks of this, pale grey or dark grey in colour, were seen at the ponds 600 yds. N.W. of Arley Castle, where the bed was found in place by T. C. Cantrill,[2] and surface fragments indicate the outcrop of the limestone between the ponds and the stream that forms the county boundary north-west of Hexton's Farm. In this stream large blocks of limestone occur, and similar fragments are numerous between here and Little London Coppice, where the remains of a kiln show that it was formerly burnt for lime. In a stream in Hallclose Coppice 770 yds. W.S.W. of Hallclose, the limestone is about 3 ft. thick ; the upper part dark grey and the lower pale brownish in colour, separated by a breccia in which the two types are mingled. *Spirorbis pusillus* (Martin) is common in both types. A thin coal is recorded by Cantrill[3] in the same stream, below the limestone. The limestone can be traced for about half a mile to the north. In Doggetts Batch, about 500 yds. S.E. of Hampton Loade ferry, large blocks of limestone probably derived from this bed were found in the stream.

Above the marl-belt with limestone come sandstones in two main beds. These have been worked in a large quarry in the upper bed near Hexton's Farm and smaller ones respectively 800 and 1,000 yds. to the north-west. It was in the last-mentioned one that Dr. Raw found the footprints referred to on p. 83. The quarry near Hexton's Farm shows some 30 ft. of red sandstone, with some grey-green beds, in rather thick beds with partings of thin-bedded material. The well marked joints trend N. 20° W. Just south of Little London Coppice the lower bed of sandstone wedges out ; the higher bed thins rapidly northwards and seems to die out in the coppice about 530 yds. south of Hadleys, near which point another bed sets in at a somewhat higher level.

The sandstone of Hexton's Farm is succeeded by beds consisting mainly of marls, on the outcrop of which are traces of what seems to be a second bed of *Spirorbis* limestone. In the streams north and north-west of Hexton's Farm there occur blocks of limestone that seem too large and numerous to have been derived from the higher bed (of Baynham's Cottage, etc.) presently to be described (p. 88). Moreover, in a shallow excavation 330 yds. S.E. of the farm, lumps of partly burnt limestone were found, suggesting that a bed of it was here formerly worked. Another shallow pit, with limestone lumps on the surface, lies about 250 yds. N.N.E. of Hexton's Farm. The limestone that was formerly

---

[1] See also, p. 88, footnote 3, below.

[2] 'Contribution to the Geology of the Wyre Forest Coalfield,' Kidderminster, 1895, p. 26.

[3] *Quart. Journ. Geol. Soc.*, vol. li, 1895, pp. 537, 540, where the locality is described as Little London Brook. Cantrill also records the finding of a coal seam, possibly the same one, in a well at the Butts, 400 yds. W.S.W. of May House.

worked and burnt near Little London (600 yds. S.W. of Hallclose), as described by Cantrill,[1] may belong to this second bed.

The sandstone of Nash End (p. 86) can be followed to a point about 380 yds. S.W. of Shropshire Farm (where a dip-fault shifts the outcrop slightly to west-south-west) and thence northward west of Mayhouse and beyond.

Near Baynham's Cottage, 500 yds. S.W. of Lowe Farm, where it was worked for roadstone,[2] another *Spirorbis* limestone occurs in the marls above the Nash End sandstone. It is recorded near Nash End by Cantrill, and is probably the source of the loose blocks in the stream running southward from Nash End to Arley Mill. North-west of Baynham's Cottage the outcrop is presumably shifted by the dip-fault already mentioned but it can be traced by fragments and shallow workings to May House and beyond. Thus between the sandstone on which May House stands and the base of the Keele Group near the river opposite Stanley we have definite evidence of two *Spirorbis* limestones, with the possibility of a third between them, in what seems clearly to be an unbroken and unrepeated series of strata.

The escarpment of the Nash End sandstone becomes more prominent northwards and, near Alveley, forms a feature about 50 ft. high. North of Alveley the outcrop broadens very considerably owing to a general coincidence of the direction of dip with the slope of the ground; but some 500 yds N.W. of Alveley a marl parting develops and divides the sandstone into two parts. The principal grindstone quarry of Alveley (Plate VA) about 100 yds. N.N.W. of the church, is in the upper part, and in 1931 showed: Dull red sandstones, in fairly thick beds, about 10 ft.; purple, grey and lavender sandstones, in beds up to 2 ft., about 20 ft.; parting of pellety rock; green sandstone, hard, seen for 2 ft. Nearly spherical concretions up to 9 in. in diameter occur sporadically in these sandstones. They are harder than the surrounding rock, being more highly calcareous as well as ferruginous, and frequently weather out as balls. Plant remains seem to have been first noted in the Alveley grindstone beds by C. E. Roberts, ' The Geologist,' 1858, p. 253. A specimen referred by R. Kidston to *Sigillaria brardi* var. *denudata* Göpp. pro sp. was obtained in this quarry by Dr. J. Pringle in 1906 (see T. C. Cantrill, *Trans. Roy. Soc. Edinb.*, vol. li, 1917, p. 1,018).

In a quarry in the lower part of the sandstone, 400 yds. E.N.E. of Moor House, the rock consists largely of ' pellet rock ' and cornstone, and is so calcareous that it was formerly burnt for lime, the remains of a kiln being still visible. Both parts of the sandstone[3] are well exposed in Gorten's Dingle north of Lake House, and in the dingle that runs northward from Alveley. The sandstones give rise to small waterfalls and, owing to the calcareous nature of the beds, there is usually a thick encrustation of tufa at such places. This is particularly well seen at Gorten's Mill, 200 yds. E.N.E. of Lake House. In Gorten's Dingle, at points 200 yds. and 250 yds. N.E. of Lake House, fragments of *Spirorbis* limestone were found below the lower part of the sandstone. If these were derived from a nearby outcrop the bed would correspond in position to the middle limestone of the area near Hexton's Farm and May House (p. 87).

To return to the neighbourhood of Nash End: the marl belt above the Nash End sandstone is succeeded by another sandstone of which the most southerly exposure is in Nash Elm Wood, 750 yds. S.S.W. of Hillfield House, beyond which place the outcrop is cut off by the cross-fault referred to on p. 87. North-west of Nash Elm Wood this sandstone makes a bold feature which becomes less prominent northward, but is, nevertheless, readily traceable to the dip fault near Lowe Farm. Beyond this fault the outcrop broadens, chiefly owing to the form of the ground, but the sandstone certainly becomes thicker. It is exposed in

---

[1] *Op. cit.*, p. 537.

[2] Cantrill, T. C., *op. cit.*, pp. 536, 537.

[3] Inasmuch as the lower part of this sandstone is connected by a practically continuous feature with the sandstone of Nash End, the section in Gorten's Dingle and Doggetts Batch provides additional evidence that the Nash End sandstone is not the basement bed of the Keele Group.

PLATE V

Geology of Dudley and Bridgnorth (*Mem. Geol. Surv.*)

A 2198

B.—Bunter Pebble Beds on Lower Mottled Sandstone; Ridge Sandpit near Stourbridge.

A 6871

A.—Alveley grindstone and building stone quarry.

quarries on both sides of the road 300 yds. and 400 yds. N.N.W. of Shropshire Farm. May House stands on a lens of this sandstone nearly detached from the main mass by a wedge of marl. Near Hallclose the sandstone, as a whole, makes a prominent bluff. A disused quarry just south of Hallclose gave the following section :—Thick-bedded sandstones, with obscure partings, strongly false-bedded lenses of 'pellet' rock in the bottom part, base irregular, 10 ft. ; soft lavender sandstones in thin bands, 1 ft. 6 in. ; dull red sandstone in thick irregular beds, split into slabs of 6 in. to 1 ft. at the south end by partings of soft lavender sandstone, 9 ft. ; massive sandstone in beds up to 3 ft. thick, seen for 7 ft. North of Hallclose the outcrop broadens and the greater part of Alveley village is built upon it. It can be traced to Gorten's Dingle, where exposures of it can be seen just west of the Bridgnorth road.

East of Nash End and Alveley other sandstones crop out above that described in the preceding paragraph ; but they are less persistent than the lower beds. North of Doggetts Batch and Gorten's Dingle the sandstones thin out rapidly. A small grindstone quarry, 550 yds. W.N.W. of Lake House, is in beds equivalent to the lower part of the Nash End sandstone (p. 88). Another disused quarry has been excavated in beds corresponding to the sandstone of Alveley village (p. 88), at about 350 yds. north of Lake House ; it shows about 4 ft. of 'pellet rock' with marl partings on 6 ft. of thick-bedded sandstones. This sandstone can be followed, diminishing in thickness, to the Pattingham Fault (p. 129) which it meets some 800 yds. S.S.E. of Quatt. Above it the Keele Group includes only a few impersistent lenses of sandstone and the same is true of the outcrop to the north-east, near Quatt Farm and Wooton.

T.H.W.

West of the River Severn.—The outcrop of the Keele Group on the west of the Severn extends from Astley Abbots southwards through Oldbury, Eardington, Chelmarsh and Sutton to Hazelwells, near Highley. The surface is in general a reddish marly clay, due to the preponderance of marl over sandstone. Although the sandstones are very evident in stream sections they do not, in this area, make clear features at the surface because the beds dip eastward with the general slope of the ground towards the valley of the Severn, whereas on the eastern side of the river below Hampton Loade they crop out in the steep westward-facing slope of the valley.

North of Binnal, near Astley Abbots, the outcrop is only half a mile wide, due to a westward encroachment of the Lower Mottled Sandstone of the Trias. At Binnal a medium-grained, brownish-red calcareous sandstone, from 15 to 20 ft. thick, is exposed and trends southwards towards Dunval. The Keele outcrop increases to a mile in width near Crosslanehead where red-brown and grey, speckled sandstone is exposed at the Rectory.

Between Crosslanehead and Cantern Brook the surface is red clay, generally stiff but loamy in places. The deep cut across the outcrop of the Keele Group made by the brook provides some good exposures. At the waterfall 450 yds. upstream from the Ironbridge Road 25 ft. of calcareous sandstone rest on some 15 ft. of lavender and grey marl with a thin band of nodular limestone resembling *Spirorbis* limestone. Another section 300 yds. above the bridge shows :—

	Ft.	In.
Red marl, at the top of the bank	2	0
Lavender-coloured calcareous sandstone, shaly and friable with scattered calcareous nodules becoming more frequent towards the base	4	0
Blocky calcareous sandstone, lavender and pink-coloured, in part a sandy limestone	2	6
Purple and lavender sand		6
Grey, purple and lavender mottled calcareous marl with nodules of limestone (? *Spirorbis*)	1	6
Purple calcareous marl.		

Blocks of limestone, with *Spirorbis*, up to 2 ft. or more in diameter lie in the stream.

A well marked fault, with a throw of 10 ft. to the west, crosses the stream at the waterfall and throws out a good spring which has deposited much tufa.

West of Bridgnorth the outcrop is less than half a mile in width and encroaches on that of the Highley Group down to the horizon of the *Spirorbis* limestone of Tasley ; it is possible that this apparent unconformity may be due to a north-north-easterly fault.

A strong spring in the valley about 700 yds. west of Bridgnorth Station was, until recently, the main supply of the town. It may be on the southward continuation of the fault seen at the waterfall in Cantern Brook. Another strong spring, known as Potseething Spring, in the coppice of that name 700 yds. south of Oldbury Church, seems to be connected with the same line of fracture.

Alternate bands of red marl and dark red sandstone, with a dip of 10° to 14° E.N.E., crop out in the banks of the stream which flows north-eastwards by Daniels Mill to the Severn.

Near the Punch Bowl Inn, west of Oldbury, there are exposures of a peculiar brown-speckled grey sandstone near the base of the group.

At Knowlsands, a large brick pit, the Bridgnorth Brick Works (Survey Photograph 6879), shows about 25 ft. of chocolate-coloured Keele marl with some marly sandstone bands dipping about 10° to E. 30° N. The top 6 ft. of the section is full of limestone nodules. In the southern corner of the excavation the Lower Mottled Sandstone can be seen resting on an uneven surface of the dark red Keele marl.

South and south-west of Eardington the Mor Brook has cut a deep channel across the Keele outcrop from Thatchers Wood to the Upper Forge, Eardington. A section at Marlbrook, 100 yds. south of the bridge on the main road, shows 2 ft. of mottled marl on 2 ft. 6 in. of grey shaly marl overlying 7 ft. of massive grey sandstone, the surface of which is coated with a layer of calcite. Marls and flaggy sandstones are exposed at intervals in the banks of the brook below the bridge.

The nearby Eardington Deep Pit (p. 74) passed through about 120 ft. of beds which can be assigned to the Keele Group, into beds of the Highley Group. A strong spring breaks out in the south side of the valley, 100 yds. south of the mill, from massive red-brown and grey sandstone and seems to be connected with a north and south fault passing just east of the mill.

By the weir 300 yds. N.E. of Astbury Hall, the lowest bed of the Trias, a dark red sandstone with conglomeratic base rests on purple Keele marl.

From the Mor Brook valley southward to Uplands, Chelmarsh and Sutton the Keele beds form a long ridge of high ground (Survey Photographs 6877-8) which extends to the Brock Hall Fault. The outcrop is here about a mile wide but there are few good sections. Some small pits near Uplands expose dark red and purple sandstones with bands of purple and grey marl and another pit, 400 yds. W. of Chelmarsh Smithy, shows 4 ft. of soft false-bedded yellowish sandstone with brown calcareous bands on 2 ft. of hard grey calcareous sandstone with traces of copper staining.

In the lane south-east of Chelmarsh Hall red marl and soft red-brown sandstone rest on hard flaggy sandstone and dip to the north-east at about 12°. At the school and about 300 yds. to the south-east dark purplish sandstones with calcareous bands and lenticular cornstones are seen near the base of the group.

In the small stream in Spadeley Rough, three quarters of a mile south-east of Chelmarsh Church, tumbled blocks of dark red sandstone occur and in this stream near its junction with another stream from Chelmarsh an outcrop of a 3-in. band of dark grey to black *Spirorbis* limestone dips to the east at about 8°. The ground here is very swampy and much tufa has been deposited. This limestone or a similar one crops out in a small stream flowing from Sutton at a point 200 yds. W. of Hempton Mill, one mile north-east of Sutton, and here again there are considerable deposits of tufa.

A boring commencing near the top of the Keele Group was put down some years ago about two thirds of a mile north of Hampton, on the edge of the Severn alluvium. It reached a depth of 420 ft., probably all in Keele beds. The record shows red and purple marl with subordinate sandstone bands to 272 ft., below which level red, purple and grey sandstones are more in evidence. Two bands of *Spirorbis* limestone were met with, one at 169 ft., and another at 271 ft., while limestone balls occurred at some level between 336 and 366 ft.

The boring at Ingram Lane (p. 74) began in red clay with a grey limestone of '*Spirorbis*' type and passed through about 130 ft. of Keele strata, mainly red and mottled marls and brown or purplish sandstones, into beds of the Highley Group. The limestone is exposed in an old pit about 50 yds. west of the reservoir.

The Brock Hall Fault crosses Ingram Lane about 500 yds. N.E. of Brock Hall and carries the base of the Keele Group south-westwards nearly to Borle Brook. Fine sections of massive red and lavender sandstone at, or near, the base are to be seen in deep gullies in the left bank of the Borle Brook valley.

Sandstones at, or near, the base are well exposed, south-east of the Kinlet Hall Fault, near Green Hall and at the Hay, half a mile west of the Vicarage, and again between the Tiphouse and Highley-Kinlet faults north-north-west of Highley Church.

In Londonderry Coppice half a mile south-south-west of Hampton Loade Station a strong spring breaks out, apparently on the line of the Tiphouse Fault, and is pumped to the reservoir to supply Highley. The overflow has deposited a remarkable series of terraces of tufa.

A small area of Keele Beds is faulted in, between the Station Fault and a branch to the east, north of Highley Station and the red marls have been worked for brick-making.

The small patch of red marly ground at the Bind near New England, coloured as Keele beds on the map, should perhaps be referred to an outcrop of red marls in the Highley Group.

R.W.P.

ENVILLE BEDS

GENERAL ACCOUNT

Introduction.—About 800 ft. above their base, on the average, the Keele beds pass up into a group that consists of highly calcareous sandstones and conglomerates separated by less calcareous or non-calcareous sandstones and marls. It is possible that these beds do not everywhere set in at exactly the same horizon, a matter which, in the absence of precise stratigraphical datum planes, it is impossible to test; but collectively they form a fairly constant belt and give rise to well-featured ground usually sharply marked off from the flatter outcrop of the Keele strata.

Lithological characters, Calcareous Conglomerate Group.—The calcareous sandstones are individually impersistent, and at outcrop are often found to wedge out suddenly or to split. They are more or less uniformly distributed throughout the group, and it is difficult to recognize the three definite and persistent calcareous " zones " described by Mr.

W. W. King.[1] The conglomerates occur as sporadic lenses in the sandstones, and rarely extend for more than a quarter of a mile at outcrop. There is a certain tendency for conglomerate lenses to recur in nearly the same vertical plane, as if the channels along which coarse material was drifted had repeatedly followed approximately the same course. The characteristic pebbles, massed in a conglomerate, scattered in pebbly sandstones, or isolated, are of red or yellow chert; but limestones, both Silurian and Carboniferous, are abundant in some localities, and pebbles of Llandovery and other sandstones, quartz, and other rocks occur. The pebbles are imperfectly rounded, often sub-angular. The matrix of the conglomerates is everywhere highly calcareous. Near Quatt lenses of breccia occur; this rock in its angular character and red colour resembles the Clent Breccia presently to be described; it differs, however, in having a calcareous matrix and in the nature of its rock-fragments (p. 100).

Mr. King[2] concluded that the materials of the conglomerates indicate a drift from south and east, there being known sources in these directions for all the constituents except the Lower Carboniferous rocks. He suggests that the latter may have come from outcrops that existed at the time of the formation of the conglomerates, and in later papers[3] adduces evidence that part of the Avonian Series may have been laid down in synclinal areas of the South Staffordshire region.

Lithological characters, Breccia Group.—The Calcareous Conglomerate Group, described above, is succeeded near Upper Penn and Bobbington by red marls and sandstones, with relatively thin beds of fine breccia at or near their base. The marls and sandstones are very similar to those of the Keele Group, but the sandstones exhibit some differences in their heavy mineral content.[4] Farther south the breccias become increasingly coarse and thick until, in the Clent Hills, the group is represented entirely by breccia. These breccias (Survey Photographs 2212-3) consist of angular rock-fragments which, in the Clent Hills, range from a foot or more in length to the size of a pea. The larger stones are set in a matrix of smaller ones graduating down to a paste of sandy marl or sand-rock; the majority consist of igneous or pyroclastic rocks resembling the Uriconian rocks of Shropshire, and the Caldecote and Barnt Green rocks of the Midlands. In addition there are purple sandstones and grits, like those of the Longmynd; quartzite, like that of the Lickey Hills; Llandovery sandstone (often fossiliferous); limestone and other rocks. The nature of the constituents varies somewhat from place to place, but they seem to have been derived mainly from areas of lower Palaeozoic and

---

[1] 'The Permian Conglomerates of the Lower Severn Basin,' *Quart. Journ. Geol. Soc.*, vol. lv, 1899, p. 97 (see p. 101).

[2] *Op. cit.*, p. 123.

[3] King, W. W., and W. J. Lewis, 'The (Lower?) Carboniferous Grits of Lye in South Staffordshire,' *Rep. Brit. Assoc.* for 1913 (1914), p. 482; and W. W. King, 'The Plexography of South Staffordshire in Avonian Time,' *Trans. Inst. Min. Eng.*, vol. lxi, 1921, p. 165.

[4] See W. F. Fleet, 'The Heavy Minerals of the Keele, Enville, " Permian " and Lower Triassic Rocks of the Midlands and the Correlation of these Strata.' *Proc. Geol. Assoc.*, vol. xxxiii, 1927, p. 1 (see pp. 8 and 11).

Geology of Dudley and Bridgnorth (*Mem. Geol. Surv.*)   PLATE VI

A 2211
A.—CLENT HILLS, CLENT; VIEW LOOKING WEST-SOUTH-WEST.

A 6889
B.—KINVER EDGE FROM NEAR SHATTERFORD.

pre-Cambrian rocks that were not available as sources of supply when the underlying Calcareous Conglomerate Group was formed.[1] The nature of the heavy-mineral grains found in these beds supports this conclusion.[2] Wherever a clear section is available the breccias are found to be well bedded.[3] When weathered they break down into a loose rubble and their permeable character is indicated by the waterless condition of the combes found in high-level parts of the outcrop and by the presence of springs at lower levels and at the base of the deposit. On the other hand, in an unweathered condition, as found in wells and fresh excavations, the rock is hard and compact.[4]

The abundance of igneous material earned for the deposit the name "trappoid breccia," and caused its origin to be ascribed to igneous activity.[5] The resemblance of many of the constituents to rocks that occur in Shropshire together with the presence of occasional striated fragments, led Ramsay[6] to invoke a Permian ice-age to explain their transportation and accumulation. Mr. King has shown that the occasional striations may be due to the grinding effects of earth-movements,[7] and that, as many of the constituents of the breccias resemble rocks occurring in place in the vicinity, it is not necessary to suppose that the materials are far-travelled. The breccias probably represent the waste products of arid upland regions washed down by intermittent torrents and deposited in overlapping fans opposite the mouths of cañons and gullies. The marls and sandstones of such localities as Bobbington and Upper Penn represent the finer materials carried farther from the hills and deposited in the plains and valleys. The Clent breccias are thus analogous to the piedmont gravels and breccias now in process of formation in many desert or semi-desert regions of Asia and the Western United States.

Relation of Breccia Group to underlying beds.—The present district affords clear evidence in support of Prof. Boulton's[8] contention that there is an unconformity at the base of these "trappoid" breccias. It has already been mentioned (p. 83) that in Hagley Park the Clent Breccia probably rests upon the upper part of the Keele Group and not upon the Calcareous Conglomerate Group. In the Enville area there is an overstep

---

[1] For details of the constituents of the breccias see W. W. King, *op. cit.*, 1899, pp. 112–118, and *idem*, 'Clent Hills Breccia,' *Mid. Nat.*, vol. xvi, 1893, p. 25.

[2] See W. F. Fleet, *op. cit.*, p. 38.

[3] The bedded character of this deposit was early recognized by Buckland; see his 'Description of the Quartz Rock of the Lickey Hill in Worcestershire and of the Strata immediately surrounding it,' *Trans. Geol. Soc.*, vol. v, 1821, p. 506.

[4] See W. W. King, *Mid. Nat.*, vol. xvi, 1893, p. 26.

[5] Murchison, R. I., 'Silurian System,' 1839, pp. 138, 496.

[6] 'On the Occurrence of Angular, Subangular, Polished and Striated Fragments and Boulders in the Permian Breccia of Shropshire, Worcestershire, etc.,' *Quart. Journ. Geol. Soc.*, vol. xi, 1855, p. 185.

[7] *Mid. Nat.*, vol. xvi, 1893, pp. 33, 34. See also T. G. Bonney, *Quart. Journ. Geol. Soc.*, vol. lviii, 1902, p. 188.

[8] 'On a Recently Discovered Breccia-Bed underlying Nechells (Birmingham) and its Relation to the Red Rocks of the District,' *Quart. Journ. Geol. Soc.*, vol. lxxx, 1924, p. 343.

on the part of the breccias towards the axis of the Trimpley Anticline, so that, at the outcrop south of Enville Sheepwalks, the thickness of the Calcareous Conglomerate Group is reduced to about half of that at Four Ashes, some two miles to the north-west. Again, three outliers of the breccia near Bowhills, Astley and Coton Hall (p. 104), appear to rest respectively on different horizons of the Calcareous Conglomerate Group. The unconformity is most distinct where the breccias are thickest and coarsest and cannot be detected where they become thin and are replaced by marls and sandstones.

Nomenclature and classification.—The term Enville Beds, to include all the beds with calcareous conglomerates and " trappoid " breccias forming the middle and upper divisions of Hull's " Salopian Permian,"[1] was first proposed (in the form " Enville Series ") by E. A. N. Arber.[2] For the lower part of the Enville Series, the Calcareous Conglomerate Group, Arber[3] proposed the name " Romsley Group." This choice was, however, unfortunate, for at Romsley, near Halesowen (Worcestershire), the locality he evidently had in mind, the group is not at all typically developed, and may even, according to some views, be absent. Prof. Boulton[4] proposed to use the term " Corley Beds " for the Calcareous Conglomerate Group. The name Corley Group, adapted from Mr. R. D. Vernon's " Corley Conglomerates,"[5] was used by T. Eastwood as a local synonym for " Enville Group," in the memoir on the Coventry district,[6] to cover " all the red pre-Triassic beds that succeed the Keele Group." It was then, however, by no means certain that the beds in question did not include some stratigraphical equivalent of the breccias of the Clent Hills ; and subsequently, Prof. F. W. Shotton[7] definitely included in the Corley Group beds that he considers should be correlated with those breccias, namely the Kenilworth Breccias. To attempt to restrict the term Corley Group to the Calcareous Conglomerate Group might thus lead to confusion. The term " Hamstead Group," used in our Lichfield Memoir (1919), suffers from a similar ambiguity[8] and is, besides, preoccupied for beds of Oligocene age. If, then, a geographical synonym for the Calcareous Conglomerate Group is desired, a new one must be selected. For this " Bowhills Group " seems very appropriate, since the beds in question are typically developed and well exposed in the vicinity

---

[1] ' Triassic and Permian Rocks of the Midland Counties of England ' (*Mem. Geol. Surv.*), 1869, pp. 13, 14.

[2] ' The Structure of the South Staffordshire Coalfield . . . ,' *Trans. Inst. Min. Eng.*, vol. lii, 1916, p. 35 (see p. 38). See also, T. H. Whitehead, ' The Subdivisions of the Red Rocks formerly classed as Permian in South Staffordshire and the neighbouring Counties,' in ' Summary of Progress ' for 1921 (*Mem. Geol. Surv.*), 1922, Appendix VII, p. 169.

[3] *Op. cit.*, p. 39.

[4] *Op. cit.*, p. 364.

[5] ' The Geology and Palaeontology of the Warwickshire Coalfield,' *Quart. Journ. Geol. Soc.*, vol. lxviii, 1912, p. 587 (see p. 603).

[6] (*Mem. Geol. Surv.*), 1923, p. 77.

[7] ' The Geology of the Country around Kenilworth (Warwickshire),' *Quart. Journ. Geol. Soc.*, vol. lxxxv, 1929, p. 167 (see p. 200).

[8] See T. H. Whitehead, *op. cit.*, p. 171.

of Bowhills, near Romsley in Shropshire, where they have been well known to geologists since the time of Murchison.[1]

For the Breccia Group Prof. Boulton[2] adopts the term " Clent Beds," also proposed (as Clent Group) by Arber.[3] This can be accepted, but it must be borne in mind that the development of the group at Clent is hardly representative, for it there consists entirely of breccias, whereas over a wider area marls and sandstones predominate.

As already mentioned (p. 34), a Permian age has been claimed for beds belonging to the Calcareous Conglomerate (Bowhills) Group on the evidence of certain fossils, consisting of reptilian or amphibian footprints and a few plant remains, found at Hamstead (Sheet 168).[4] Arber[5] was of opinion that this assemblage was more likely to be due to the conditions under which the beds were deposited than to their geological age, and considered that his Enville Series probably represented a desert facies of part of the highest Coal Measures (Stephanian) of the Continent.

Prof. Boulton,[6] whilst agreeing that the Calcareous Conglomerate Group may reasonably be grouped with the Carboniferous System, considers that the Breccia or Clent Group, on account of the unconformity at its base, should remain in the Permian System in which it, along with the underlying portions of the "Lower New Red Sandstone" (i.e., the Keele and Bowhills Groups), was placed by Murchison.[7] Prof. Boulton suggests that it may be in part a time-equivalent of the Magnesian Limestone Series of the north-east of England. There is no palaeontological evidence for the correlation of the Breccia or Clent Group with any part of the Permian System. The only fossils (other than those derived from older formations) hitherto found in the group consist of some undeterminable plant remains, some of which appear to be casts of large fern-like petioles, discovered by Mr. W. W. King in a borehole near Stourbridge.[8] The absence of palaeontological evidence, of course, equally prevents definite correlation with any part of the Carboniferous System. Much depends upon the importance attached to the unconformity at the base of the Breccia Group. Prof. Boulton[9] considers that in the Birmingham district the break between the breccias and the overlying Trias is generally not so pronounced as that at their base ; but does not claim that, in the

---

[1] 'Silurian System,' 1839, p. 48. In this work the name is spelt ' Bowells,' and on the Old Series Geological Map (Sheet 61 S.E.), it is ' Bowels,' spellings that reflect a local pronunciation that still survives.

[2] *Loc. cit.*

[3] *Loc. cit.*

[4] Hardaker, W. H., ' On the Discovery of a Fossil-bearing Horizon in the "Permian" Rocks of Hamstead Quarries, near Birmingham,' *Quart. Journ. Geol. Soc.*, vol. lxviii, 1912, p. 639, and E. Dix, ' Note on the Flora of the Highest " Coal Measures " of Warwickshire,' *Geol. Mag.*, 1935, p. 555 (see p. 557).

[5] *Op. cit.*, p. 41.

[6] *Op. cit.*, p. 365.

[7] *Phil. Mag.*, n.s., vol. xix, 1841, p. 419. See also ' Silurian System,' 1839, p. 54.

[8] See Arber, *op. cit.*, pp. 39, 40.

[9] ' The Rocks between the Carboniferous and Trias in the Birmingham District,' *Quart. Journ. Geol. Soc.*, vol. lxxxix, 1933, p. 53 (see p. 79) ; and *Geol. Mag.*, 1933, p. 560.

Midlands generally, the unconformity at the base of the Clent Group is greater than that at the base of the Bunter. In the present district the break at the base of the Bunter seems undoubtedly to be greater than that at the base of the Breccia Group (see, e.g., pp. 93, 122). The latter, indeed, cannot be detected where the breccias themselves are thin and the group is mainly represented by marls and sandstones ; and in such places conditions may be similar to those in the Kenilworth area, where Prof. F. W. Shotton[1] can find no unconformity between the equivalent of the Clent Beds and the underlying Calcareous Conglomerate Group. As Prof. Boulton has pointed out,[2] movements occurred in the Midlands intermittently throughout Upper Carboniferous times. It is doubtful (for example) whether those which resulted in the unconformity between the Breccia Group and the Calcareous Conglomerate Group were as intense and widespread as those which, in early Staffordian times, produced the "Symon Fault" in Coalbrookdale and the unconformity at the base of the Halesowen Group of South Staffordshire and of the Highley (Sulphur Coal) Group of the Wyre Forest. It is thus difficult to say what should be regarded as the "main" Hercynian movements in the Midlands of England ; but it is at least clear that these movements were not completed before the deposition of the Clent Beds. These beds, especially where they are represented by marls and sandstones, seem in their lithological characters to be linked to the underlying red Upper Carboniferous rocks ; but it appears reasonable to interpret the "trappoid" breccias as the products of rapid mechanical denudation resulting from elevations produced during a certain phase of the Hercynian movements. The relative importance attached to historical, stratigraphical, tectonic and lithological considerations will, however, inevitably vary with different observers, and, in the absence of palaeontological evidence, the age of the Clent Beds, whether Carboniferous or Permian, must remain in doubt.

Distribution.—The Enville beds crop out at various places along the western border of the South Staffordshire Coalfield. Of these that in the south-western corner of the district includes Clent and Walton Hills (Survey Photographs 2030, 2210-1, 2214-7), and Wychbury Hill (Survey Photographs 2202, 2204-5) near Pedmore. This is the type area of the Clent Breccia where that deposit has its coarsest and thickest known development. The Bowhills Group seems to be represented by some highly calcareous sandstones to the north-east and north of Walton and Clent hills ; but there are no calcareous conglomerates. Other outcrops of the Enville beds occur, between branches of the Western Boundary Fault, near Old Swinford, Amblecote and Kingswinford. A larger area is found west of Sedgley (Survey Photographs 1977-80), near Baggeridge Wood and Upper Penn.

The most extensive outcrop of the Enville beds occurs in the middle of the district, near Enville and Bobbington. Smaller outcrops are found

---

[1] 'New Evidence on the Origin of Breccias and Conglomerates in the Warwickshire Coalfield,' *Geol. Mag.*, 1933, p. 466 (see p. 474).

[2] *Quart. Journ. Geol. Soc.*, vol. lxxxix, 1933, p. 80.

UPPER COAL MEASURES: ENVILLE BEDS.

along the line of the Enville Fault (p. 128) near Compton and Kingsford, and south-east of Trimpley. To the west of the Romsley Fault (p. 129) an outcrop extends from Kingsnordley to the neighbourhood of Romsley. Finally, the Bowhills Group crops out on the north-west side of the Brockhall and Pattingham faults (pp. 129, 130), near Quatt.

DETAILS

CALCAREOUS CONGLOMERATE OR BOWHILLS GROUP

Details of this group in the eastern part of the district will be found in ' The Geology of the Southern Part of the South Staffordshire Coalfield ' (*Mem. Geol. Surv.*, 1927) as follows :—Clent Hills area, pp. 141-2, Brettel Lane and Kingswinford, pp. 142-3, Sandyfields, Baggeridge Wood, Goldthorn Hill, etc., pp. 144-5.

Enville, Four Ashes, Gatacre, etc.—To the south of Enville Sheepwalks a quarry at Compton Court Farm, 500 yds. N.W. of Compton, shows about 16 ft. of rather coarse calcareous conglomerate with pebbles of limestone and chert up to 4 or 5 in. long. The rock is slickensided and jointed, the joints running N.20°E., approximately parallel to the Enville Fault which passes about 150 yds. east of the quarry. The dip is to north-east, and may be as high as 20°. The flatter, clayey outcrop of the Keele Group begins a few yards south of the quarry, and the base of the Breccia Group lies some 160 to 170 yds. north of it ; so that, with a dip of 20°, the breadth of outcrop leaves room for a thickness of not more than 175 ft. in the Bowhills Group. The lens of conglomerate dies out rapidly both to east and west, though west of the quarry a few scattered pebbles occur in sandstones. About 450 yds. W. by S. of the quarry another lens of conglomerate sets in. This underlies dull red coarse sandstones, with a few scattered pebbles, exposed in a small quarry about 650 yds. W. by S. of that at Compton Court Farm. The dip here is not more than 10°, and there is still room for no more than 175 ft. in the Bowhills Group. Farther to west-south-west the dip decreases rapidly and the outcrop broadens. Near No Man's Green three ' ribs ' due to hard beds can be distinguished, but in none is a conglomerate developed, although the sandstones in an old gravel pit 200 yds. south of No Man's Green are coarse and pebbly.

Between No Man's Green and Gatacre the outcrop is traversed by several faults, most of them small, and it is frequently impossible to identify individual lenticular beds of sandstone on opposite sides of these faults.

From No Man's Green to the moat half a mile south-west of Cox Green the lowest beds of the Bowhills Group make a bold bluff, but no conglomerates are developed. In a higher sandstone belt, however, a considerable lens of conglomerate occurs and is exposed in a gravel pit about 500 yds. S.W. of Cox Green. The pit showed (in 1929) flaggy lenticular sandstone up to about 2 ft. thick, in the middle of 12 ft. of conglomerate. It is estimated that the beds from the base of the Bowhills Group, near the moat, to the top of this conglomerate lens correspond approximately to that part of the group which crops out near Compton Court Farm. Near the Hollies and Cox Green, however, higher beds have come in, and, in a sandstone belt amongst these, a quarry at Cox Green shows about 10 ft. of brown, false-bedded sandstones without pebbles.

In the neighbourhood of Four Ashes the outcrop shows what is probably the maximum thickness of the Bowhills Group. With a dip of 5° this would amount to about 400 ft. Dips of 5° or 6° have been noted in places ; but in false-bedded rocks such observations are not reliable and the general dip may be less. Still, the thickness is probably well over 300 ft.[1] Hereabouts conglom-

---
[1] W. W. King (*Quart. Journ. Geol. Soc.*, vol. lv, 1899, Table I, facing p. 108), estimates it at 325 ft.

erates do not appear in the lower part of the group, but several lenses occur in the upper part. One of these is exposed in a quarry about 600 yds. E.S.E. of Four Ashes Hall, on the south side of the Enville road. South-east of the quarry a fault is indicated by interruption of the features and by a strong spring known as White Well. The shift of the outcrops is south-westward on the south-east side of the fault, but where the conglomerate would be expected to appear on this side there are only flaggy calcareous sandstones to be seen. Another large lens of conglomerate crops out in the grounds of Four Ashes Hall and is exposed in the cellars.

Near Tuckhill two conglomerate lenses occur in the top of the lowest belt of calcareous sandstones, which hereabouts make a very bold feature, rising abruptly from the flat clayey ground of the Keele Group. One of these conglomerates crops out at Dudhill, indicated by numerous pebbles on the surface and by an exposure of conglomerate resting on reddish-brown sandstone near the house. This lens dies out rapidly northward, for the road section north-east of the house shows only coarse sandstones, with scattered pebbles and chips of chert, on brown-red calcareous sandstones with partings of chocolate marl. The second lens of conglomerate is at Tuckhill and there is a quarry in it, now much overgrown, east of the church. Another quarry, just north of the church, shows calcareous conglomerate on sandstone, and conglomerate is exposed, north of the church, in the road to Six Ashes. The lens, its outcrop slightly shifted by a fault, extends northward to Cherry Orchard Farm, 400 yds. N. by W. of the church. Near Tuckhill and Six Ashes seven or eight lenticular beds of more or less calcareous sandstone can be distinguished. North-east of Six Ashes three lenses of calcareous conglomerate occur in the upper part of the group, nearly one above the other.

Near Broad Oak and Gatacre Park there is no evidence of any conglomerate, but scattered pebbles of red and yellow chert occur in the sandstones in places. The Bridgnorth road, 550 yds. N.E. of Broad Lanes, exposes brown sandstones, in part coarse, but without pebbles. About two thirds of a mile north of Broad Lanes the base of the Bowhills Group meets the Pattingham Fault (p. 129). The beds of the higher part of the group, which crop out farther north, include a large lens of conglomerate at the Castle, 730 yds. N.W. of Gatacre. A quarry near the Castle shows about 12 ft. of calcareous conglomerate with pebbles, up to 3 in. long, of limestone, chert, quartzite and quartz ; a few appear to be of volcanic rock. About 500 yds. north of the Castle the outcrops are shifted westward by a cross-fault. On the north side of this fault dull red or reddish-brown sandstones, in part coarse-grained, crop out north-east of Lower Beobridge and are exposed, in a broken and slickensided condition, along the road to Beobridge. These are unlike Triassic sandstones and are probably part of the highest beds of the Bowhills Group, brought against the south-east side of the Pattingham Fault.

In the Claverley borehole 287 ft. of beds, beginning at 185 ft. 6 in. from the surface, can be placed in the Bowhills Group ('Middle Permian').[1] They consist of red, brown and grey calcareous sandstones, with beds of crimson marl and one or two of conglomerate.

Kingsnordley, Bowhills and Romsley.—At Kingsnordley the Bowhills Group crops out, with a north-west strike, on the west side of the Romsley Fault. The beds consist mainly of sandstones, with relatively thin marl-partings ; but two lenses of conglomerate are indicated by abundant pebbles on the surface respectively 500 yds. and 950 yds. S.W. of Kingsnordley. They are both in the same belt of sandstones, but seem to be unconnected. A much larger lens, in the next sandstone belt above, extends from 800 yds. S.W. of Kingsnordley southwards to Coton Farm. This, again, is shown by many pebbles, chiefly of chert, on the fields. A gravel pit 350 yds. south of Coton Farm in a lower lens shows

---

[1] See W. Gibson, *Trans. Inst. Min. Eng.*, vol. xlv, 1913, p. 35.

about 20 ft. of calcareous conglomerate with pebbles up to 6 in. long, mostly of chert and limestone, with some of quartzite.

The lowest beds of the Calcareous Conglomerate (Bowhills) Group make a feature near Green House. They consist of sandstones, pebbly in places, but without actual conglomerates. The ground near Green House and Coton Hall illustrates very well the discontinuous character of the hard beds in the Bowhills Group by the manner in which the features to which they give rise die out rapidly or bifurcate. The dingle south of Coton Hall exposes the lower part of the group, though rather overgrown and obstructed by fallen timber; the lowest beds crop out 630 yards S.W. of Coton Hall and consist of calcareous sandstone and cornstone, while similar beds, separated by marls, crop out higher up the dingle, and, at a small waterfall 400 yds. south of Coton Hall, some 10 ft. of flaggy sandstones rest on 15 ft. or more of calcareous conglomerate.

Conglomerate extends from about 400 yds. S.E. of Coton Hall southwestward to the road, probably as two distinct lenses. The lower one has been quarried south of the road, 700 yds. west of Astley, where about 15 ft. of rather coarse conglomerate is shown (Survey Photographs 6891-2). Derived Lower Carboniferous fossils, in pebbles, or detached in the matrix, are fairly common here, and the following forms have been collected[1] :—

Crinoid fragments
Caninia ?
Koninckophyllum ?
? Lithostrotion
   (Diphyphyllum) sp.
Michelinia tenuisepta (*Phillips*)
Michelinia sp.
Syringopora sp.

Zaphrentis delanouei ?
   Edwards & Haime
Z. konincki ? *Edwards & Haime*
Zaphrentis ?
Fenestella cf. membranacea
   (*Phillips*)
Cleiothyridina royssii (*Davidson non Léveillé*)
Spirifer ?

The fossils are fragmentary and Dr. Stubblefield considers that the species recognized are not sufficiently diagnostic for a reliable opinion to be given either of age or of likely area of derivation.

Bowhills Dingle affords further exposures in the lower part of the Bowhills Group, the base of which crosses it 650 yds. S.W. of Astley. Higher up the dingle the proximity of an outcrop of *Spirorbis* limestone is indicated by nodules and fragments. The dingle is in the Bowhills Group up to where the Romsley Fault crosses it, 700 yds. a little east of south of Astley, and if the outcrop of the limestone is below this point it would provide the first instance of a Spirorbis limestone in the Calcareous Conglomerate Group of the Enville Beds in this or any other district with the exception of the Warwickshire Coalfield, where the Astley Court Limestone lies in the corresponding group.[2] It is, however, quite possible that the fragments of limestone have been carried down from one of the Keele outcrops in Perryhouse Dingle and the upper part of Bowhills Dingle mentioned on p. 85.

Near Bowhills the outcrop of the Bowhills Group forms a broad ridge rising above the relatively low-lying ground to the west and to the east on the upthrow side of the Romsley Fault. Some of the characteristic reddish-brown calcareous sandstones are exposed in the lanes south-east of Birds Green. A quarry 680 yds. E.N.E. of Dodds Green is in sandstone and a lenticular conglomerate that dies out a short distance to the south-east, but continues for about half a mile to the north-west. Bowhills itself is situated on the largest outcrop of calcareous conglomerate in the district, probably in two superimposed lenses

---

[1] Named by Dr. Stanley Smith (corals) and Dr. C. J. Stubblefield. Reg. Nos. T.W. 409–21, T.W. 481–96. See also W. W. King, *op. cit.*, Table IV, p. 125.

[2] Eastwood, T., in 'The Geology of the Country around Coventry' (*Mem. Geol. Surv.*), 1923, pp. 80, 90.

without any perceptible parting of sandstone or marl. Quarries in the upper and lower lenses show rather coarse calcareous conglomerate, with pebbles up to 6 in. long. An east to west fault crosses the outcrops about 150 yds. south of Bowhills; but the conglomerate continues to the south of it, and is exposed in an old quarry about 650 yds. south of Bowhills. The pebbles here again range up to 6 in. in length and include, besides limestone and chert, some of purple sandstone.

Farther south no more conglomerates appear, but some of the sandstones are pebbly. Beds near the base of the Bowhills Group, consisting of highly false-bedded sandstones, on sandstones in thicker beds, can be seen in a quarry 400 yds. W. of Romsley. Pebbly sandstones are exposed in another old quarry 400 yds. south of Poolhouse Farm.

Quatt.—South-west of Quatt a belt of sandstones forms a steep bluff in the Long Cover, on the east bank of the Severn. These sandstones, though there is some discontinuity of individual components, resolve themselves into three main bands too thin to be shown separately on the one-inch map. They, and higher beds, are well exposed near the Mill Pool, 500 yds. S.W. of Dudmaston Hall and in the dingle south of the Hall. At the outlet of the Mill Pool the following section from above downwards, in beds forming part of the lowest sandstone band, was noted :—Conglomerate, at least 15 ft.; sandstone, red, with green fish-eyes (this appears to pass, a few yards northward, into greenish and pink-mottled hard sandstone), 3 ft.; red marl, which passes northward into soft sandstone, and then wedges out, 0 to 4 ft.; sandstone, brown and purplish, hard, calcareous, 4 ft. The conglomerate is fairly coarse, with pebbles up to 6 in. long; all the largest are sub-angular and most are imperfectly rounded, some of the limestone pebbles are discoidal. The pebbles are of crystalline limestones, like those of the Carboniferous Limestone, purple compact limestone, grey compact limestone of ' *Spirorbis* ' type, purple quartzite and sandstone, and highly calcareous sandstone like that at the base of the section. The beds above the main sandstone belt, in the dingle south of Dudmaston Hall, include calcareous sandstones, red and green non-calcareous sandstones, and red marls.

An old quarry at the northern end of the Long Cover shows about 25 ft. of the sandstones in the highest band of the main belt. These are mostly brown in colour, but parts are red and dark purple. Most of them are calcareous. At the top are lenses of highly calcareous pellet rock with chips of purple marl, and of yellow material that is possibly chert. The bottom bed in the quarry contains a few sub-angular fragments of grey or yellow chert. The dingle about a quarter of a mile north of Lye Hall affords a good section in these and higher beds. Perhaps the most interesting exposure is at a 10-ft. waterfall about 280 yds. east of the mouth of the dingle. The rocks at the fall itself are thickly encrusted with tufa, but the banks show the following section from above downwards :— sandstone, flaggy, false-bedded, 6 or 8 ft.; marl, sandy, red, 10 ft.; sandstone, soft, marly, red with green blotches, 3 ft.; breccia, red with green streaks, soft, marly, small fragments, interbedded with sandstone and passing down into conglomerate, 8 to 10 ft.; sandstone, hard, calcareous, purple (forms ledge of fall), 4 ft.; marl, sandy, red, about 2 ft.; sandstone, calcareous, thick-bedded, probably 4-6 ft. The breccia resembles the finer varieties of the Clent Breccia, but differs in having a calcareous matrix; its constituents are similar to those in the conglomerate at the Mill Pool. The conglomerate into which the breccia passes down contains larger fragments, most of which are imperfectly rounded.

Conglomerate and fine breccias crop out a few yards above the fall; and similar breccias were found in the branch dingle commencing about 450 yds. N.E. of Lye Hall. Here the beds dip westward owing to folding in the vicinity of the Brock Hall Fault (p. 130). On the south-east side of this fault, higher up the same branch dingle, purple and green breccia with lavender and greenish sandstones dip steeply north-west, and are evidently close to the Kinlet Hall Fault (p. 130).

The Brock Hall Fault shifts the outcrops south-westward, but the main belt of sandstones can be recognized on its south-west (downthrow) side. Exposures in them, including fine breccias and pebbly sandstones, can be seen at various points south of Lye Hall.

The strata near Quatt, described in the preceding paragraphs, were referred tentatively to the Keele Group in the ' Summary of Progress ' for 1930. It is now possible to relate them to the beds penetrated in a borehole about two thirds of a mile north of Hampton (p. 91), and to estimate that the base of the main belt of sandstones in Long Cover lies about 850 ft. above the base of the Keele Group. This gives a thickness for the latter rather less than it seems to have near Alveley (p. 84) but still above the average for the district as a whole ; so that there need be no hesitation in referring the beds near Quatt, with their calcareous sandstones, conglomerates and breccias, to the Bowhills Group.

Trimpley.—About half a mile south-east of Trimpley church the Bowhills Group crops out from beneath the Clent Breccia near Jacobs Ladder. There are now no exposures in the present district, but farther south, near Warshill Farm (Sheet 182), the characteristic calcareous sandstones and conglomerate can be seen.

Breccia or Clent Group

For details of this group in the eastern part of the district the following pages of ' The Geology of the Southern Part of the South Staffordshire Coalfield,' (*Mem. Geol. Surv.*, 1927) should be consulted :—Clent Hills, Hagley Park and Wychbury Hill, pp. 148-9, Stourbridge, Amblecote and Kingswinford, pp. 149-50, Penn Common, etc., p. 150.

Enville, Gatacre and Bobbington.—In this area the lowest beds of the Breccia and Clent Group consist of a belt of breccias that can be traced north-westward from Enville Sheepwalks to Gatacre. The outcrop that forms the high ground of the Sheepwalks and extends into the grounds of Enville Hall is comparable in size with that of Clent itself. Its breadth is greater than one would expect from the thickness (which cannot greatly exceed 120 ft. though estimated by W.W. King as 225 ft.) ; but this is largely due to changes of dip near a small N.—S. fault and to dip-slope topography.

This outcrop affords very few exposures of the breccia in place ; but debris allows its general character to be deduced. It is, in general, coarse, though not uniformly so, and nowhere so coarse as on the Clent Hills. The bank of a lane 400 yds. E.N.E. of No Man's Green shows a few feet of coarse breccia with fragments up to 7 in. long. On the other hand an exposure 150 yds. S.W. of Leigh House Farm shows about 6 ft. of soft marly breccia with lenses of rather fine-grained sandstone. In general it may be said that in the Sheepwalks area few of the fragments exceed 4 in. in length. In the steep bank about 600 yds. S.W. of Leigh House Farm the coarse debris from rabbit burrows includes many examples of volcanic rocks like those of the Uriconian in Shropshire and of purple grit and sandstone like the Longmynd rocks. A discontinuous section in the upper part of this belt provided by a road-bank just south-east of Gilbert's Cross shows breccias varying from fine marly to fairly coarse with stones up to 2 in. long, interbedded with marls and coarse, gritty sandstones containing scattered sub-angular pebbles.

North-west of Gilbert's Cross the breccia-belt decreases steadily in thickness and in the size of its fragments. Near Four Ashes it is probably less than 80 ft. thick and few fragments are more than an inch or two in length. The breccia-belt can be traced, its boundaries becoming increasingly indefinite, by debris to Gatacre, where it can hardly be much over 20 ft. thick. The banks of a pond 470 yds. N.W. of Gatacre show red clay with angular fragments up to 2 in. long, but mostly much smaller.[1]

---

[1] W. W. King (*op. cit.*, p. 105) records fragments at Gatacre up to 6 in. in length.

The beds between this and a higher belt of breccias at Enville consist mainly of marls, some of which can be seen near the western end of Church Gorse. These are about 40·ft. thick, while the upper belt of breccias appears to be between 40 and 50 ft. thick. Farther north-west, as the breccia-belts become thinner, the separation between them increases. These upper breccias seem everywhere to be less coarse than the lower ones. Near Church Gorse fragments up to 3 in. long were observed, and Mr. King records some up to 6 in. Just north of Church Gorse a lens of breccia separates from the upper part of the belt and dies out in about 450 yds. About 550 yds. N.W. of Church Gorse the outcrop of the main part of the belt is shifted by a fault. Beyond the fault it can be followed by debris and a distinct ridge to which it gives rise, but the thickness of the breccias and the size of the constituents diminish rapidly until, about half a mile south-east of Bobbington, it becomes impossible to separate the breccias from the marls in which they lie.

The beds above the breccias consist of red marls or clays and sandstones in which there are very few exposures. North of Church Gorse red clays above the upper breccias have been dug for brick-making. In the banks of Philley Brook, from Poolhouse Farm eastward, red sandstones, generally rather fine and angular in grain, and purple and green sandy marls can be seen in places. Small sub-angular rock-fragments occur in some of the red clays. Farther north the outcrop becomes increasingly obscured by drift and drift debris. In a brook just north of Bobbington, red and green sandy marls with lenticular sandstones must lie only a short distance above the horizon of the upper breccias ; but here, and farther north, where these breccias can no longer be traced, the position of beds in the sequence is doubtful. Thus, some fine-grained red and green sandstones, with red sandy marls, exposed in the roadside three quarters of a mile south-east of Claverley, may be below the horizon of the upper breccia belt.

A better section, in the road about 700 yds. W. by S. of Aston, shows, from above downwards : purple and green marly clay, 3 ft. or more ; dull red sandy marl in thin beds, 4 ft. ; dull red flaggy micaceous sandstone, 3 ft. ; red, rather hard sandstone 2 ft. 6 in. ; soft sandstone, 3 ft. ; harder sandstone, 6 or 8 in. ; sandy marl, 6 in. ; hard sandstone, 10 in. ; sandy marl, 0 to 6 in. ; brown-red sandstone, seen for 4 ft. All these strata are false-bedded. They dip westward at about 20° and appear to be somewhat disturbed.

In the Claverley borehole the beds from the surface to 185 ft. 6 in. were placed by W. Gibson[1] in the ' Upper Permian ' (i.e., Clent Group). They consist of red marls and sandstones with two beds of breccia, one, 6½ ft. thick, at the base and the other, 16 ft. thick, with its base at 146 ft. from the surface. Another borehole,[2] put down between 1855 and 1865 behind the White Cross Schools, Bobbington, penetrated Enville beds beneath 20 ft. or more of drift. The former consist, judging by the record, of rather thick red, purple and mottled marls, with thinner sandstones and two beds described as ' conglomerate.' These may be breccia,[3] and the bottom of the Breccia Group may, perhaps, be placed at the base of the lower one, 474 ft. 9 in. from the surface. If these beds belong to the Clent Group an interesting feature is the apparent presence of traces of coal,[4] represented by ' black smuts ' in a hard mottled rock at 290 ft. 6 in.

---

[1] *Trans. Inst. Min. Eng.*, vol. xlv, 1913, p. 35.

[2] Section communicated by Mr. W. W. King. Mr. King (in MS. on section) gives a somewhat different interpretation of the record from that adopted herein. He regards some of the highest beds as Bunter and places the base of the Breccia Group (' Upper Permian ') at 228 ft. from the surface.

[3] The breccias in the Claverley borehole were described by the sinkers as ' conglomerates ' ; see W. Gibson, *loc. cit.*

[4] Two ' smuts ' above 20 ft. were probably carbonaceous streaks in the drift, and a 2-in. coal which T. C. Cantrill (*Quart. Journ. Geol. Soc.*, vol. li, 1895, p. 541) was informed was found in two wells at White Cross may have been the same. ' Light gobstone ' recorded in the White Cross borehole at 456 ft. is believed by Mr. King to indicate a bad coal.

UPPER COAL MEASURES : ENVILLE BEDS.

A note on the section states that ' in a lane near the Bank a thin Permian coal crops out.' The Bank is half a mile north-west of White Cross, and about 700 yds. N.E. of the site of the Claverley borehole, and there is hardly room for doubt that the beds that crop out there belong to the Breccia or Clent Group.

A third borehole that presumably penetrated the Clent Group was put down in the years 1857-8 at Highgate Common (or Forest).[1] The total depth was 565 ft. Below 12 ft. or more of drift the strata down to 150 ft. are, with little doubt, Lower Mottled Sandstone. The remainder consist of red sandstones and marls, with one bed 8 ft. thick described as conglomerate, at 281 ft., and another 2 ft. thick called ' rocky marl with pebbles,' at 467 ft. The description of the latter certainly suggests a breccia of the Clent Group, and the whole of the strata below the Bunter may belong to that group ; though some of the lowest may belong to the Bowhills Group[2] (see below, p. 123).

Compton, Kingsford, etc.—Four detached outcrops of the Breccia Group occur along the line of the Enville Fault between Compton and the Kidderminster-Bridgnorth road. They lie on the downthrow (east) side of the fault, their boundaries with older rocks to the west being sharply defined and comparatively straight. On the other hand, the boundaries with the Lower Mottled Sandstone to the east are, except in one case, sinuous and evidently controlled by the contours of the ground, and with the exception mentioned are considered to be unconformable junctions. On Sheets 61 S.E. and 55 N.E. of the Old Series map the boundaries on both sides are shown as faults.

All these outcrops, with the exception of the southernmost, give rise to marked features. The most northerly, which begins at Pigeon House Farm, forms a hill known locally as ' Bladders Bank,' near the Lydiates. There are no sections, but breccia debris includes fragments up to 5 in. long. The second outcrop which begins about 350 yds. S.W. of the Lydiates extends south-south-westward for over a mile, and it is possible to distinguish three ridges of relatively coarse breccia separated by hollows consisting mainly of marl with, perhaps, some finer breccia. The Lower Mottled Sandstone transgresses these bands, at both ends of their outcrops, in a manner suggestive of overstep rather than of faulting. The breccias of the lowest (westernmost) band are exposed in the roadside 450 yds. S.W. of the Lydiates, where they are coarse, with fragments up to 8 in. long. Finer breccias in the same band may be seen in the brook 450 yds. S.S.E. of Greyfields Court. The dip here is 35° to W. by S. A small quarry 200 yds. N.E. of Castle Hill shows a few feet of breccia in the middle band, with fragments up to 4 in. long. No exposures were found in the highest band, but the debris seems to be finer.

The third outcrop lies north-west of Horseleyhills Farm and is recognizable by its feature and by the presence of angular fragments of tuffaceous sandstone, quartzite and other rocks on the surface. It is possible that the junction with the Lower Mottled Sandstone in this case may be a fault. In the case of the fourth outcrop, on the Bridgnorth road, 630 yds. S.W. of Horseleyhills Farm, the junction with the Bunter is certainly a fault, for the breccia is in contact with the Pebble Beds. This outcrop is only a few yards broad and is too small to be shown on the one-inch map. The breccia was temporarily exposed in a pipetrench in the winter of 1931-32, when it was seen by Mr. King. Moderately coarse debris was still visible later in 1932.

Trimpley.—Farther to south-south-west, near Jacob's Ladder, the Breccia Group again crops out, faulted against lower measures on the west, and against

---

[1] Section in J. B. Jukes's MS. collection. Another version, differing in unimportant details, was communicated by T. C. Cantrill, who received it from Mr. W. W. King in 1895. From local information, obtained in 1929, the site appears to be about 450 yds. nearly due north of Highgate Farm.

[2] Mr. W. W. King (*Trans. Inst. Min. Eng.*, vol. lxi, 1921, p. 167) considers that the strata below the Bunter, to 311 ft., are " Permian " conglomerates (Bowhills Group) and those below Keele Beds.

Bunter Pebble Beds and Lower Mottled Sandstone to the east, but overlying the Bowhills Group to the south (p. 101). The outcrop seems to consist entirely of breccia ; this is moderately coarse and includes fragments of purple sandstone and igneous rocks.

Romsley, Astley and Coton Hall.—About 500 yds. south of Bowhills an outlier of Clent Breccia rests directly upon a calcareous conglomerate. The outcrop is on arable land and there is considerable mingling of the respective constituents on the surface, but the boundary of the breccia is partly defined by a small feature. Another outcrop, indicated by similar evidence, lies just northwest of Astley, and is bounded on the east by the Romsley Fault. A third outcrop, north of Coton Hall, forms a broad band and gives rise to a slight ridge. The breccia here rests for the most part upon marls and sandstones in the Bowhills Group but, near Coton Farm, it is in contact with one of the calcareous conglomerates and there is, again, much mingling of the respective constituents. The breccia, judging from soil indications, is overlain by marls with some sandstone. The breccias of the three outcrops described in this paragraph are fairly coarse, but less so than those in the Sheepwalks area.

T.H.W.

CHAPTER VI

TRIASSIC ROCKS

GENERAL ACCOUNT

The present district, with part of that to the north,[1] has good claims to be considered the type area of the English Trias.[2] The Keuper Marl does not occur in the area represented on Sheet 167, but the Lower Keuper Sandstone and all the members of the Bunter Series are well and typically displayed, and occupy, either at the surface or immediately beneath the superficial deposits, rather more than two-fifths of the area.

Broadly speaking, the Triassic rocks form part of the great syncline separating the older strata of Shropshire from those of the South Staffordshire Coalfield ; but in the present district that syncline is split in two by the pre-Triassic rocks of the Trimpley, Enville and Bobbington Anticline. The structure is further modified by faults which replace, wholly or in part, the eastern limbs of the two subsidiary synclines, and, as a result, the dip of the Triassic rocks is predominantly eastward over the whole district. Moreover, repetitions produced by these strike faults increase the breadth of outcrop and affect the scenery by giving rise to three bold escarpments of the Bunter Pebble Beds and four of the Lower Keuper Sandstone. The Triassic rocks have suffered less disturbance than the older strata, upon which they rest unconformably ; but their overstep in the present district is not very striking, as the rocks upon which they lie range only from Keele Group in the west to Clent Beds in the centre and east.

The succession and limits of thickness of the several subdivisions of the Triassic rocks will be found in the table of formations, Chapter I (p. 3).

Lower Mottled Sandstone.—The lowest member, the Lower Mottled Sandstone, is generally brick-red in colour, ranging to reddish-brown or buff, with irregular blotches or streaks of a pale greenish colour. It is usually soft enough to be worked with a pick, though sufficiently coherent to stand in vertical faces, thus making possible the excavation of "rock-houses" near Bridgnorth and Kinver Edge. Where protected from denudation by overlying Pebble Beds or by even a thin covering of superficial gravel the Lower Mottled Sandstone will stand up as a steep escarpment or steep-sided bluffs, ridges and knolls. Where not so protected it is reduced to relatively low-lying, gently sloping ground.

The Lower Mottled Sandstone is of medium to rather fine grain, the

---

[1] See 'The Country between Wolverhampton and Oakengates' (*Mem. Geol. Surv.*), 1928, Chap. VII.

[2] See E. Hull, 'Triassic and Permian Rocks of the Midland Counties of England' (*Mem. Geol. Surv.*), 1859, p. 32.

larger grains being well rounded, but the smaller ones imperfectly so. In places the grains are particularly well rounded and of approximately uniform size and the rock approaches the "millet seed" type. In the present district the formation is devoid of pebbles except rarely at its summit, and at its base near Eardington (p. 110), where a conglomerate occurs.

"False" bedding is a conspicuous feature and the laminae frequently slope at high angles. This circumstance, together with the roundness of grain, suggests that the Lower Mottled Sandstone is of aeolian formation. This mode of origin now seems to be conclusively proved by the valuable studies of the Lapworth Club of Birmingham University, an account of which has recently been published,[1] which show that in range of grain-size and degree of sorting the Lower Mottled Sandstone agrees closely with dune and desert sands. The cross-lamination may dip in any direction, lenticular units with the dip in different directions, or at different angles, being superimposed. A westerly direction of dip is, however, distinctly prevalent, especially in the more highly inclined laminae which Prof. Shotton interprets as representing the leeward slopes of barchan dunes. In the north-western part of the district the direction tends to be between north and west; in the south-eastern part it is more commonly between south and west.

The thickness of the Lower Mottled Sandstone varies from about 600 ft. in the Bridgnorth area to, possibly, 400 or 500 ft. near Highgate Common and over 825 ft. at Kinver. Near Stourbridge it is a little over 250 ft. It is absent in the neighbourhood of Clent and of Penn Common.

Bunter Pebble Beds.—Owing to the repetition due to strike faults the Pebble Beds crop out in three principal areas. The central one, with an escarpment extending from Abbot's Castle Hill to the neighbourhood of Wolverley, may be considered the main outcrop. The eastern outcrop is more broken, but includes an escarpment in and to the north of Himley Park, and another extending from Ridgehill Wood, near Kingswinford, to Bunker's Hill. Smaller outcrops, detached by faulting, occur between Himley and Kingswinford, and others near Penn Common, at Hagley and on the Clent Hills. The western outcrop is that of the Bridgnorth area, with its bold escarpment to the east of the Severn.

The Bunter Pebble Beds, with a thickness of from 370 to 400 ft., are sharply marked off from the Lower Mottled Sandstone and generally rest upon an eroded and pot-holed surface of the latter (Plates V B and VII A). As a whole they consist of sandstones and pebbly sandstones of coarser and more irregular grain than the Lower Mottled Sandstone, and of a duller, brownish-red colour. The basement bed, in the Bridgnorth area and the main escarpment, is a breccia or coarse grit, with sub-angular fragments and some part-rounded pebbles. Near Bridgnorth and on Abbot's Castle Hill the matrix is calcareous, and the basement bed stands in a bold cliff above the Lower Mottled Sandstone, or even overhangs it. In the main escarpment, south of Abbot's Castle Hill, the matrix of the

---

[1] Shotton, F. W., 'The Lower Bunter Sandstone of North Worcestershire and East Shropshire,' *Geol. Mag.*, 1937, pp. 534–53.

A 6841

A.—QUEEN'S PARLOR, BRIDGNORTH; PEBBLE BEDS ON LOWER MOTTLED SANDSTONE.

A 6838

B.—LOWER MOTTLED SANDSTONE, FALSE-BEDDED; BRIDGNORTH.

basement bed becomes in places soft and sandy or even marly, and decreases in thickness until, at Kinver Edge, where it caps the escarpment, it is but 2 ft. thick. In the eastern outcrops no distinct basement breccia occurs, but the lowest beds there include, in places, lenses of calcareous breccia. The actual basement bed is often a non-calcareous conglomerate though there may be, as at Wollaston, conglomerate with a calcareous matrix a few feet above.

The pebble beds vary from compact, well cemented pebbly sandstones and conglomerates, in places calcareous and hard, to nearly incoherent gravel or shingle, with but a small amount of sandy matrix. On the whole the former type prevails except near Clent and Penn Common. It is noteworthy that, in this and other districts, the incoherent type tends to occur where the Lower Mottled Sandstone is absent.

While it is generally agreed that the Bunter Pebble Beds are of fluviatile origin, there is less agreement as to the source of the contained pebbles. In the present district, as in other areas, quartzites of various types are the commonest constituents, but do not, perhaps, predominate so overwhelmingly as in some localities. Pebbles of Carboniferous limestone and chert, of volcanic rocks resembling those of the Uriconian and of purple tuffaceous grits like those of the Longmyndian are fairly common, and fragments of fossiliferous Llandovery sandstone and of cornstone and sandstone like those of the " Old Red Sandstone " occur. The pebbles of Carboniferous rock are most abundant in the main and western outcrops, and, since some of this material is of larger size than that in the Bowhills Group, neither a distant nor second-hand source need be sought. The rocks of Longmyndian and Uriconian type, on the other hand, may well have been derived from the Clent Breccia, as the average size of such pebbles is less than that attained in the coarser parts of that formation. The Llandovery sandstone and the " Old Red " rocks could probably have been derived from a Midland source, either directly or, in the case of the former, through the Clent Breccia. Some of the quartzite pebbles also could conceivably have originated in the Midlands. The suggestion[1] that they may have been derived, ready made, from Longmyndian conglomerates could, however, hardly apply to any large proportion, since, on the average, the pebbles in these conglomerates are distinctly smaller than those in the Bunter Pebble Beds. Still, the common purple or " liver-coloured " type is known in the Cambrian Quartzite formation of the Midlands. It would seem that the only extensive exposures of Cambrian and pre-Cambrian rocks in the Midlands in Bunter times were to the east of the present district.[2] Dr. W. F. Fleet[3] concludes that the heavy mineral suites of the Bunter, especially

---

[1] Sherlock, R. L., 'A Correlation of the British Permo-Triassic Rocks,' *Proc. Geol. Assoc.*, vol. xxxvii, 1926, p. 1 (see p. 8).

[2] See L. J. Wills, 'An Outline of the Palaeogeography of the Birmingham Country,' *Proc. Geol. Assoc.*, vol. xlvi, 1935, p. 238, and Plate 16. (Note, the title of this plate has been interchanged with that of Plate 17.)

[3] 'The Heavy Minerals of the Keele, Enville, " Permian " and Lower Triassic Rocks of the Midlands, and the Correlation of these Strata,' *Proc. Geol. Assoc.*, vol. xxxviii, 1927, p. 1 (see p. 34); and ' Petrography of the Upper Bunter Sandstone of the Midlands,' *Proc. Birm. Nat. Hist. and Phil. Soc.*, vol. xv, 1929, p. 213 (see p. 214).

of the Pebble Beds, indicate a " supply from freshly-denuded crystalline or metamorphic rocks." The newly-derived minerals may have come from distant sources, and much of the quartzitic, igneous and metamorphic material in the Bunter Pebble Beds probably had an origin outside the Midland area. Though quartzite pebbles containing Ordovician and Devonian fossils, like those in the Budleigh Salterton Pebble Bed, have not yet been found in the present district, prolonged search might disclose them, as they have been recorded from several other Midland localities.[1]

The conglomerates and shingle-beds lie in the lower portion of the Bunter Pebble Beds. Individual lenses are very impersistent, as is strikingly shown by the fact that a borehole at Prestwood,[2] the site of which is surrounded by outcrops of conglomerate, passed through 626 ft. of sandstones, of which about 370 ft. must be assigned to the Bunter Pebble Beds, without meeting a single pebble. In the southern part of the main outcrop, near Wolverley and Cookley, pebbly sandstones and coarse pebbly grits, in part calcareous, range nearly to the top of the formation, but conglomerates of massed pebbles are rare. Elsewhere only occasional layers of sub-angular fragments are found in the higher beds. Near Bunker's Hill highly micaceous sandstone occurs near the top of the formation.

The boundary between the Bunter Pebble Beds and the Upper Mottled Sandstone is vague at the outcrop ; but in many borehole sections a bed of marl $1\frac{1}{2}$ to 4 ft. thick is recorded at what may be considered to be the junction.

Upper Mottled Sandstone.—The Upper Mottled Sandstone differs but little from the Lower Mottled Sandstone in colour and texture. In general the grains are more rounded, and in parts the interstitial material is marly, giving a " loamy " character to the rock which renders it suitable for moulding sand. Thin beds of red marl and of micaceous sandstone occur. Although false-bedded, like the rest of the Bunter Series (see Survey Photograph 6873), the laminae in the Upper Mottled Sandstone are generally less steeply inclined than those in the Lower Mottled Sandstone, and parallel laminae persist for greater horizontal distances and through greater thicknesses of rock. From this it may perhaps be concluded that the formation was, in the main, water-deposited. Its thickness, in the eastern part of the district, ranges from 500 ft. or over in the north to about 200 ft. in the south ; probably between 300 and 400 ft. near Stourbridge. In the western part of the district the outcrop is too much interrupted by faults to allow of a reliable computation of thickness, in the absence of data from boreholes.

Lower Keuper Sandstone.—The outcrop of the Lower Keuper Sandstone is divided, as a result of the folding and faulting, into four principal

---

[1] See T. G. Bonney, 'The Bunter Pebble-beds of the Midlands and the Source of their Materials,' *Quart. Journ. Geol. Soc.*, vol. lvi, 1900, p. 287 (p. 290) ; and C. A. Matley, ' The Source of the Pebbles of the Bunter Pebble-beds of the English Midlands,' *Geol. Mag.*, 1914, p. 211.

[2] Section communicated by the South Staffordshire Waterworks Company ; cores examined by Mr. Whitehead.

areas; of these two are continuations of those in the Wolverhampton Sheet 153. An outlier near Lower Penn and Oreton to the west of the Stapenhill Fault (p. 127) forms a third, and the fourth, and largest, extends from Stourbridge to the southern border of the district near Broom. In addition, a small outlier of the lowest beds, on the downthrow side of the Stapenhill Fault west of Wollaston, is of interest.

The outcrop of the Lower Keuper Sandstone is, topographically, well marked off from that of the Upper Mottled Sandstone by an escarpment only less prominent than that of the Bunter Pebble Beds. In detail, however, the line of demarcation is less distinct. The coarse basement beds of the Lower Keuper Sandstone rest in most places upon a bed of marl or sandy marl, generally from 1 to 3 ft. thick, and frequently taken as the base of the Keuper. Between it and the typical Upper Mottled Sandstone, however, there occur in the northern parts of the district some 30 to 40 ft. of sandstones, in part calcareous, some coarse in grain and a few pebbly, with thin beds of micaceous and sandy marl. Some of these sandstones contain " millet-seed " grains. In the northern margin of the present district and in the adjacent areas to the north (Sheet 153) the lowest bed of calcareous sandstone with " millet-seed " grains has been taken as the base of the Keuper. Even in the Stourbridge area sandstone not of Upper Mottled Sandstone character occurs in places between the typical beds of that formation and the coarse Keuper "basement beds." The most marked break occurs at the base of these coarse beds, where there are usually signs of erosion. These, however, are not more conspicuous than those at the base of the Bunter Pebble Beds; and there is no evidence in the area under review of any general unconformability of the Lower Keuper Sandstone.

In this district the Lower Keuper Sandstone cannot be divided into Basement Beds, Building Stones and Waterstones. The basal beds (above the marl bed) consist of coarse pebbly sandstones and lenses of calcareous grit with marl pellets (" cat brain "). The majority of the pebbles are of quartz and quartzite, but limestone pebbles are rather common in the western outcrops. The average size is smaller than in the Bunter Pebble Beds. In the Stourbridge and Broom area the pebbly beds range well above the middle of the formation. Some non-pebbly sandstones, dull or brown-red in colour and angular in grain, are interbedded with them, but there is no continuous group of non-pebbly sandstones, analogous to the Building Stones of other areas. The highest beds lie outside the district, but both to north and to south the transition to the Keuper Marl is rapid.

DETAILS

LOWER MOTTLED SANDSTONE

Wordesley, Wombourne, Himley and Wollaston.—At Wordesley a small area of Lower Mottled Sandstone crops out on the east side of the Stourbridge syncline (p. 124). About 50 ft. of false-bedded red sandstone are exposed, beneath the Pebble Beds, in a quarry 300 yds. N.E. of the church.

On the east side of the Stapenhill Fault (p. 127), near Wombourne, red and yellowish-mottled sandstone appears immediately below the Pebble Beds behind the Wodehouse. In an exposure at Himley Hall, in the continuation of the same outcrop, the highest part of the Lower Mottled Sandstone is mainly soft red sandstone, but has gritty lenses and a few very small pebbles.

Many exposures of the Lower Mottled Sandstone can be seen in the banks of the canal and mill-leat south of the River Stour where the latter breaks through the Bunter Pebble Bed escarpment near Audnam. A considerable face of Lower Mottled Sandstone, well displaying the false bedding, was formerly to be seen at the Ridge Sand Mine (Survey Photograph 2196), north of the road about half a mile W. by N. of Wollaston church.

Abbot's Castle Hill, Kinver, etc.—In places along the Abbot's Castle Hill escarpment the Pebble Beds can be seen resting upon the false-bedded, red, Lower Mottled Sandstone. Farther south there are few exposures of the top part of the latter formation but several small ones in lower portions occur on Highgate Common. Numerous small sections can be seen between Enville and Stourton and in the banks of the canal near the Hyde, and in some of these the steeply inclined false-bedding laminae are striking. The inclination is usually to south of west. The best exposures, however, are to be found on Kinver Edge (Survey Photographs 6894-5), Blakeshall Common and near Drakelow, where many crags of Lower Mottled Sandstone display the false bedding to advantage. In one of these crags, Holy Austin Rock (Survey Photograph 6893), a 'rockhouse,' still inhabited, has been made in the soft sandstone, and other such excavations, now abandoned, may be seen at Nanny's Rock and near Baxter's Monument. In some of the beds, about 100 ft. below the top of the formation, lenses with well rounded and polished grains of 'millet-seed' type occur. Beds near the base of the formation can be seen in the roadside about 250 yds. S.W. of the Lydiates.

T.H.W.

Bridgnorth, Quatford and Quatt.—North of Bridgnorth the Lower Mottled Sandstone is well exposed on both sides of the Severn valley. Its general dip is here about 7° to the east and its total thickness may be estimated at about 550 ft. On the west it overlies the Keele Group, as can be clearly seen near Stanley Hall and Hoards Park, while on the east it forms the striking features of Pendlestone Rock and High Rock (Survey Photographs 6844-5) overlooking the river.

Bridgnorth High Town is built upon a mass of the sandstone that rises 100 ft. or more above the river and shows fine sections of the false-bedded rock (Survey Photographs 6837-9, 6842). On the east side of the Severn there are good exposures in the cutting of the Wolverhampton road below the Hermitage.

South of the town, near Oldbury, the outcrop is confined to the east side of the valley but at Knowlesands it crosses again to the west side and is well exposed in the road and railway cutting, while its base can be seen resting on Keele beds in the pit of the Bridgnorth Brickworks.

There are many outcrops round Eardington (Survey Photographs 6862-4) and at the old weir north-east of Astbury Hall the sandstone rests, with a conglomeratic base, on purple marl of the Keele Beds. At Quatford the river cliff and road cuttings (Survey Photographs 6854-5) show excellent sections and in the highest beds around Gags Hill nodules of the sandstone cemented by barium sulphate are abundant.

R.W.P.

Sections in the Bridgnorth road (Survey Photographs 6856-7) north-west of Dudmaston Hall show the red sandstone with false bedding steeply inclined, mainly in a north-westward direction. Other exposures occur in the east bank of the Severn, near Lodge Farm, and along the stream east of the Holt, Quatt.

BUNTER PEBBLE BEDS

Clent and Hagley.—On the south-west flanks of the Clent Hills and near Hagley, the Pebble Beds, resting directly upon the Clent Breccia, are of the shingle-bed type with some lenses of red or brown sand or sand-rock. In Great Farley Wood white vein quartz was observed by Mr. Eastwood to be a fairly common constituent of the pebble-beds.

Upper Penn and Wombourne.—On Penn Common the Pebble Beds are, again, of the semi-coherent, shingle type and here also rest directly upon the Clent Beds. In a gravel pit on Colton Hills Dr. Robertson noted brown sand with the shingle and some beds of hard sand-rock. A fair proportion of the pebbles are more than 6 in. long. They include, besides the usual quartzites, grits, rotten volcanic rocks, much silicified limestone and many lumps of marl like that in the pit of Penn Brickworks (in Clent Beds) close by.

At the Wodehouse, Wombourne, the basal beds of the Pebble Beds consist, in descending order, of : pebbly sandstone, 12 ft. or more ; soft, thin-bedded marly sandstone, 6-8 in. ; calcareous breccia, about 3 ft., separated by a marly parting of about 1 ft. from the Lower Mottled Sandstone. Near the Foxhills the Pebble Beds form a double escarpment owing to a bed of hard conglomerate, with a calcareous matrix, about 50 ft. above the base.

Kingswinford, Wordesley, Wollaston, etc.—In the escarpment (Survey Photographs 2199, 6899-6901) extending from Ridgehill Wood, Kingswinford, to Bunker's Hill, the lowest beds of the Bunter Pebble Beds appear everywhere to be non-calcareous conglomerates. In a borehole at the Stourbridge Waterworks pumping station, 120 yds. south of Ashwood Farm (the Tack, see p. 189), Mr. W. W. King noted that the lowest beds of the formation consist, in descending order, of : micaceous dull red sandstone, 13 ft. ; conglomerate, 20 ft. 6 in. ; dull red sandstone, 3 ft. ; conglomerate 13 ft. 6 in. In places, however, hard conglomerate with a calcareous cement occurs a short distance above the base. Such is exposed in the road (Survey Photograph 2195) 900 yds. west of Wollaston Church, and, again, 850 yds. south by west, in another road. About 50 yds. west of the first exposure, the basement beds, consisting of non-calcareous pebble-beds, rest on an eroded and pot-holed surface of Lower Mottled Sandstone (Plate V B).

The upper part of the Bunter Pebble Beds can be seen in the left bank of the Stour. An exposure about 500 yds. S.W. of Audnam shows, in descending order : brownish-pink sand-rock with scattered pebbles and angular fragments, over 20 ft. ; marly sand-rock, 3 ft., on brick-red sand-rock. A sand pit 500 yds. north of the cross-roads at Norton Covert showed some 15 ft. of red sand-rock, with pebbles averaging about one per cubic foot of rock, the top part of which is highly micaceous.

Abbot's Castle Hill, Kinver, Wolverley, etc.—On Abbot's Castle Hill, from Hill End to Tinker's Castle (where the Seisdon road crosses the escarpment) the basement bed of the Bunter Pebble Beds consists of coarse calcareous grit with sub-angular fragments. Pebbles of Carboniferous limestone and chert are numerous. The best exposure is in a crag (Survey Photographs 6849-51) about 250 yds. E. by S. of Hill End, where these basement grits rest on an eroded and pockety surface of the Lower Mottled Sandstone. Another exposure (Survey Photograph 6852) 640 yds. farther east by south shows about 4 ft. of calcareous conglomerate and pebbly grit, the upper part with pebbles up to 2 in. or more in length, isolated or in lenses, the lower 2 ft. with pebbles mostly less than 1 in. in length and closely massed. At the east end of the crag a wedge of rather coarse sandstone, up to 1 ft. thick, intervenes between the conglomerate and the normal Lower Mottled Sandstone, the junction with the latter being fairly even and not pockety.

South of Tinker's Castle the basal beds of the Bunter Pebble Beds are not hard and calcareous, but consist of semi-coherent shingle. From Blackhills

Plantation southwards the basement bed is a red sandy marl with sub-angular pebbles, which can be traced along the escarpment near Rumford Hill and the Hampton Valley. Higher beds are exposed in several places near Rumford Hill and Stourton Field. A typical section in an excavation 300 yds. south of Checkhill Farm shows, in descending order : gravel with lenses of coarse sand, 6 ft. ; reddish-brown marl, 0 to 4 in. ; brown coarse sand-rock, 2½ ft. ; coarse green grit, 6 in. ; green and reddish-brown marl, 2 ft. Another exposure, about 700 yds. N.E. of Stourton Hall, shows : coarse gravel with a clayey matrix containing pebbles up to 8 in. long, about 4 ft. ; calcareous conglomerate, lenticular, up to 1 ft. ; clayey gravel with smaller pebbles than the upper bed. The pebbles are limestones and cherts, some crinoidal, purple tuffs resembling Uriconian types and purple grits, like Longmyndian rocks. Similar pebbles were seen in other exposures. Carboniferous limestones and cherts are common in this part of the outcrop.

Exposures near the top of the Bunter Pebble Beds in the east bank of the canal near Prestwood House consist of coarse reddish brown sandstones with, in places, small subangular pebbles. Calcareous conglomerate crops out about 180 yds. east of the Prestwood Pumping Station (situated 850 yds. W.S.W. of Prestwood House) of the South Staffordshire Waterworks, and gritty, pebbly sandstones in the west bank of Smestow Brook, to the south. There are many exposures of conglomerate and pebbly sandstone to the west of the pumping station, but, nevertheless, as stated on p. 108, the cores of the boring there showed not a single pebble. Those parts which must be assigned to the Bunter Pebble Beds consist of dull red or reddish-brown sandstones of medium or moderately coarse grain.

There are good exposures of conglomerate and pebbly sandstone, near the base of the Pebble Beds, in the banks of the canal north of Dunsley Hall, and, again, south of Dunsley, but near here the marly basement bed was not seen.

The road south-east of Kinver Church affords a fine section in the lower part of the Bunter Pebble Beds, though the actual base is not exposed. From above downwards the section shows :—coarse conglomerate, with pebbles up to 6 in. long in a matrix of red micaceous sandstone, about 15 ft. ; thin-bedded marly sandstone, about 20 ft. ; red, non-pebbly, fine-grained sandstone, wedging out eastward, up to 5 ft. ; coarse conglomerate, with many pebbles 6 in. long, 6 in. to 3 ft. ; red sandstone, rather coarse-grained, 3 ft. ; irregular parting of marl with pebbles, 6 in. ; dull red, medium-grained sandstone, 12 ft. or more ; marly conglomerate with pebbles up to 4 in. long, thickness up to 4 ft. ; hard calcareous conglomerate, tufa-covered, with lenses of grit with calcite cement, and many limestone pebbles, 12 to 15 ft. In the road to Cookley, 240 yds. S.E. of Kinver Church, conglomerate containing many limestone pebbles is exposed. A specimen stated by Dr. J. Pringle to be comparable with *Schellwienella crenistria* (Phillips) was found in one such pebble. There are also cherts, including yellow cherts like those in the Bowhills Beds (p. 92), and volcanic rocks resembling those of the Uriconian. This conglomerate rests on about 20 ft. of non-pebbly sandstone, with another conglomerate below.

On Kinver Edge (Survey Photographs 6896-8, 6902) the basement bed can be seen to consist of marly or sandy breccia with, in places, coarse pebbly grit above. Shingle beds higher in the sequence make a feature parallel to and south-east of the Edge, but this dies out near the Lodge. A calcareous grit or breccia forms the basement bed along the remainder of the escarpment to where it meets the Enville Fault near Horseleyhills Farm. Near Drakelow it appears to be about 2 ft. thick, but lenses of similar calcareous breccia occur in places above the actual base. Calcareous grit and conglomerate near the base of the Pebble Beds are exposed in the Kidderminster to Bridgnorth road about 650 yds. E.N.E. of Lower Barns Farm, and here, again, there are many pebbles of limestone and chert. In an exposure 530 yds. S.W. of Horseleyhills Farm, showing about 12 ft. of calcareous conglomerate and breccia resting on the Lower Mottled Sandstone, the breccia contains large angular stones about 1 ft. above the base. Amongst them was noted a fragment, 10 in. long, of purple grit, like those of the Longmyndian, and also fragments of quartzite, rhyolite and limestone.

There are many exposures of higher beds near Blakeshall, Wolverley and Cookley. Red sandstone with calcareous grit and breccia, a little above the middle of the Pebble Beds, appears in the sides of a valley about 900 yds. N.W. of Cookley Church, while a lane near Wolverley Church shows up to 50 ft. of pebbly sandstones.

Pebbly sandstones not far below the top of the formation can be seen alongside the canal about half a mile S.S.W. of Whittington. The south bank of the canal, 200 yds. N. of Austcliff, is an overhanging cliff about 25 ft. high, showing false-bedded pebbly sandstones overlain by coarse grit with imperfectly rounded pebbles and many quite angular rock-fragments. These beds, again, are not far below the top of the formation.

Several large exposures, some forming cliffs 50 ft. high, occur in the valleys north-west and west-north-west of Franche, and show sandstones with few or many pebbles and conglomerates. The basement beds cap the Lower Mottled Sandstone cliff of Ridgestone Rock and Jacob's Ladder, at the north end of Habberley Valley.

T.H.W.

Bridgnorth.—The Pebble Beds crop out over a considerable area extending from Newton, about 3 miles N.N.E. of Bridgnorth, southwards to the Hermitage, on the east side of the Severn opposite Bridgnorth, and thence south-eastwards to Burf Castle and Gags Hill (Survey Photographs 6846, 6869). Exposures are very numerous, especially in the lower beds, and the junction with the Lower Mottled Sandstone is visible at many points.

The River Worfe cuts right across the outcrop from near Burcote to Worfe Bridge on the Shifnal road (Survey Photographs 6882-3). Close to the bridge, on both sides of the river, there are good exposures of the basal conglomerate and breccia resting on strongly false-bedded Lower Mottled Sandstone (Survey Photograph 6874). The dip of the Pebble Beds at the junction appears to be about 8° N.E. Upstream pebbly sandstone is exposed at Rindleford Mill and on the right bank 500 yds. above the mill a band 2 ft. thick of very coarse conglomerate is composed of pebbles, up to 9 in. in length, of quartzite, quartz and sandstone. The left bank shows a fine section over a distance of about 400 yds., the general dip being about 4° E. Good exposures occur again near Burcote Mill and Burcote House.

The base of the Pebble Beds is well defined from Worfe Bridge southwards along the crest of the precipitous east side of the Severn valley by Pendlestone Rock to the Hermitage (Survey Photographs 6840-1, 6843). At many points the basement conglomerate is highly calcareous, apparently owing to the presence of pebbles of Carboniferous limestone. Sections just north of the Wolverhampton road show some 30 ft. of conglomerate with pebbles of crinoidal limestone up to 6 in. in length, purple grit, chert and dark red calcareous sandstone up to 10 in. in length. In the road cutting 6 ft. of pebbly sandstone on 12 ft. of very coarse conglomerate with pebbles of quartzite, limestone and other rocks overlie red, false-bedded, Lower Mottled Sandstone.

Another good section is the cutting on the Stourbridge road (Survey photograph 6884) about half a mile south of Hermitage Farm, which shows some 30 ft. of pebbly sandstone and coarse conglomerate with pebbles of limestone, quartzite and quartz. There are many other exposures to the south-east as far as Gags Hill and Burf Castle where the Pebble Beds are brought into contact with the Upper Mottled Sandstone by a northward-trending fault. East of this fault a small area of Pebble Beds occurs at Roundabout Coppice south-east of Roughton.

R.W.P.

UPPER MOTTLED SANDSTONE

Wollaston, Stourbridge and Churchill.—A sand-pit 170 yds. W.N.W. of Wollaston Church showed (in 1922), in descending order : thin-bedded sand-rock, 8 ft. ; massive sand-rock, 10 ft. ; soft sand-rock with thin green beds, upper surface irregular, 3 ft. ; marly sand-rock, 4 ft. ; soft flaggy sand-rock, lower part micaceous, 8 ft. or more. At Holloway End the Upper Mottled Sandstone has been extensively worked for moulding sand (Survey Photograph 2200).

A good section in the upper beds of the Upper Mottled Sandstone, with the overlying Lower Keuper Sandstone (p. 115), is afforded by the railway cutting south of Stourbridge Junction Station and others occur in the road just south of Stourbridge and 450 yds. north of St. Mary's Church, Old Swinford. These beds are soft brick-red or brownish sandstones, massive or thin-bedded, with rather marly sandstones below.

Numerous exposures of the Upper Mottled Sandstone may be found to the west of the Lower Keuper Sandstone outcrop near Churchill. A sand-pit 300 yds. south of the cross-roads at Norton Covert showed (in 1922) about 30 ft. of brick-red sand-rock with some very micaceous bands and beds of tough marl. The beds are here disturbed and the south face of the pit showed an overfold with its axis pitching northward.

Lower Penn, Wombourne, Prestwood, etc.—Near Lower Penn the outcrop of the Upper Mottled Sandstone is largely drift-covered, but exposures of red soft and marly sandstone occur in cuttings of the Wolverhampton railway and of brick-red sandstone, near the top of the formation, around the high ground formed by the Lower Keuper Sandstone. The best sections are in sand-pits near Ounsdale, of which that to the east of the canal showed (in 1929), in descending order : marly sand-rock, with seepages of water, about 25 ft. ; more 'open' sand-rock, very uniform, with false-bedding only on a small scale, about 15 ft. ; marly sand-rock in beds about 1 ft. thick. The rock is worked for moulding sand (p. 181). A pit 450 yds. south by east of Wombourne Church showed about 35 ft. of brown-red marly sand-rock, very evenly bedded, with green streaks at wide intervals. The sand-rock is overlain by drift gravel and the bedding is disturbed at the top in places. A borehole at the Bratch Pumping Station of the Bilston Waterworks (p. 187), begun at the bottom of a 140-ft. well, passed through 147 ft. of red sandstone into 4 ft. of red marl which may be regarded as the base of the Upper Mottled Sandstone.

Exposures of the lower part of the Upper Mottled Sandstone occur in the banks of the mill-stream south of the River Stour just west of the Stapenhill Fault, near Prestwood. Though there are numerous sections in the outcrop west of the Stapenhill Fault farther south, the only one that needs mention is by the roadside 150 yds. W.N.W. of Island Pool, near Cookley, where up to about 10 ft. of red soft sandstones, rather fine in grain, may be seen, not many feet above the base of the formation.

Rudge and Claverley.—A narrow outcrop of the Upper Mottled Sandstone near Rudge and Claverley is bounded on the east by the Lower Keuper Sandstone and the Pattingham Fault (p. 129) and on the west by the Patshull Fault (p. 129). Most of the northern part is covered by drift, but there are a few exposures just beneath the Lower Keuper Sandstone. In one such, in the road about a quarter of a mile W.S.W. of Shipley, the rock below the bed taken by Dr. Robertson[1] as the base of the Keuper is red and white mottled, finely grained, gently false-bedded sandstone, with small lenses of coarser, round-grained sandstone ; these lenses become more numerous towards the top and being in most cases calcareous stand out on the weathered surface. There are some 12 to 14 ft. of such beds, with fine-grained, brick-red and white mottled sandstone below. Other sections in the top part of the Upper Mottled Sandstone can be seen in roads north-east

---

[1] In 'The Country between Wolverhampton and Oakengates' (*Mem. Geol. Surv.*), 1928, p. 140.

and south-east of Ludstone. In one (see Fig. 12, p. 117) about 30 ft. of fine-grained, brick-red, false-bedded sandstone is exposed below the harder brown sandstone taken as the base of the Keuper (Fig. 12). There are sections lower in the Upper Mottled Sandstone in and near Claverley. The best, in a road-cutting south-west of Woodfield, shows 30 ft. of brick-red sandstone, rather fine in grain, on red sandstone with green spots and streaks, less markedly false-bedded than the upper part.

T.H.W.

Worfield and Roughton.—North of Worfield a broad outcrop of Upper Mottled Sandstone extends from near Hartlebury in the west to Chesterton in the east. Much of this ground is obscured by boulder clay and glacial gravels, especially west of the River Worfe.

Exposures of the soft, red, false-bedded sandstone are to be seen in the lanes round Worfield, by the side of the Wolverhampton road at Roughton, and below the escarpment of the Lower Keuper Sandstone at Hilton. Farther south round Hoccum and Barnsley the Upper Mottled Sandstone forms undulating ground with occasional exposures in the lanes. The soft red sandstone is also seen at intervals below the escarpment of Lower Keuper Sandstone trending southward from Dalicote through Upper Farmcote.

R.W.P.

LOWER KEUPER SANDSTONE

Stourbridge, Hagley and Clent.—The Lower Keuper Sandstone occupies the middle of a shallow syncline from Stourbridge southwards. The base appears in the railway cutting 440 yds. N.N.W. of Stourbridge Junction, where the section shows, from above downwards : brown-red sandstone with pebbles and marl pellets, very false-bedded, 10 ft. or more ; dull red sandy marl with thin green and red sandstones, 1 ft. 3 in. ; dull greenish sandstone with occasional lenses of greenish marl base, about 10 in., passing down into soft brick-red sand-rock (Upper Mottled Sandstone). In the cutting about 420 yds. south of Stourbridge Junction Station the lowest beds of the Lower Keuper Sandstone consist of massive pale red pebbly sandstones, dipping south-westward, overlain by calcareous marl-breccia ('catbrain'). Farther south-westward the cutting shows conglomerate, pale red coarse sandstone with marly pellets and pebbles, brown-red coarse sandstone with scattered pebbles, and dark, micaceous fissile sandstone. The last-mentioned occupies about the middle of the syncline, for, still farther south-westwards, about 100 yds. from the Hagley road-bridge, very pebbly sandstones dip north-westward. To the south, about half a mile north of Hagley Station, thin beds and pockets of highly calcareous cornstone-like material and marl occur amongst the pebbly sandstones ; two small faults cross the cutting hereabouts. Numerous roadside exposures of the Lower Keuper Sandstone can be found near Pedmore, Hagley, Lower Hagley and Broom. One north of the cross-roads at Holy Cross, Clent, shows some 35 ft. of coarse and pebbly sandstones. Here, as elsewhere, the sandstones are in many, cases pitted or weathered to a honey-comb appearance.

On the western side of the syncline the base of the Keuper Sandstone forms a prominent escarpment from Stourbridge to beyond the southern border of the district. An exposure 50 yds. N.E. of The Quarry shows calcareous breccia, 3 ft. ; dark red marl, weathering purple, 2 ft. ; on soft brick-red sandstone (Bunter). The roadsides 400 yds. south of The Quarry show coarse red sandstone, very false-bedded, overlain by a 2-ft. bed of 'catbrain.' In Pedmore Quarry, 400 yds. S.S.W. of The Quarry, the basal Keuper Sandstone is a dull red, non-pebbly sandstone, but a little farther south coarse calcareous sandstone with pellets of marl ('catbrain') was seen at or very near the base.

Lower Penn and Wombourne.—In the outcrop of Lower Keuper Sandstone to the west of the Stapenhill Fault a bed of sandy marl occurs at the base. This in places contains ' millet-seed ' grains of sand and the sandstones for a few feet above also include bands with such grains. In the eastern of the two small outliers near Oreton a section at the base of the Keuper showed rather soft brown-red sandstone on coarse gritty calcareous sandstone, or breccia, on thin-bedded marly sandstone (Bunter). A roadside exposure on Shovel Bank, 600 yds. east of Oreton, shows dull reddish and yellowish sandstones with scattered pebbles up to 1 in. in length near the base of the Keuper and, though the sandy marl is not visible here, a spring in the little valley north of the road indicates its probable presence. A quarry near the top of the escarpment about 1,000 yds. S.W. of Bearnett House shows, in descending order : rather soft, dull-red sandstone with scattered pebbles, 6 or 8 ft. ; thin-bedded sandstone with scattered pebbles, 2 ft. ; brownish sandstone with irregular strings of rather large pebbles, up to 4 in. long, 4 ft. exposed. A short distance southward, dull-red, rather coarse sandstone, with scattered small pebbles and lenticular inclusions of marl up to 6 in. long, contains a lens of black sandstone. A quarry at the south end of the outcrop, 760 yds. N. by W. of Wombourne Church, shows, in descending order : coarse, brown, pebbly sandstone, about 4 ft. ; brown, dull-red and dark-red sandstones, 6 ft. ; red, sandy micaceous marl passing down into flaggy sandstone, 2 ft. ; dull-red and greenish-brown sandstones, 3 ft. Springs, depositing tufa, are thrown out by the marl, which is regarded as the base of the Keuper.

Wollaston.—On the downthrow side of the Stapenhill Fault, west of Wollaston, the presence of a small outlier of the basal beds of the Keuper Sandstone is indicated by a feature, and by many pellets of marl, small pebbles and pieces of coarse sandstone and ' catbrain ' on the surface. Brown, coarse sandstone containing marl pellets, with brick-red sand-rock below, was seen in place 350 yds. E.N.E. of the north-east corner of Gibbet Wood.

Rudge and Claverley.—In the road section south-west of Shipley, the basal beds of the Keuper may be summarized[1] as follows : reddish-brown sandstone, fine to medium in grain, and fine to coarse conglomerate, 25 ft. ; sandy marl, micaceous, with ' millet-seed ' quartz grains, irregular, 6 in. to 3 ft. ; non-pebbly sandstone, thickly bedded, 10 ft. ; micaceous sandy marl, 6 in. to 1 ft. ; calcareous sandstone, fine-grained, 2 to 4 ft. ; pebbly sandstone, calcareous, with marl pellets, 1 ft. ; alternating sandstone and sandy marl, both with ' millet-seed ' grains, 12 to 13 ft. ; sandy marl with ' millet-seed ' grains, 1 ft. ; red mottled sandstone alternating with red to buff calcareous sandstone with ' millet-seed ' grains, 3 ft. Dr. Robertson places the base of the Keuper at the lowest definite band of the calcareous sandstone with ' millet-seed ' grains

A section in basement beds beginning 420 yds. N.E. of Ludstone Hall is illustrated by Fig. 12 (Survey Photograph 6848). A retaining wall obscures part of it and small faults render some details uncertain, but it is clear that over 40 ft. of beds intervene between the typical Upper Mottled Sandstone and the coarse conglomerate. The sandstone numbered 9 on Fig. 12, however, closely approaches the Bunter type.

Chesterton, Hopstone, Hilton and Farmcote.—Along the west bank of Stratford Brook, near the Walls, Chesterton, Dr. Robertson noted highly current-bedded sandstone with scattered pebbles. In the road 320 yds. S.W. of the Walls 4 or 5 ft. of calcareous marl conglomerate, with pebbles, mainly under 2 in. in length, were observed by him not far above the base of the Keuper Sandstone. A little higher in the sequence, *i.e.*, farther south-east, about 20 ft. of dark, reddish-brown, non-pebbly sandstones are exposed.

Sections are numerous near Hopstone. Hilton Brook, north of Hopstone, 300 to 400 yds. west of the road from Upper Ludstone, has a cliff of coarse

---

[1] For further details see T. Robertson in ' The Country between Wolverhampton and Oakengates ' (*Mem. Geol. Surv.*), 1928, p. 139. See also E. Hull, ' Triassic and Permian Rocks of the Midland Counties of England ' (*Mem. Geol. Surv.*), 1869, p. 31 and Fig. 8.

TRIAS : DETAILS.

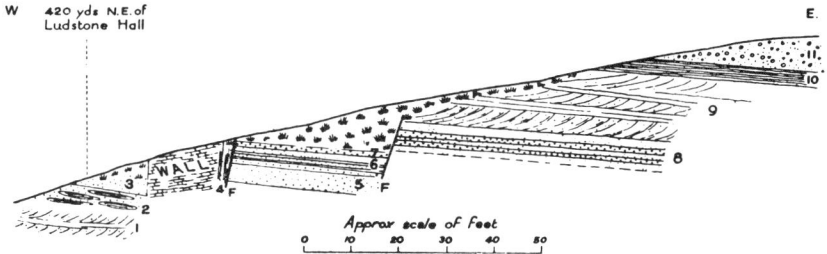

Fig. 12.—*Section of Keuper basement beds in the road north-east of Ludstone.* 1. Brick-red soft sandstone (Upper Mottled Sandstone). 2. Hard brown sandstone, lenticularly interbedded with chocolate sandy marl. 3. Hard brown sandstone, rather coarse. 4. Sandstone and marl, nearly vertical, apparently between two faults. 5. Reddish-brown sandstone. 6. Reddish-brown sandstone, thin-bedded,. with marl bands. 7. Hard, coarse gritty sandstone. 8. Red sandstone with coarse hard bands. 9. Red sandstone, false-bedded, with soft bands. 10. Dark red sandy marl. 11. Conglomerate passing up into pebbly grit, calcareous. F, fault.

brown sandstone with scattered pebbles and lenses of conglomerate. The pebbles are chiefly of quartzite, but some of grit and others apparently of volcanic rock were seen. Another long continuous section, in the roadside about 400 yds. N. by W. of Hopstone, shows red and brown sandstones of which the lower and upper parts are pebbly, the middle part containing few pebbles. The upper part of the sandstone has a curious undulating bedding. A small exposure just east of Sutton Mill, 500 yds. N.E. of Hopstone, shows coarse dull red sandstone apparently vertical, owing, probably, to the proximity of the Patshull Fault. The south side of a gully near the smithy, Hopstone, shows some 40 ft. of red coarse sandstones of which some parts have contorted bedding.

Farther south the outcrop is largely covered by superficial deposits, but coarse, red-brown sandstones with scattered pebbles can be seen in the roadside at Farmcote Hall.

Though there is no exposure of rock in place, a feature indicates the position of the base of the Keuper Sandstone northeast of Bine Farm, and the appearance of pieces of coarse sandstone, small pebbles and sub-angular fragments in the soil closely defines the position of the Pattingham Fault thereabouts.

T.H.W.

The Lower Keuper Sandstone gives rise to the steep bank east of Hilton which is continued south of Hilton Brook in the sharp feature trending southward from Dalicote. In the roadside 150 yds. S.S.W. of Dalicote a 10-ft. section shows coarse yellowish current-bedded sandstone and coarse lenticular conglomerate with pebbles of quartzite and quartz up to 6 in. in length ; a similar conglomerate is present near the foot of the Lower Keuper Sandstone feature farther south.

The base of the sandstone is thrown westwards by a fault trending southward by the Cross and a good section, in the lane 300 yds. N.W. of the Cross, shows a coarse conglomerate with large pebbles of Carboniferous limestone overlying the soft red Upper Mottled Sandstone. This conglomerate can be traced along the edge of the high ground to near Morfevalley Farm. Just south of the road west of Upper Farmcote a section shows 20 ft. of conglomerate and in the lane 400 yds. S.E. of Morfevalley as much as 35 ft. of the conglomerate are exposed.

R.W.P.

CHAPTER VII

FOLDS AND FAULTS; EXTENSIONS OF THE COALFIELDS; IGNEOUS ROCKS

Folds and Faults

GENERAL ACCOUNT

INTRODUCTION

The present district is dominated by what may be called the west Staffordshire and east Shropshire, or Stafford, Syncline (Fig. 13). This is separated on the north-west by the broken, complex Longmynd-Wrekin Anticline[1] from the Cheshire or Prees Syncline.[2] On the south-east the South Staffordshire Coalfield, considered as a whole and in its relation to the Trias outcrops, appears as an anticlinal block with an axial trend parallel to that of the Stafford Syncline, but this general form is much disguised by subsidiary structures, especially within the present district. The coalfield serves, nevertheless, to separate the Stafford Syncline from another which may be called the Birmingham Syncline. The axial direction of these broad folds is in general from north-north-east to south-south-west ("Caledonoid"), and the pitch of the synclines is north-north-eastward. The faults show the same general "Caledonoid" trend, but, more particularly in the east, a north-north-west to south-south-east ("Charnoid") direction appears in places, whilst in the south a tendency to swing to a north and south course seems to indicate the influence of the Malvern axis.

This broad structural pattern, due largely to post-Triassic movements, is modified by folds of shorter wave-length but usually greater amplitude that are in the main pre-Triassic in age.

TECTONIC HISTORY

Pre-Carboniferous movements, in their effects, are not easy to distinguish from those of post-Avonian but pre-Coal Measures date. The unconformity between the Upper and Lower Old Red Sandstone must be ascribed to the Caledonian folding and such folding may have given rise to separate basins of deposition for the Lower Carboniferous rocks of Coalbrookdale[3] and of Titterstone Clee.[4] At least it prevented the

---

[1] 'The Shrewsbury District' (*Mem. Geol. Surv.*), 1938, p. 169.

[2] 'The Geology of the Country around Wem' (*Mem. Geol. Surv.*), 1925, p. 54, and *op. jam cit.*, p. 178.

[3] Dixon, E. E. L., in 'The Country between Wolverhampton and Oakengates' (*Mem. Geol. Surv.*), 1928, p. 41 and Table, p. 42.

[4] *Idem*, 'The Geology of the Titterstone Clee Hill Coalfield,' *Proc. Roy. Soc. Edinb.*, vol. li, 1917, p. 1,064. See also L. J. Wills, 'An Outline of the Palaeogeography of the Birmingham Country,' *Proc. Geol. Assoc.*, vol. xlvi, 1935, pp. 221, 223.

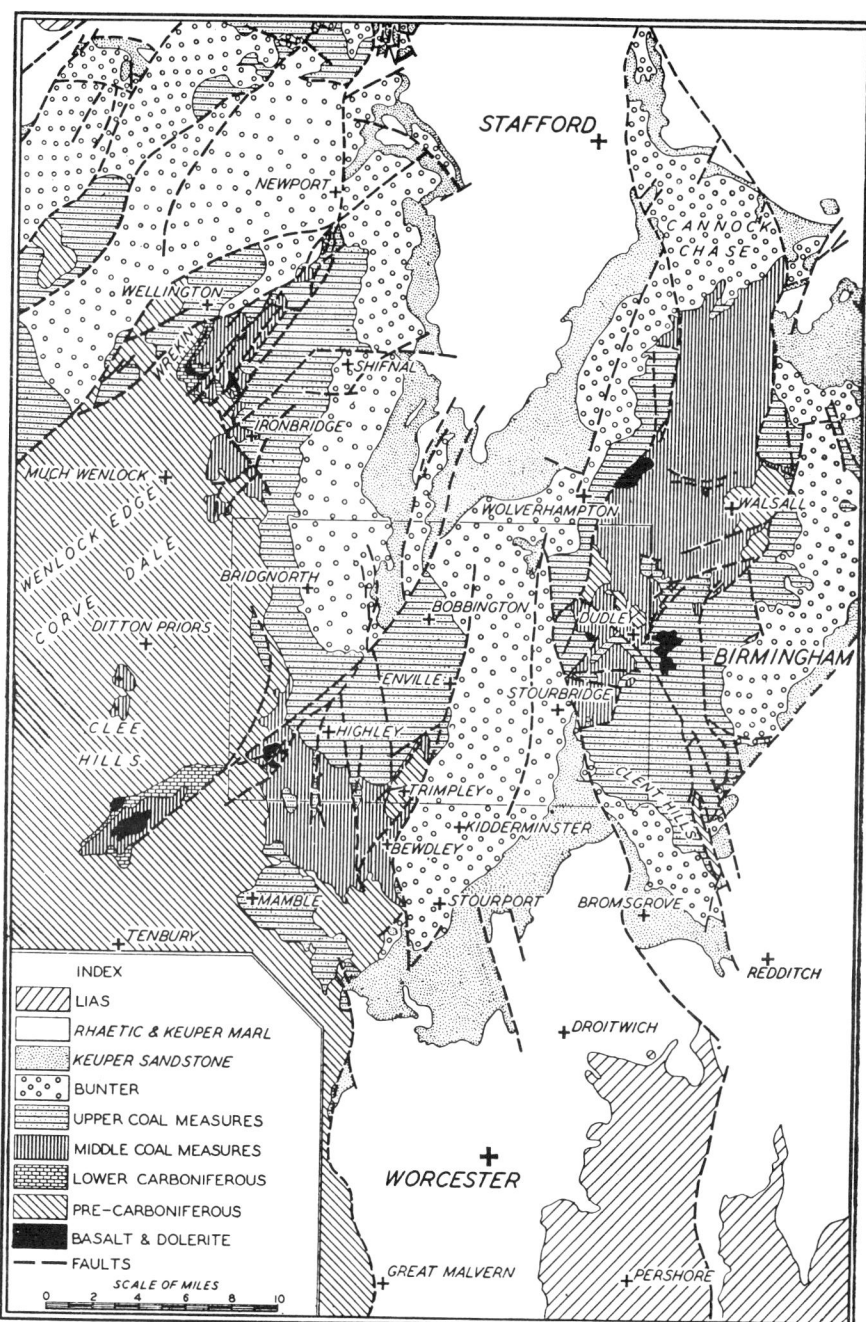

Fig. 13.—*Map to illustrate the Geological Structure of the Dudley and Bridgnorth District in relation to surrounding areas.*

Tournaisian sea of the south-west Province from reaching the former area. The Clee Hill Syncline was perhaps initiated by this Caledonian folding.

Post-Avonian folding separated the Lower Carboniferous rocks of Coalbrookdale from those of the Clee Hill area, if they were not separated *ab initio*, and caused the 'Middle' (Yorkian) Coal Measures to be deposited unconformably upon them in both areas. Uplifts of the pre-Carboniferous rocks at this time may perhaps have caused the deposition of the 'Middle' Coal Measures to fail in the Bridgnorth area (whilst taking place to a thickness of about 1,400 ft. in the parallel Clee Hill Syncline), with the result that the Upper Coal Measures subsequently overlapped onto the pre-Carboniferous rocks.[1] The Linley Anticline, the axis of which crosses the extreme north-west corner of the district, is one of the folds across which such overlap takes place, and seems clearly to be in part pre-Carboniferous, or at least pre-Yorkian, in date.[2]

Pre-Coal Measures, if not pre-Carboniferous, movement must have produced the main folding and much of the faulting in the Downtonian and Lower Old Red Sandstone rocks of the Trimpley inlier, for, as pointed out on p. 12,[3] their structure, unlike that of the Coal Measures, is anticlinal only in part, and the lowest beds actually appear in the northern synclinal portion. This implies a large amount of pre-Coal Measures denudation, and the greater thickness of the Kinlet Beds at Shatterford, as compared with Highley (pp. 45, 56), suggests that no upstanding obstacle to deposition survived in Yorkian and early Staffordian times.

In the South Staffordshire Coalfield pre-Coal Measures folding of the Silurian rocks is obvious from the disposition of their various members beneath the Coal Measures. Even in the case of folds by which the Coal Measures themselves were affected, earlier movement is shown by the overstep of those measures towards the anticlinal axes, by the greater prevalence of coarse basal conglomerates to the Coal Measures around the anticlines than in the synclines, and, in the case of the Sedgley and Dudley Anticline, by differences between some of the lower measures on either side,[4] suggestive of a barrier during deposition. In some of the faults also pre-Coal Measures movement took place.[5] There is, however, no evidence of such movement along the Western Boundary Fault, for where, as near Sedgley (p. 126), the throw in the Coal Measures and in the Silurian rocks can be separately estimated, it appears to be about the same in both.

Then the presence of beds of Wenlock age beneath the Coal Measures, as proved in the Claverley and Smestow boreholes (pp. 39 and 195), whilst

---

[1] But see below, p. 121.

[2] Robertson, T., in 'The Country between Wolverhampton and Oakengates' (*Mem. Geol. Surv.*), 1928, p. 161.

[3] See also W. W. King, *Trans. Worc. Nat. Club*, vol. vii, 1921, p. 320.

[4] See T. H. Whitehead in 'The Geology of the Southern Part of the South Staffordshire Coalfield' (*Mem. Geol. Surv.*), 1927, p. 23.

[5] See *e.g.*, W. W. King, 'The Plexography of South Staffordshire in Avonian Time,' *Trans. Inst. Min. Eng.*, vol. lxi, 1921, p. 157.

Downtonian rocks form the floor to west and east, indicates a generally anticlinal disposition of the pre-Carboniferous rocks, masked by the syncline of Coal Measures and Trias.[1]

Early Staffordian movement gave rise to the unconformity between the ' Middle ' Coal Measures and the Coalport Beds (the " Symon Fault ") in Coalbrookdale,[2] between the Kinlet and Highley Groups in Wyre Forest and between the Etruria Marl and the Halesowen Group in South Staffordshire. In the last-named area the lower part (perhaps 200 ft.) of the Etruria Marl lies below the plane of unconformity and the upper part appears to have been deposited during the movements.[3] In the Wyre Forest Coalfield (within the present district) a thin remnant of the Etruria Marl seems to be preserved below the unconformity (p. 42) and more may have been denuded before the deposition of the Highley Group. In the Coalbrookdale Coalfield it seems probable that the equivalent of the Etruria Marl either was not deposited or was denuded during the " Symon Fault " folding and that the Coalport Group represents, in the main, the Highley and Halesowen Groups. If any remnant of the Etruria Marl does survive there then that remnant must be represented by the lowest part of the Coalport Group and lies above, instead of below, the unconformity.

Further movements along the Linley and parallel axes at this time may have caused the denudation of the ' Middle ' Coal Measures (if they were ever deposited) from areas where the Upper Coal Measures now rest upon pre-Carboniferous rocks.[4] The Trimpley Anticline may have been initiated by early Staffordian movements,[5] though there is no evidence of a greater degree of overstep here than elsewhere on the part of the Highley Group (see above, p. 120).

These early Staffordian movements seem to have followed preexisting lines, mainly " Caledonoid " in the west, but " Charnoid " in the south-east, and to have accentuated already existing structural features. In places, for example at Halesowen (Sheet 168), deposition seems to have kept pace with the deepening of troughs. The tendency to follow pre-existing lines of movement is shown by the circumstance that the Etruria Marl and the ' Middle ' Coal Measures are commonly below the average thickness in the same localities.

Post-Staffordian movements resulted in the unconformity at the base of the Clent Beds. To them must be attributed some of the folding of the Trimpley Anticline,[6] for the Clent Beds overstep the underlying Bowhills Beds towards the axis (p. 93). This was the period of the main folding along the Malvern and Abberley axis, but the Trimpley Anticline

---

[1] See L. J. Wills, *op. cit.*, Fig. 28, p. 227.

[2] Pocock, R. W. in ' The Country between Wolverhampton and Oakengates ' (*Mem. Geol. Surv.*), 1928, p. 158.

[2] See T. H. Whitehead, *op. cit.*, pp. 87 and 164.

[4] See L. J. Wills, *op. cit.*, p. 232.

[5] Cantrill, T. C., ' A Contribution to the Geology of the Wyre Forest Coalfield,' Kidderminster, 1895, p. 36.

[6] Cantrill, T. C., *loc. cit.*

retains a "Caledonoid" trend probably impressed upon it by earlier movements. By analogy with the Trimpley area it would seem not improbable that some of the folding in the South Staffordshire Coal Measures took place at this time,[1] but at present there appears to be no hope of distinguishing such folding from the effects of post-Clent Beds and pre-Triassic movement.

It is possible that, at this time, movement was localized in the neighbourhood of pre-existing axes, such as those of Malvern, Trimpley, the Lickey Hills and the now-buried "Mercian Highlands." The effects of the movements are not apparent where, as near Bobbington, breccias form an inconsiderable proportion of the Clent Beds; but possibly, at a distance from the highlands whose wastage provided the material for the breccias, deposition more or less kept pace with movement.[2]

Post-Clent Beds and pre-Triassic movements are indicated by the unconformable relation of the Trias to the Clent Beds. This is illustrated most clearly in the south-east of the present district where, between Clent and Hagley Park, the Bunter oversteps the whole local thickness (about 450 ft.) of the Clent Breccia.

Some post-Clent Beds accentuation of the Trimpley axis may have taken place, for the breccia of Enville Sheepwalks shows a slightly anticlinal disposition (p. 101). It is possible, however, that this is due to post-Triassic movement.

Post-Triassic movements gave rise to the broad fold pattern described on p. 118. To them must be ascribed the principal uplift of the South Staffordshire Coalfield and a considerable proportion of the folding and faulting within it, including further movement along pre-existing faults, such, for example, as the Western Boundary Fault.

Post-Triassic folding along the Trimpley axis or one parallel and near to it is indicated by the curved outcrop, concave westward, of the Bunter from Wolverley to Abbot's Castle Hill. This was presumably contemporaneous with the post-Triassic throw of the Enville Fault, which largely replaces the eastern limb of the Trimpley Anticline.

The Clee Hill Syncline, the influence of which upon the present district is indicated by the occurrence of Upper Old Red Sandstone south of Stottesdon (p. 24), probably owes its present elevation to post-Triassic movement. It now seems to constitute a south-westward culmination of the Stafford Syncline, to which it stands in a somewhat similar relation as the Long Mountain Syncline does to that of Prees.[3]

---

[1] *Cf.* L. J. Wills, *op. cit.*, p. 238.

[2] *Cf.* F. W. Shotton, ' New Evidence on the Origin of Breccias and Conglomerates in the Warwickshire Coalfield . . .," *Geol. Mag.*, 1933, p. 466 (see p. 474), and T. Eastwood in ' The Country around Birmingham ' (*Mem. Geol. Surv.*), 1925, p. 58.

[3] Wedd, C. B., ' The Principles of Palaeozoic and Later Tectonic Structure between the Longmynd and the Berwyns,' ' Summary of Progress ' for 1931, Part II (*Mem. Geol. Surv.*), 1932, p. 6. See also *idem* in ' The Shrewsbury District ' (*Mem. Geol. Surv.*), 1938, p. 178.

The South Staffordshire Coalfield

The principal structural element of that part of the South Staffordshire Coalfield which lies within the present district[1] is the composite Sedgley and Dudley Anticline, with a north-north-west to south-south-east axial trend. It is complicated by the subsidiary periclines of the Castle Hill, Dudley, the Wren's Nest and Hurst Hill, the axes of which trend more nearly north and south. At Sedgley a shallow, southward pitching syncline has a nearly parallel axis.

Parallel to the Sedgley and Dudley Anticline, on the north-east, is the Coseley Syncline,[2] in which a part of the Etruria Marl is preserved. On its south-west side the anticline is flanked, and south-east of Dudley largely replaced, by the Russell's Hall Fault, a south-westward downthrow ranging, in the Coal Measures, from about 120 ft. near Gornal to about 300 ft. where it leaves the district near Windmill End.

The area to the south-west of the Sedgley and Dudley Anticline is characterized by folds of north-north-east to south-south-west trend. Of these the most important is the Netherton Anticline, which brings up Silurian rocks on its axis near Netherton (p. 17) and Wollescote (p. 20). The anticline pitches north-north-eastward and the outcrops of the Coal Measures curve round the end. At its south-western end the fold is truncated by the nearly east-to-west Wollescote Fault. Traces of an anticlinal disposition of the Upper Coal Measures to the east of Hagley[3] may conceivably be due to a displaced continuation of the Netherton axis.

Two other nearly east-and-west faults form the Brierley Hill Trough, in which a wedge of Etruria Marl is thrown down. The northern of these two faults crosses the Netherton Anticline.

To the west of the Netherton Anticline, in the "Pensnett Basin," the measures are stepped northwards by a series of offshoots from the Western Boundary Fault (the Brockmoor, Corbyn's Hall and Shut End faults) with west-north-westward downthrows. Farther north a much broken anticline, with its axis in part parallel to that of Netherton, is shown by the Silurian outcrops of Turner's Hill and Ellowes Park (p. 17), affecting also the Coal Measures near Gornal and Straits Green. Northwards the anticlinal axis seems to curve round nearly into parallelism with those of the Castle Hill, Wren's Nest and Hurst Hill, until it is cut off by a cross-fault at Sedgley, which separates it from the shallow syncline mentioned above.

The Main Syncline

The eastern limb of the Stafford Syncline is marked by the outcrop of the Bunter beds from Cannock Chase (Fig. 13, p. 119) to Upper Penn in the present district. Farther south it is lost, owing to interruption by

---

[1] For details of the folds and faults of this part of the coalfield the reader is referred to 'The Geology of the Southern Part of the South Staffordshire Coalfield' (*Mem. Geol. Surv.*), 1927, pp. 160, 165-71 (*passim*), 173-4.

[2] Part of the 'Tipton to Oldbury Syncline' of W. W. King (*Trans. Inst. Min. Eng.*, vol. lxi, 1921, p. 56).

[3] 'The Geology of the Southern Part of South Staffordshire Coalfield,' (*Mem. Geol. Surv.*), 1927, p. 169.

the Western Boundary Fault; unless the outcrops of Upper Coal Measures along the western side of the South Staffordshire Coalfield and small patches of westward-dipping Bunter near Kingswinford and Wordesley can be regarded as fragments of it.

The western limb of the syncline is clearly indicated by the Bunter outcrop from Newport to the present district near Bridgnorth, to the south-east of which it is interrupted by the approximately axial Pattingham Fault (p. 129).

The long and broad outcrop of Bunter deposits from north of Seisdon to beyond the southern border of the district is due to the distortion of this western limb by the subsidiary Enville and Bobbington Anticline, and to repetition by the Enville and Stapenhill faults.

The Stourbridge Syncline

The Keuper Sandstone near Stourbridge, flanked on both sides by the Bunter, occupies a shallow syncline which seems like a portion of the Birmingham Syncline (p. 118) displaced north-north-westward by the Western Boundary Fault. It is, however, perhaps better regarded as a fold complementary to the post-Triassic anticline of Enville and Bobbington.

The Trimpley Anticline

The movements by which the Coal Measures of the Trimpley Anticline were folded would appear to have raised the pre-Carboniferous rocks as a block. In the northern, synclinal, part of the Trimpley Inlier these rocks seem to have been thrust west-north-westward over the Coal Measures[1] without themselves suffering much further folding. In the southern part, where the structure was already anticlinal, accentuation of the folds in the pre-Carboniferous rocks took place, with over-folding, involving the Coal Measures, towards the west-north-west (pp. 55, 129).

The outcrops of the Coal Measures show that the Trimpley Anticline pitches north-north-east. Some continuation of the structure in that direction under the Triassic rocks might be expected, but little evidence of it is at present available. The axis would presumably pass between the sites of the Claverley and Smestow boreholes. If the dip of the Coal Measures continues to the east of the Claverley borehole at much the same angle as between that borehole and the outcrops of the Keele and Enville beds to the west, then, allowing a moderate downthrow for the Enville Fault, the various measures would be found at about the levels at which the Smestow borehole appears to have proved them; *i.e.*, no assumption of an anticlinal fold is required.[2] On the other hand, Mr. W. W. King's view (p. 103) that the beds under the Bunter in the Highgate Common borehole belong to the Bowhills Group, with Keele beds below, involves a rise of the measures eastward from the Claverley borehole (Fig. 14).

---

[1] See T. C. Cantrill, *op. cit.*, Plate facing table of contents.
[2] The horizontal section at the foot of the one-inch map (Sheet 167) has been drawn on this basis. *Cf.* Fig. 14.

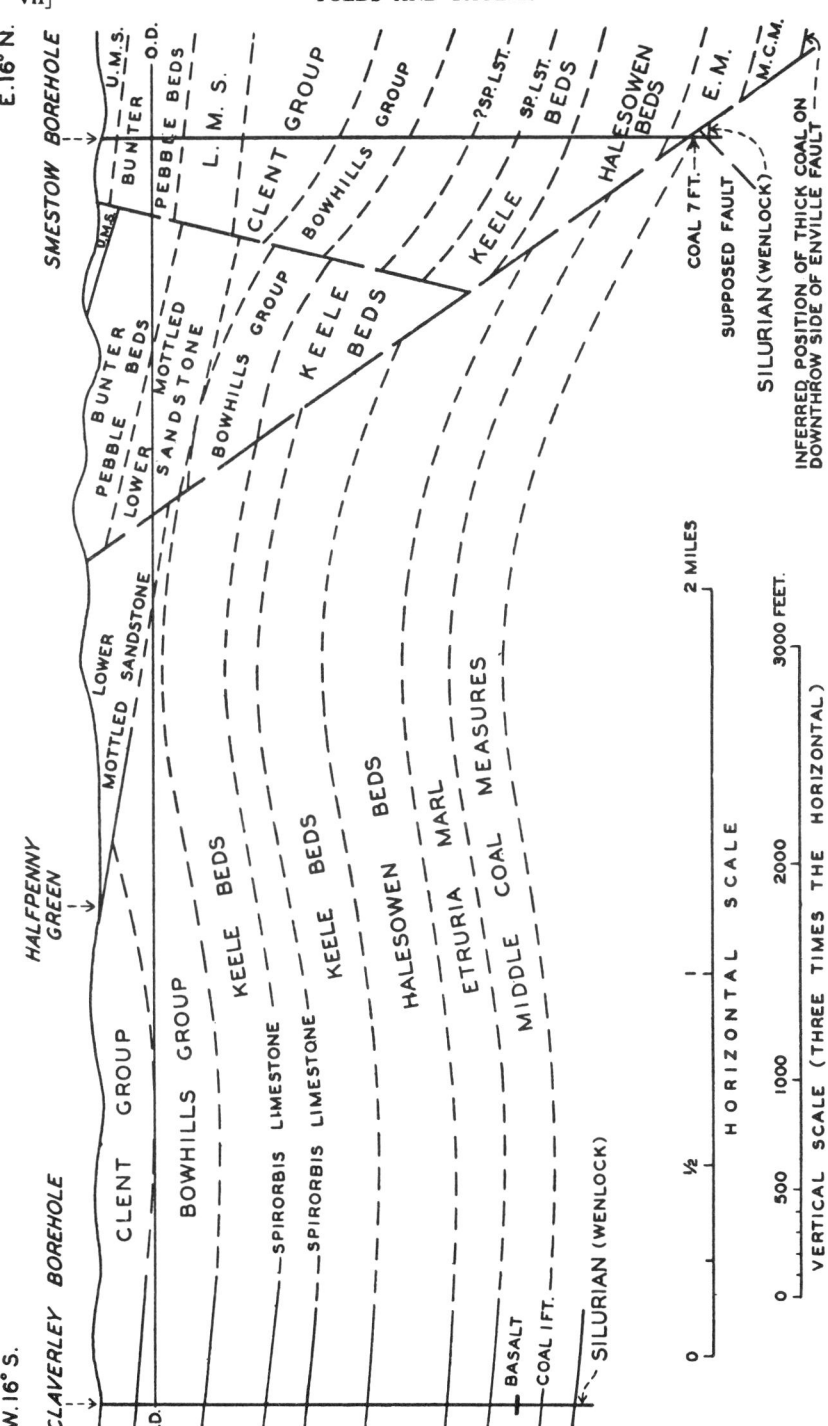

Fig. 14.—*Hypothetical Section from the Claverley to the Smestow Borehole.*

This would require a greater pre-Triassic throw for the Enville Fault, accompanied, perhaps, by eastward downfolding, to account for the levels in the Smestow borehole. There might thus be an anticlinal fold, broken by the Enville Fault, at about the position where a continuation of the Trimpley Anticline would be looked for, but, unlike that anticline, with its eastern limb steeper than its western one.

T.H.W.

Western Area

The post-Avonian folding in the western area appears to have been intense and in the succeeding period of denudation the Lower Carboniferous and older rocks were reduced to an approximately level surface prior to the deposition of the 'Middle' (Yorkian) Coal Measures. That these deposits were laid down over an extensive and for the most part level surface including Coalbrookdale, the Clee Hills and Wyre Forest is indicated by the general similarity both in lithology and horizon of their earliest formed strata; especially in regard to the ganister rock near the base (p. 44) which has been recognized in the Coalbrookdale area, at Brown and Titterstone Clee, in the Wyre Forest Coalfield, and again in the very small outliers of basal Middle Coal Measures west of Deuxhill and at Cleobury Mortimer (Sheet 182).

Movement during the deposition of the Yorkian strata is indicated by the variation in total thickness of these beds.

The early Staffordian movement which produced the strong folding and some faulting below the " Symon Fault " in the Coalbrookdale Coalfield and the upcast of the Old Red Sandstone beds on the west in the Clee Hill area and to the south of the district at Mamble and Heightington was followed again by extensive denudation to an approximately level surface on which the lowest beds of the Highley Group were deposited unconformably on the Kinlet Group and with overlap onto the Old Red Sandstone to west and south.

The essential horizontality of the surface on which deposition of these Upper Coal Measures commenced is shown by the widespread similarity of their lowest beds, but again it is clear from the variation in thickness of the group that movements, which seem to have followed pre-existing lines, took place during its deposition.

R.W.P.

DETAILS

The Western Boundary Fault.—The Western Boundary Fault enters the district near Blakenhall and throws at first 'Middle' Coal Measures and then Silurian (Lower Ludlow) rocks against Enville Beds. Near Sedgley the fault splits into three branches, which in a short distance re-unite, whereby strips of Middle Coal Measures and of Keele Beds are brought in. Between Sedgley and Cotwall End the westward downthrow of the main or westernmost branch may be estimated as between 1,700 and 1,800 ft. both in the Coal Measures and in the Silurian rocks. Farther south-west, in Himley Wood, where the fault, here single, brings Keele beds against the lower part of the Etruria Marl, the throw appears to be about 900 ft.

South-west of Himley Wood the fault gives off a branch, the Lloyd House Fault running a little west of north. From this point to Kingswinford the Western Boundary Fault throws down Bunter rocks against Etruria Marl. The Round Hill Pits,[1] 650 yds. E.S.E. of Holbeache House, began in the Bunter and passed through the fault into the Coal Measures. This part of the fault is in general very steeply inclined, its inclination from the vertical (hade) being in places apparently only 5°, though just south of the Round Hill Pits it appears to be as much as 40°.

From Kingswinford south-south-eastward the Western Boundary Fault is generally in two branches, enclosing strips of Enville or Keele beds between them. Boreholes at Wordesley (p. 189) and Amblecote (p. 190) have proved Clent beds under the Trias to the west of the western branch. The combined throw of the branches from Kingswinford to Brettel Lane, where the Brockmoor Fault is given off to the north-east, must be between 1,500 and 1,600 ft. Near Amblecote and Stourbridge, where measures below the Thick Coal abut against the eastern branch, the aggregate downthrow must be about 1,900 ft.

South-east of Stourbridge the throw probably diminishes rapidly as successively higher members of the Upper Coal Measures appear on the upthrow side, but want of data as to what underlies the Trias on the downthrow side prevents any reliable estimate. At Hagley, Bunter Pebble Beds appear on the upthrow side of the fault and from Clent to the southern border of the district its course at the surface is wholly in the Triassic rocks.

The Lloyd House Fault, etc.—The Lloyd House Fault branches from the Western Boundary Fault (p. 126), and runs to Earlswood, where it appears to meet another fault. For the greater part of its course it throws Bunter beds on the west against Keele or Enville beds on the east, but near Earlswood the Bunter Pebble Beds are on both sides. The throw in the Triassic rock appears to be about 300 ft.; that in the Coal Measures is difficult to estimate as the rocks beneath the Bunter to the west are unproved. What appears to be the Lloyd House Fault was met with underground north of the Wodehouse, about 250 yds. west of its surface position, indicating a hade of about 20° From information supplied by Mr. L. Holland, a heading and horizontal borehole driven from the Thick Coal into the fault apparently penetrated Etruria Marl. This would suggest a throw of between 270 ft. and 470 ft.; i.e., probably about the same in the Coal Measures as it is in the Trias.

Between the Lloyd House Fault and the Western Boundary Fault the most important dislocations have a north-north-eastward course and throw down to east-south-east. About 500 yds. east of the shafts of Baggeridge Colliery a gate road encountered a westward downthrow which brought up Downtonian beds to face the Thick Coal (p. 10, and Fig. 2, p. 16). Farther east, one of the east-south-eastward downthrows brought in the Thick Coal again at about the same level.

The Stapenhill Fault.—This fault seems to begin to the east of Lower Penn, but is possibly a continuation of the disturbances associated with the Bushbury Fault (Wolverhampton Memoir, p. 163). Near Wombourne its westward downthrow in the Triassic rocks is about 750 ft. Continuing southwards the fault crosses the River Stour near Stapenhill Farm, from which it was named.[2] About 100 yds. N.W. of this farm a small hill, Primrose Hill, is formed of an outlier of Bunter Pebble Beds, dipping west-north-west, against the east side of the fault. About one and a quarter miles farther south a similar outlier of Pebble Beds forms Round Hill, and some 550 yds. north of this, on the downthrow side of the fault, is the small outlier of Keuper Sandstone mentioned on p. 116. The throw of the fault here can thus be estimated as about 550 ft.; farther south it

---

[1] 'Vert. Sects.' (*Geol. Surv.*), Sheet 26, No. 48.

[2] It is believed by Charles Lapworth, though not apparently in any publication. Stapenhill Farm, not named on the one-inch map, is about 450 yds. west by north of the north-west corner of New Wood.

seems to diminish. Near Iverley House Farm the fault abruptly truncates the Bunter Pebble Beds escarpment of Bunkers Hill. It then continues in a south-south-westward direction, with Upper Mottled Sandstones on the downthrow (western) side and Bunter Pebble Beds on the upthrow side nearly to the southern border of the district. This it leaves in Hurcott Wood, its throw there being probably not much more than 150 ft.

The Enville Fault.—The Enville Fault is first traceable in the Bunter outcrops west of Trysull. It crosses the main Bunter Pebble Beds escarpment north of Blackhills Plantation, where its throw in the Triassic rocks can hardly be more than 50 ft. If, as suggested below (p. 132), it is this fault that was passed through near the bottom of the Smestow borehole, its throw in the Coal Measures must be considerably greater and may reach 500 ft., though if, as discussed on p. 124, the fault is accompanied by·folding in the Coal Measures, the throw will be variable (Fig. 14, p. 125).

From Highgate Common to Enville Sheepwalks, Clent Beds, on the west (upthrow) side of the fault, abut against Lower Mottled Sandstone. From Compton south-south-westward the fault flanks and largely replaces the eastern limb of the Trimpley Anticline. South of Compton, as described on p. 103, breccias of the Clent Group on the east are thrown down against Coal Measures to the west and in the stream about 400 yds. S.E. of Greyfields Court they can be seen within a few yards of one another. Farther south at first Lower Mottled Sandstone and then Bunter Pebble Beds are thrown against the pre-Carboniferous rocks of the Trimpley inlier. The stream in Parkatt Wood, about 300 yds. N.N.E. of Lower Barns Farm, exposes Pebble Beds in close proximity to Downtonian cornstone and calcareous sandstone dipping very steeply east-south-east. A short distance east of the fault the Pebble Beds dip west by south, and it is mainly to this slight roll in the Bunter that the curious inlier of Lower Mottled Sandstone to the east of Lower Barns Farm is due.

Near Trimpley the Enville Fault splits and strips of Kinlet, Clent and Bow-hills beds appear between the pre-Carboniferous rocks and the Bunter; though at one place the westernmost branch has pre-Carboniferous rocks on both sides but with opposing dips.

The Trimpley Anticline and associated Faults.—In the synclinal portion of the pre-Carboniferous inlier a group of more or less radially disposed faults, centred approximately on the Kidderminster road, may represent adjustments to the post-Coal Measures movements. One of them affects the little outlier of Coal Measures in Birch Wood (p. 56).

The axis of the anticline south of the Park Attwood Fault seems to lie west of Mary Moors, between the two outcrops of *Psammosteus* Limestones (p. 26), which dip away from one another. North of Holbeache the axis is replaced by faults, but the eastern limb of the fold is traceable by the *Psammosteus* Limestone outcrop near Littlegains Farm, by eastward-dipping higher beds near Lower Barns Farm and perhaps by those cropping out east of Birchwood (near Bodenham Farm, p. 25), which are separated from the synclinal area by a fault. This would appear to imply relative south-eastward displacement on the south-west side of this fault and eastward displacement along the north side of the Park Attwood Fault, the syncline having been thus pushed, wedgewise, into an originally parallel anticline, and also raised relatively to it. The greater part of this movement must have been pre-Carboniferous, for, as already pointed out, the syncline and anticline must have been very much in their present relative positions when the Coal Measures were deposited.

Part also of the distortion and fracturing of the anticline may have been the result of pre-Carboniferous movements, but part may have accompanied the post-Coal Measures movements. Most of the faults in the pre-Carboniferous rocks south of the Park Attwood Fault are probably at least in part of post-Coal Measures date, for some of them appear to affect the Coal Measures, for example, south of Folly Point and in North Wood (Sheet 182). The accentuation, with

FOLDS AND FAULTS.

westward overfolding, of the anticlinal structure probably resulted in further displacement along the Park Attwood Fault.

The fault between the pre-Carboniferous rocks and the Coal Measures on the west side of the inlier is, with little doubt, a reversed one, though direct proof of this is wanting. Its straight course, however, forbids its being regarded as a low-angle 'thrust.' For a short distance south of the Park Attwood Fault the overfolding appears to be accompanied by faulting at or near the junction between the Coal Measures and the Lower Old Red Sandstone. Farther south, however, this does not appear to be the case. In the bank of the railway near Hill Farm (Sheet 182) Coal Measures and Old Red marls and sandstones can be seen in close proximity, inverted but apparently unfaulted.[1]

At the northern end of the Trimpley Anticline faults more or less longitudinal in direction help to bring the outcropping Coal Measures round the 'nose' of the fold. As mentioned on p. 86, some strike faulting appears to occur along the western limb of the fold.

The Romsley Fault.—This begins at the Pattingham Fault near Wooton and at first throws the highest beds of the Keele Group, on the west, against sandstones on about the horizon of the Alveley grindstone beds. This indicates an eastward downthrow of about 500 ft. The maximum throw is probably near Bowhills, where beds high in the Bowhills Group are brought against a lower horizon of the Keele beds than that near Wooton, and may be nearly 1,000 ft. Farther south the throw diminishes rapidly, but the continuance of the fault beyond where the base of the Bowhills Group meets it, at the Kidderminster and Bridgnorth road, is shown by abrupt changes of strike in the Keele and lower beds. The outcrop of the Shatterford basalt is slightly shifted by the Romsley Fault, which appears to terminate at the fault along the west flank of the Trimpley Anticline.

The Pattingham and Patshull Faults.—The Pattingham Fault enters the district from the north, near Rudge, and for about two miles throws Lower Keuper Sandstone on the west against Bunter on the east. It is responsible for the abrupt termination of the Abbot's Castle Hill escarpment at Hilend. At Lower Beobridge it is joined by the Patshull Fault, also a westward downthrow. South of Quatt the Brock Hall Fault (p. 130) meets the Pattingham Fault and from this point the latter is continued as the Kinlet Hall Fault (p. 130).

The Arley Park Fault.—Some displacement of the outcrops near Alveley appears to be due to a fault of eastward downthrow which may arise at the Pattingham Fault near Quatt Farm, where, however, it must be very small. A short distance south-east of Alveley the direction of downthrow changes to west, and here the Arley Park Fault may be considered to begin. Its westward downthrow near Hall Close is probably less than 100 ft. but increases near Pickards Farm to, probably, over 200 ft.

Crossing the Severn west of Upper Arley the fault continues, though with diminished throw, affecting the outcrops of the Kinlet and Highley beds as mentioned on pp. 56 and 76, until it meets the Station Fault near Woodhouse Farm.

The Station Fault.—Arising at the Kinlet Hall Fault (p. 130) near Hampton Loade, this fault at first has Keele beds on both sides. After about a mile beds high in the Highley Group appear on the west side and abut against Keele beds well above the base of the group on the east or downthrow side. About 450 yds. S.W. of Hadleys the fault splits and both branches cross the Severn, preserving a block of Keele beds in a trough between them.

T.H.W.

The two branches reunite about 300 yds. S.S.E. of Highley Station and the position of the fault can be closely fixed on the railway, a third of a mile south of

---

[1] Cantrill, T. C., *op. cit.*, p. 14.

the station, where red Keele beds with *Spirorbis* limestone are thrown down on the east against grey sandstones high in the Highley Group. From this point the fault continues southwards and crosses Borle Brook just east of the railway bridge.

R.W.P.

Under the River Severn, nearly opposite the mouth of Borle Brook, an eastward downthrow has been proved in colliery workings. A short distance farther south workings ended just east of the river, at what is believed to be a continuation of the same fault. The surface position of the fault near here appears to be along the river, which it leaves where the latter begins to turn south-eastward towards Upper Arley. Farther south, as already mentioned (p. 76), the Station Fault throws down Highley beds on the east against Kinlet beds on the west. It leaves the district near the Woodhouse Farm.

T.H.W.

The Highley-Kinlet Fault. This fault, so named because it separates the area of coal worked from the Highley pits from that worked from the Kinlet pits, has approximately the same point of origin from the Kinlet Hall Fault as the Station Fault. With a south-easterly trend it crosses the Severn half a mile below the ferry at Hampton Loade, throwing down Keele beds on its western side against Highley beds. West of the river it curves towards the south through Hazelwells.

South of Hazelwells the throw is about 200 ft., and the fault apparently passes beneath St. Mary's Church, Highley. Thence, in Highley beds, it crosses Borle Brook a third of a mile west of the railway bridge and continuing towards Severn Lodge may unite with the Station Fault.

The Tiphouse Fault.—Arising also from the Kinlet Hall Fault, but at a point west of the River Severn, this fault crosses Borle Brook about half a mile northwest of Netherton and then turns almost due south towards Button Bridge. West of Netherton the downthrow is about 160 ft. to the west. No serious attempt appears to have been made to work coal to the west of the Tiphouse Fault.

R.W.P.

Near Button Bridge an eastward downthrow leaves the Tiphouse Fault on the west side and the two faults let down a wedge of Highley beds between them (see p. 77), which has been traced beyond the southern border of the district.

T.H.W.

The Kinlet Hall Fault.—This may be regarded as the continuation of the Pattingham Fault from the point south of Quatt where it meets the Brock Hall Fault. East of the Severn it throws Enville beds on the north-west against the Keele beds north of Butter Cross. West of the river it brings Keele Beds down against Highley beds, north of New England, near where it crosses Borle Brook. In its south-westward course from this point, although its position across Kinlet Park can be traced with considerable accuracy, it is difficult to determine the amount and direction of throw ; indeed, to the south of the Park the outcrops clearly indicate a downthrow to the south-east. It may be that in this region the structure is complicated by a certain amount of horizontal movement such as is indicated by slickensided surfaces of basalt in Raggits Quarry and elsewhere.

Minor faults in Kinlet Park affect the outcrop of the basalt and associated Coal Measures but appear to be of local significance only.

The Brock Hall or Billingsley Fault.—This important fault leaves the Pattingham-Kinlet Hall Fault south of Quatt, crosses the River Severn at Hampton Loade and, with its downthrow to the south-east, appears to balance to a great extent the effect of the Kinlet Hall Fault and its branches, the Highley-Kinlet and the Tiphouse faults. At Brock Hall it brings Keele beds on the south-east against Highley beds ; the base of the Keele Group is carried forward nearly to Borle Brook between it and the Kinlet Hall Fault. The workings of Billingsley Colliery were limited by it towards the south-east. At a point 550 yds. south of the Cape of Good Hope the fault bifurcates, the two branches reuniting two and a half miles to the south-west near Walton. The old Harcourt Pit is situated in the trough between these faults. It is noteworthy that there is no evidence of the presence of basalt in the Billingsley area to the north-west of the Brock Hall or Billingsley Fault.

The Deuxhill Fault.—This fault appears to arise in the Downtonian ground near Harpswood and its downthrow to the west is responsible for the Deuxhill outlier of Upper Coal Measures (Highley Beds) to the west of the main outcrop at Woodlands and The Hill. It crosses Borle Brook 150 yds. S.E. of the Ford at Eudon Mill and can here be located accurately for about 100 yds. The fault line is well defined where it crosses the small brook from Eudon George and again in two small streams half a mile west of Glazeley Church. It passes southwards just east of Deuxhill Farm at the southern end of the outlier of Coal Measures. In the Old Red Sandstone ground to the south it is not easily traceable but disturbances of the beds in Horsford Brook appear to be on the fault line. A strongly marked fault-face with vertical slickensides exposed in the brook 80 yds. S.W. of Plym Hall is probably on the line of this fault and west of Chorley it appears to be responsible for the small outlier of Kinlet beds beyond the main outcrop.

R.W.P.

Extensions of the Coalfields

The Productive Coal Measures have now been proved, and largely worked, over nearly all the area under the Keele and Enville beds to the west of the Western Boundary Fault near Sedgley. The coals should also lie at a similar moderate depth beneath the same Upper Coal Measures between the branches of the Western Boundary Fault from Kingswinford to Stourbridge ; but they are likely to be too disturbed to be worth working in such narrow areas, and in places the Thick Coal may actually be cut out by the meeting of the two faults.[1]

Further extensions of the South Staffordshire Coalfield must be sought to the west of the Lloyd House Fault and of the Western Boundary Fault south of where the former meets it. As mentioned on p. 127, the westward downthrow of the Lloyd House Fault appears to be about 300 ft. in the Triassic rocks. If it is the same in the Coal Measures the Thick Coal may be expected at an average level of about 1,600 ft. below Ordnance Datum on the downthrow side; rising from about $-1,900$ ft. near Bearnett House to perhaps $-1,500$ or $-1,400$ ft. at the north end of Himley Park, if the direction and amount of dip are similar to those on the east side of the fault. This area would be bounded on the west by the Stapenhill Fault, the intersection of which with the Coal Measures will presumably be somewhat to the west of its surface outcrop.

Farther south the South Staffordshire seams are likely to persist for some distance between the Western Boundary Fault and the Stapenhill Fault, the average depth of the Thick Coal being probably about 2,300 ft.

---

[1] See T. H. Whitehead in ' The Southern Part of South Staffordshire Coalfield ' (*Mem. Geol. Surv.*), 1927, p. 177.

The seams may possibly deteriorate in a southward direction, as they do in the " old " coalfield to the east of the Boundary Fault, and, judging from the result of the Wassel Grove sinking,[1] are not likely to be present in a workable condition much south of Stourbridge.

The first direct evidence concerning the Coal Measures to the west of the Stapenhill Fault is provided by the Smestow borehole (p. 39). The record of this is far from satisfactory and it is difficult to locate with certainty the base of the Trias and the boundaries between some members of the Upper Coal Measures. One horizon that does seem to be clearly recognizable, however, is the base of the Halesowen Group, at a depth of 2,660 ft.[2] Below this much core was lost and an abnormally small thickness of Etruria Marl was passed through, so that there seems little reason to doubt that a fault was encountered at a depth of about 2,700 ft. This fault would appear to have an eastward downthrow and, though the Enville Fault (p. 128) would require a hade of about 60° to pass through the borehole at this depth, no other is known at the surface that is likely to have done so. It is possible, however, that north of Blackhills Plantation the Enville Fault has a more north-easterly course than that drawn on the map (Sheet 167), which would bring its surface position nearer to the site of the Smestow borehole.

The thickness of the grey ' Middle ' Coal Measures in the borehole is also very small, and this could be explained by the presence of a second fault, throwing down westward and bringing up the basal part of the Middle Coal Measures, with Silurian rocks, in a horst between itself and the first fault (see Fig. 14, p. 125). It may be remarked that the coarse grit resting on the Silurian shales is evidently the basement bed of the Coal Measures, so that there is no fault actually between the Coal Measures. and the Silurian rocks.

If the 7-ft. coal seam recorded in the Smestow borehole at a depth of 2,725 ft. represents the Thick Coal it is clearly nearer to the base of the Halesowen Group than is normal, no doubt because the eastward downthrow passes through the borehole between those horizons ; *i.e.*, the coal is on the upthrow side of what is presumed to be the Enville Fault. Judging from the position of the base of the Halesowen Group, that of the Thick Coal on the downthrow side of the fault should be at a level of about — 3,000 ft. O.D. (Fig. 14, p. 125). It seems likely that the measures lie in a shallow syncline between the Enville and Stapenhill faults. They probably rise westward to the former, though the extent to which they do so depends somewhat upon the interpretations of the structure discussed on p. 124. In the middle of the syncline the Thick Coal may well be at a greater depth than 3,000 ft. below O.D., and minor complications may exist to which the Triassic rocks at the surface afford no clue. The Coal

---

[1] ' The Southern Part of South Staffordshire Coalfield ' (*Mem. Geol. Surv.*), 1927, pp. 68, 176. The " Swinford Basin " of Mr. W. W. King (*Trans. Inst. Min. Eng.*, vol. lxi, 1921, p. 153) might, however, carry the seams somewhat farther south than their limit on the east of the Western Boundary Fault.

[2] *i.e.*, according to the section communicated by Mr. J. Bloomer ; W. Gibson puts it at 2,650 ft.

Measures may perhaps also rise eastward to the Stapenhill Fault, as does the Keuper Sandstone north of Wombourne, in which case the actual throw of that fault would be proportionately reduced, but it may very well be over 1,000 ft. in the Coal Measures.

There is, then, some evidence of a basin of coal at a not unworkable depth between the Stapenhill and Enville faults in the latitude of Smestow; but the unsatisfactory record of the Smestow boring and the unfortunate circumstances that it appears to have passed through a fault before reaching the coal renders it desirable that this area should be further tested by one or more boreholes before any attempt is made to develop it. Boreholes are also highly desirable farther south, between the same two faults, if only to ascertain whether, as seems probable on general grounds, the Coal Measures tend to rise in that direction. The southern limit of workable coal may probably be regarded as lying in about the same latitude as to the east of the Stapenhill Fault (p. 132), say in the neighbourhood of Kinver. North of Smestow the depth to the coal would tend to increase, as the dip is probably northwards and the Triassic cover thickens.

To the west of the Enville Fault, though the ' Middle ' Coal Measures are at a less depth than to the east, the results of the Claverley borehole (p. 39) suggest that the prospects are not promising. The 1-ft. coal seam, the thickest met with, at a depth of 2,044 ft. 6 in., might conceivably represent an attenuated remnant of the South Staffordshire Thick Coal, for it lies at about the same distance below the base of the Etruria Marl. Moreover, at 800 yards due west of the shafts of Baggeridge Colliery the average thickness of the Thick Coal is 22 ft. 3 in. and if the 7-ft. coal in the Smestow borehole represents the Thick Coal this implies a reduction in thickness of about 15 ft. in a little over 2 miles. At the same rate of westward attenuation the Thick Coal would thin out altogether before reaching the site of the Claverley borehole, so that there is no inherent improbability in its being represented there by a 1-ft. seam. Thus, until the area west of the Enville Fault has been further tested by a borehole nearer to that fault than the Claverley borehole it would seem wiser to discount it as a potentially productive field. The Highgate Common (or Forest) borehole was in the desired position, but did not go deep enough to provide the required information and, moreover, was made so long ago (in 1857-8) that its interpretation is now a matter of difficulty (see p. 103).

Approaching the same area from the west, the Highley-Brooch Coal has been followed under the Severn to the west of Alveley and in 1935 the Alveley shaft was sunk, to meet the workings, at a point 800 yds. S.W. of Alveley Church. Already (1945) the workings have reached points some distance north, east and south of the shaft. A fault trending north-north-west has been located extending through the western margin of Alveley village with an easterly downthrow of 66 ft. Evidence farther north suggests an increasing throw in that direction. A considerable area of workable coal may be anticipated south of the Pattingham-Kinlet Hall Fault. It is known that, west of the Severn, the Highley and Kinlet seams thin out southward (p. 45) and, in view of the barren character of the Kinlet beds at Shatterford (p. 55), their southern limit east of the river is hardly likely to be much south of a west-to-east line opposite the mouth of Borle Brook.

The Romsley Fault would form the natural eastern boundary of this field, but whether the seams will persist in a workable condition as far east as this is uncertain, having regard to the evidence from the Claverley borehole and the Shatterford sinking. The presence in the Claverley borehole of marine bands that can probably be correlated with similar bands in South Staffordshire and the northern part of the Wyre Forest Coalfield (p. 40) certainly suggests continuity of deposition between the two fields. Such continuity may, however, have been intermittent, as far as the present district is concerned, and the connexion indirect. On the whole the available evidence does not suggest that the area between the Pattingham and Romsley faults, on the west, and the Enville Fault, on the east, is worthy of practical consideration until other more promising areas have been developed.

Prospects would appear equally uncertain in the area to north-west of the Pattingham Fault. Though the connecting link between the Wyre Forest and South Staffordshire coalfields may have lain in this area there is a probability that in the western part of it the 'Middle' Coal Measures (Kinlet Group) may be absent as at Eardington (p. 44) and near Bridgnorth, owing to Upper Coal Measures resting directly upon the pre-Carboniferous rocks. Furthermore, the presence of Triassic rocks at the surface would increase the depth of the coals, if they exist, as compared with that in the area south-east of the Pattingham Fault.

T.H.W.

IGNEOUS ROCKS

The only igneous rocks in the district are the olivine basalts and dolerites that occur amongst the Coal Measures of the South Staffordshire Coalfield, of Shatterford and Kinlet, and have also been proved underground at Highley and in the Claverley borehole. A few miles to the west of the district the well-known basalt of the Clee Hills belongs to the same group.

The igneous rocks in South Staffordshire were studied by J. B. Jukes[1] and S. Allport,[2] who both considered them to be of Carboniferous age. Later their general resemblance to the Tertiary basalts of Scotland and to the Butterton and Swynnerton dyke in Staffordshire, which is intrusive into Keuper Marl, led to the suggestion that the basalts in the Coal Measures might be of Tertiary age.[3] In 1931 Dr. R. W. Pocock[4] claimed the extrusive character of some of these rocks, and that in places they

---

[1] 'The South Staffordshire Coalfield' (*Mem. Geol. Surv.*), edit. 2, 1859, p. 117.

[2] 'On the Basaltic Rocks of the Midland Coalfields,' *Geol. Mag.*, 1870, p. 159 ; and 'On the Microscopic Structure and Composition of British Carboniferous Dolerites,' *Quart. Journ. Geol. Soc.*, vol. xxx, 1874, p. 529.

[3] Watts, W. W., *Proc. Geol. Assoc.*, vol. xv, 1898, p. 399, and vol. xix, 1905, p. 179.

[4] 'The Age of the Midland Basalts,' *Quart. Journ. Geol. Soc.*, vol. lxxxvii, 1931, p. 1.

suffered denudation in Coal Measures times (pp. 143 and 144, below). The evidence advanced, together with the fact that they belong to the alkaline suite of igneous rocks,[1] which distinguishes them petrographically from both the Scottish Tertiary basalts and from the Butterton and Swynnerton dyke, was considered sufficient to establish their Carboniferous age. This view has been strengthened by the work of W. D. Urry,[2] who has determined the age of the Clee Hill basalt, by the helium method, to be 135 million years, a figure which places it in the Upper Carboniferous on his revised Helium Method Time-scale.

Recently the relations of the Midland basalts to the associated sediments have been studied by Dr. C. E. Marshall,[3] who, while regarding them as dominantly intrusive, apparently accepts their general Carboniferous age.

It is probable that igneous activity occurred intermittently in the district. The Little Wenlock basalt (Sheets 152-3) is Lower Carboniferous, other of the basic rocks may be of Middle Coal Measures (Yorkian) age, but in South Staffordshire the main Rowley Regis mass overlies a considerable thickness of Etruria Marl (p. 136), so that its age is presumably Staffordian. Since none of the igneous rocks has been found to penetrate or overlie beds of the Halesowen or Highley groups it would seem that the latest phase of the igneous activity was associated with the early Staffordian earth-movements (p. 121), though other phases preceded those movements.

T.H.W., R.W.P.

SOUTH STAFFORDSHIRE COALFIELD

The greater part of the basalt and dolerite of the Rowley Regis area lies outside the present district,[4] only extending into it near Kates Hill, Dudley, where it is not well exposed and is to a considerable extent built upon. Its relation to known horizons in the Coal Measures, *e.g.*, the Thick Coal, underground clearly shows that it transgresses these measures in a west-north-westward direction. In the Twin Quarry, near Oakham (Sheet 168) the basalt is overlain by a small thickness of Etruria Marl, and shows a thin chilled edge at the contact, the marl being but little altered.[5] The thickness of igneous rock in this quarry must be nearly 100 ft., but in No. 2 Shaft of Grace Mary Colliery, about 350 yds. to the south-east, only about 12 ft. is present. This rapid change in thickness, together with the transgressive relation of the basalt to the Coal Measures

---

[1] See T. Robertson in discussion on R. W. Pocock, *op. cit.*, p. 12.

[2] 'Age determinations of Carboniferous basic rocks of Shropshire and Colonsay,' *Geol. Mag.*, 1941, pp. 45-61.

[3] 'Field relations of certain of the Basic Igneous Rocks associated with the Carboniferous strata of the Midland Counties,' *Quart. Journ. Geol. Soc.*, vol. xcviii, 1942, p. 1.

[4] See T. H. Whitehead in ' The Geology of the Country around Birmingham ' (*Mem. Geol. Surv.*), 1925, p. 102.

[5] Dr. F. Raw suggests that this is due to the anhydrous and limeless nature of the Etruria Marl in contact with the dolerite (in W. S. Boulton, 'Report on Week-End Field Meeting in the Birmingham District,' *Proc. Geol. Assoc.*, vol. xlii, 1931, p. 300).

and its chilled contact edge, is strong evidence that the Rowley mass is intrusive.[1]

A quarry on Tansley Hill showed (in 1920) the edge of the main intrusion, consisting of two comparatively thin sills separated by Etruria Marl with espley grits. About 350 yds. to the south-west a clay pit (Survey Photograph 1936) near the Dudley and Birmingham road exposes purple marls with espley rocks, nearly vertical, and amongst them basalt with its top portion very scoriaceous and amygdaloidal. This suggests that it is a lava flow, but in that case it must be of appreciably earlier date than the main Rowley Regis mass, for it lies in the lower part of the Etruria Marl, probably not much over 100 ft. above the base, whilst the Rowley basalt, in the Grace Mary pit already mentioned, overlies about 550 ft. of that formation.

Dr. J. Phemister describes the rock of Tansley Hill as follows :—

" A fresh specimen (E.12475) is an olivine-basalt containing numerous microphenocrysts of olivine which is replaced entirely by aggregates of brown fibrous serpentine or of serpentine and pale green mica-like material. Microphenocrysts of labradorite, containing about 65 per cent. An, grade in size downwards to the laths which form the groundmass of the rock. The plagioclase shows normal zoning to oligoclase at the terminations of the prisms. Small serpentinized crystals of olivine and thick plates of iron ore are numerous throughout the groundmass, and prismatic grains of greenish and faintly purple augite are occasionally subophitic to feldspar but have in general normal intersertal disposition between the plagioclase laths. Interstitial material includes aggregates, occasionally spherulitic, of serpentine, a very small amount of alkali feldspar, some analcite and calcite. Apatite occurs as swarms of fine needles. Petrographically the rock belongs to the Dalmeny type of olivine-basalt. Other specimens are vesicular and amygdaloidal varieties of the same type. They are greatly decomposed, the ferro-magnesian minerals being replaced by carbonates, and the feldspar by calcite and analcite (E. 12478, 14805)."

Igneous rock has been encountered intrusive into the ' Middle ' Coal Measures at a depth of about 380 ft., in Warren's Hall Colliery,[2] between Tansley Hill and Bumble Hole. Here the comparatively unaltered part of the rock is a micro-ophitic olivine basalt, with labradorite feldspar, augite of a deep purplish-brown colour and olivine replaced by serpentine. In parts of the intrusion farther from the centre the feldspar is partly replaced by calcite and a micaceous or kaolinitic aggregate, and in the outer portions the rock is a " white trap,"[3] with the feldspar nearly completely altered as described and the other minerals also largely replaced by calcite and a skeleton aggregate of iron ores. Calcite is present also as veins and coarsely crystalline aggregates filling cavities.

A small basalt intrusion occurs near Brewin's Bridge, Netherton, in the Downtonian rocks (Fig. 3, p. 18). It appears to be later than the fault there exposed,[4] for in the canal section it has run up the fault-plane

---

[1] See discussion on R. W. Pocock, *op. cit.*, p. 11.

[2] The description is from slides lent by the late H. W. Hughes.

[3] See J. S. Flett, in ' The Geology of the neighbourhood of Edinburgh ' (*Mem. Geol. Surv.*), 1910, p. 311 ; also T. C. Day, ' Chemical Analyses of White Trap from Dalmeny,. Granton, Weak Law and North Berwick,' *Trans. Edinb. Geol. Soc.*, vol. xii, 1930, p. 189.

[4] See also W. W. King and W. J. Lewis, *Geol. Mag.*, 1912, p. 490.

where the massive upper part of the Downton Castle Sandstone abuts against it. A few yards south, where flaggy beds lower in the Downton Castle Sandstone come against the fault, the basalt has crossed the latter and intruded itself as thin sills amongst the flags. The igneous rock has been described by Mr. W. J. Lewis[1] as an ophitic olivine dolerite, containing fresh olivine, augite highly decomposed into chloritic products, feldspar intermediate between labradorite and andesine and a small quantity of magnetite. The rock is entirely free from glassy material, the residual magma having given rise to a second generation of feldspars and augite.

Another occurrence is at Barrow Hill, near Pensnett, where it has been suggested that the basalt has a laccolitic form, intrusive in the lower part of the Etruria Marl.[2]

Dr. Phemister describes the Barrow Hill rock as of the same type as that from Tansley Hill,

" but the feldspar microphenocrysts are more definitely porphyritic ; that is, the fabric is hiatal. They may form polysomatic groups with serpentinized olivine (E.12480), or may be individually disposed in more or less parallel orientation (E.12481). The groundmass may show good fluidal texture and variation in grain-size (E.12481)."

Dr. C. E. Marshall,[3] who has recently carried out a magnetometer and electrical survey of the Barrow Hill area, concludes that the intrusion took place dominantly in a near-vertical direction from the east and south-east, accompanied by a considerable disruption of the strata. The intrusion is considered by him to represent a late Carboniferous vent or feeder, filled with an intrusion agglomerate in which local country-rock xenoliths occur in large numbers, exhibiting a wide range in degree of alteration. The igneous rock is mainly a fine-grained microtaxitic analcite-bearing olivine basalt.

Near the Graveyard, Gornalwood, an outcrop of basalt appears to be approximately on the horizon of the " Ten Foot Measures," above the Thick Coal. This occurrence displays to advantage the spheroidal type of weathering.[4] Another, exposed in a quarry immediately south of the Himley road, 400 yds. south of the Graveyard, is intruded somewhat higher in the Middle Coal Measures. Yet another intrusion lies on the upthrow side of the Russell's Hall Fault near Little London Fields, and is on a horizon a little below the Thick Coal. Between Barrow Hill and Springs Mire basalt, referred to as " green rock," has been met with underground in various measures ranging from the Whitestone to those between the Lower Heathen Coal and Gubbin Ironstone. A short distance north and west of Barrow Hill basalt was found in the Thick Coal, while near Coopers Bank it replaced the Heathen Coal. Farther west,

---

[1] In W. W. King and W. J. Lewis, *loc. cit.*

[2] See ' The Geology of the Southern Part of the South Staffordshire Coalfield' (*Mem. Geol. Surv.*), 1927, p. 184.

[3] See C. E. Marshall, ' The Barrow Hill intrusion, South Staffordshire,' *Quart. ourn. Geol. Soc.*, vol. ci, 1946, p. 177.

[4] *Op. cit.*, 1927, Plate VIIB (Survey Photographs 1,970-1).

near the Shut End Fault, " green rock " occurs in all the measures from above the Whitestone to the Thick Coal. On the downthrow side of the same fault it is found in all the measures except the fireclay there worked, which is probably that beneath the Fireclay Coal.

SHATTERFORD

The Shatterford basalt can be traced for about two and a third miles from the northern edge of Arley Wood to Eymore Wood. At the northern end of its outcrop it is probably cut off by a fault (p. 80), but at the southern end it appears to die out.

Long known as the Shatterford ' dyke,' this igneous body is certainly not a dyke, for it conforms to the bedding of the sediments in which it lies and its generally steep inclination is due to its having been folded with those sediments.[1] Dr. Pocock[2] claims it as a lava flow on the grounds of absence of alteration of the associated sediments, the presence of pipe amygdales at its base, the vesicular nature of its top and the absence of transgression of or fingering out into the sediments.[3] Whilst many of the exposures, especially those in the quarries north-north-east of Witnell's End, support this view, there is other evidence that appears to conflict with it. In one section at least, described below (see also Fig. 15, p. 139), there seems to be some alteration of the sediments above as well as below the basalt, and the top of the latter has a well-defined chilled edge of uniform thickness. If the presence of pipe amygdales can be regarded as definitely diagnostic of a lava flow, then at least those parts of the Shatterford basalt in which they occur must be lavas. Though, so far as can be judged in the absence of well-defined datum planes in the Kinlet Beds, the basalt maintains the same general horizon in the Coal Measures throughout its outcrop, the possibility of some degree of transgression cannot be ruled out, for the beds in contact with the base and top are not everywhere of the same character ; and it is by no means certain that the igneous rock occupies the same horizon in the Coal Measures underground in the Shatterford sinking as it does at the surface. It is possible that the apparently conflicting evidence can be reconciled by regarding the igneous rock as a sill inserted under a comparatively thin cover of sediments, perhaps imperfectly consolidated. It is even possible that the rock is intrusive at some parts of its course and extrusive at others. In any case, however, its age can differ but little from that of the associated sediments, which, on the evidence of plants collected from a few feet below the basalt, is Yorkian.

At the northern edge of Arley Wood, 500 yds. south of Hightrees Farm, the basalt is exposed in a quarry. It is here moderately fine-grained and shows a rude spheroidal structure. Its thickness cannot be less than 25 ft. and its feature ends abruptly in the field a few yards north

---

[1] See T. C. Cantrill, ' A Contribution to the Geology of the Wyre Forest Coalfield . . . ," 1895, p. 33.

[2] *Op. cit.*, p. 3.

[3] But see C. E. Marshall, *Quart. Journ. Geol. Soc.*, vol. xcviii, 1942, pp. 11-13.

of the quarry, where it is probably cut off by a fault. Mr. W. W. King informs us that it is reported to have been met with in a well at Starts Green, but no trace of it was found at the surface there or anywhere along the north-east and east sides of the Trimpley Anticline.

In Arley Wood, 300 yds. south of the quarry above-mentioned, the basalt forms a considerable crag formerly known as Munster's Hill. The next noteworthy exposure is in the stream between Arley and Coldridge Woods, where the basalt can be seen overlying the plant-bearing shales described on p. 56. Between the stream and Witnell's End is an almost continuous line of quarries in which, in places, the top and base of the basalt are exposed. At a point about 500 yds. N.N.E. of Witnell's End the top 2 in. of the basalt, beneath the beds described on p. 56, differs from the remainder and may represent a weathered or possibly a chilled surface. It is underlain by green vesicular basalt. The beds here dip at about 30° W.N.W., and the total thickness of the basalt appears to be about 50 ft. About 300 yds. N.N.E. of Witnell's End the base of the basalt, with pipe amygdales in the lowest 5 in., rests on rather sandy beds that do not show any obvious sign of alteration.

Dr. Phemister describes a fresh specimen of the igneous rock from north of Witnell's End as :

" a theralitic analcite-dolerite (E.14495) like that from Kinlet (p. 146). The plagioclase is extensively replaced by analcite and by chloritic material. The nature of the alkali feldspar is difficult to determine in the body of the rock. Stout prisms which surround a nucleus of zeolite prove, however, to have a negative sign and a small optic angle. This feldspar thus appears to be the same variety as at Kinlet (see p. 146)."

In the stream 300 yds. south of Witnell's End the basalt with the overlying and underlying sediments can be seen in a continuous section, the beds being inverted. The underlying beds (Fig. 15, No. 1) are sandy shales with a " dicey " or cuboidal fracture, and appear to be somewhat hardened. Between them and the igneous rock is a 1-in. band (Fig. 15,

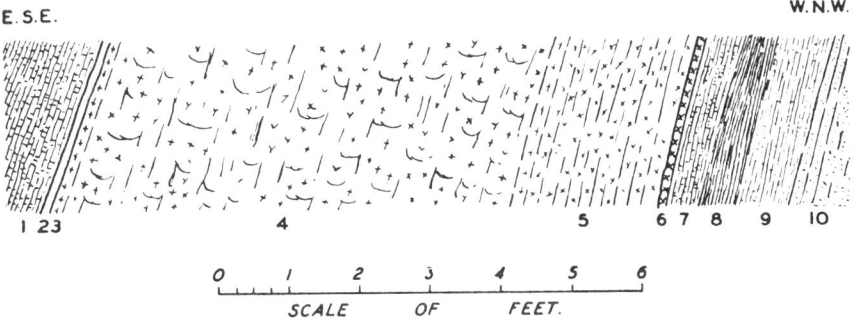

FIG. 15.—*Section of the Shatterford Basalt and Country Rock in dingle 250 yards south of Witnell's End.* 1. Dark shale with cuboidal fracture. 2. Contact-altered shale. 3. Compact fine-grained basalt. 4. Massive basalt. 5. Fine-grained basalt with platy jointing. 6. Chilled edge of basalt. 7. Dark shale with cuboidal fracture. 8. Black shale with coal streaks. 9. Dark sandy shale with cuboidal fracture. 10. Sandstone. Bedding inverted.

No. 2) of olive-green fine-grained rock. This (E.16805) is described by Dr. J. Phemister as

" composed of finely divided quartz grains, mica flakes and undetermined clayey material, the whole being stained yellow. There is an unusually large proportion of micas and the flakes show a strong tendency to parallel orientation. Many of the micas appear to be detrital and none looks like the usual contact-alteration variety of biotite."

Nevertheless, in view of its sharply defined limits and its position, it can hardly be doubted that the band is a contact-hardened sandy shale. Next to it comes about 2 in. of compact, fine-grained basalt (Fig. 15, No. 3), evidently the rapidly cooled basal portion of the rock. This is followed by about 6 ft. of massive basalt (Fig. 15, No. 4), not so fine-grained but nowhere coarse, with a rude jointing relatively widely spaced, parallel to the upper and lower contacts. This is succeeded by 2 ft. of finer-grained basalt (Fig. 15, No. 5), pale green in colour, with closely spaced joints parallel to the others. Then comes about 1 in. of compact, fine-grained, olive-green rock (Fig. 15, No. 6), which is described by Dr. Phemister as follows :

" E.16806 consists of two parts, one a vesicular basalt, greatly decomposed and stained, containing microphenocrysts of feldspar and of olivine or augite, or both. The other part is non-porphyritic and shows broad laths of yellow isotropic material probably after feldspar, and rare small serpentine pseudomorphs probably after olivine. The rock contains irregular bands and patches of radiating serpentine aggregates."

This is evidently the chilled upper surface of the basalt. It is overlain, stratigraphically, by dark shale (Fig. 15, No. 7) with a conchoidal to cuboidal fracture, which appears to be hardened to the same degree as that below the igneous rock (Fig. 15, No. 1). The junction plane of the chilled basalt with this is gently undulating, without the irregularities that would be expected at the upper surface of a lava flow. The chilled edge is of uniform thickness and shows no sign of erosion, and though vesicles can be seen in the microscope slide, none is visible to the naked eye, as is the case in the exposure north-north-east of Witnell's End described above. Some calcite is present in the compact basalt at the base, and in the " platy " basalt near the top. Although the thickness of basalt is only about 8 ft., it appears to be complete, and the section is undisturbed, apart from the inversion.

A fault, however, crosses the stream just below the above-described section, and is probably responsible for a second exposure of basalt. In view of the small thickness of basalt in the first exposure, as compared with that along the strike to north and south, it is possible that the intrusion (or flow) is here in two separated portions.

To the north of the Kidderminster road at Shatterford the rock in contact with the top of the basalt is a coarse pebbly sandstone, not shale as in the quarries near Witnell's End. A fault shifts the basalt at the road and, probably for this reason, it is not seen in the road-side. Just south of the road old workings expose the base, and here again pipe amygdales can be seen. At the Deep Pit, Shatterford, the basalt is stated to be 28 ft. thick,[1] but the boring from the bottom of the shaft ended in the rock, so the whole may not have been proved. The thickness at the outcrop near Shatterford does not, however, appear to be much more than 30 ft.

From the Kidderminster road the basalt can be followed by a marked feature and occasional exposures into Eymore Wood. In an old quarry 540 yds. S.S.W. of the road it dips to west at 75°.

About 1,100 yds. W.N.W. of Littlegains Farm two quarries in Eymore Wood are excavated along the strike and show clearly a change of align-

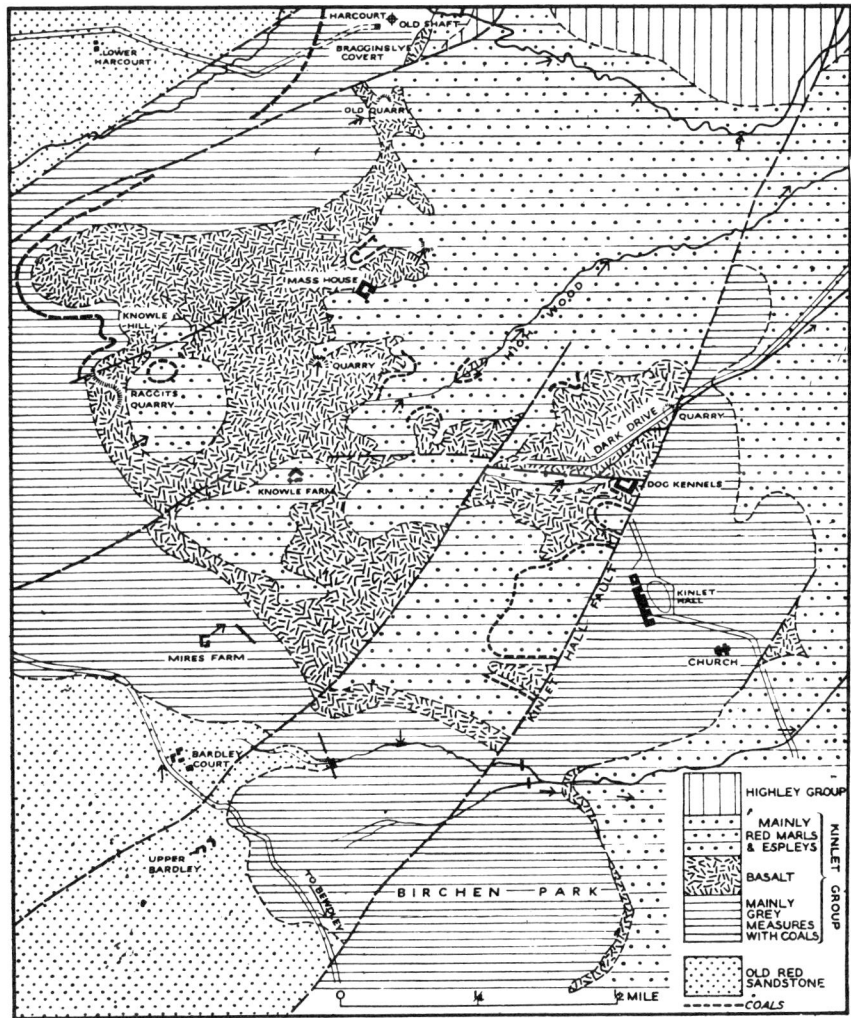

Fig. 16.—*Map of the Kinlet basalt outcrops.*

---

[1] Roberts, G. E., *The Geologist*, 1861, p. 426.

ment due probably to the shifting of the basalt by the Romsley Fault. The basalt is vesicular and shows closely spaced " platy " jointing parallel to the upper and lower surfaces. The lower surface is in contact with dark shales. The igneous rock, still making a prominent feature, can be traced to a stream some 280 yds. S.S.W., where it is exposed with a thickness of only about 8 ft. Its lower surface is in contact with hardened shale. The upper contact is not visible, but sandstones appear about 12 ft. to the west. Beyond the stream the feature dies out, and in the next stream to the south there is no sign of the basalt.

T.H.W.

KINLET AND HIGHLEY

In the Kinlet area (see Fig. 16) the basalt appears to occupy a position approximately at or just above the horizon of the Highley-Brooch Coal. The most northerly outcrop observed is in Bragginslye Covert 600 yds. S. 25° E. of the Smithy at High Green. An old quarry in the same covert 600 yds. N. 5° E. of Mass House farm shows coarse grit resting on greenish vesicular basalt weathering spheroidally. The thickness of the igneous rock does not seem to be as great as farther south where the outcrop widens between Mass House and Knowle Hill.

About 200 yds. N.W. of Mass House an isolated outcrop of indurated shale and sandstone in the basalt has a high dip towards the south.

In a small quarry, 200 yds. N.E. of Mass House, coarse ferruginous and greenish grit, dipping at 65° N.N.E., must be only a few feet above the igneous rock, since basalt debris occurs at the surface close by. The house itself and the southern side of the farmyard are on these grits, which here dip about 25° S., while the rest of the farmyard is on basalt.

A good section of the basalt and the overlying sediments is exposed in Mass House Quarry (Survey Photograph 6916) about 300 yds. S. 20° W. of Mass House.[1] It shows :—

	Ft.	In.
Sandstone, yellowish-buff, flaggy ; the lower layers are dark brown in colour and consist of quartz grains in a matrix of basaltic material with pebbles of basalt and portions of amygdales	2	0
Grit, coarse, composed of small quartz pebbles and sand with angular fragments of decomposed basalt, in a matrix of decomposed basaltic material ; it fills depressions in the surface of the underlying basalt ... ... up to		6
Basalt, weathering spheroidally, top vesicular (scoriaceous), the vesicles becoming large about 2 in. from the top surface ; the lower parts less decomposed but still vesicular, vesicles filled with secondary minerals seen to about	20	0

The interpretation placed on this section is that the basalt was a vesicular lava flow and that the grit and sandstone were deposited upon

---
[1] See R. W. Pocock, *op. cit.*, p. 2.

A.—Mass House Quarry, Kinlet; spheroidal weathering of basalt.

B.—Raggitts Quarry, Kinlet; 'Dykes' in decomposed basalt.

its uneven surface, while denudation of part of the igneous rock in the neighbourhood provided some of the material for these sediments.[1]

The base of the basalt is to be seen near the entrance to Raggits quarry on the west side of Knowle Hill. The section is in the wooded bank on the north side of the entrance, and 900 yds. W. 15° S. of Mass House. The igneous rock here rests on dark carbonaceous shale which appears to be unaltered.

Raggits quarry itself (Survey Photographs 6914-5) shows a remarkable section of over 50 ft. of basalt, much of which is highly decomposed and friable. It is traversed by a number of " dykes " of tougher doleritic material which are almost vertical and have a north-easterly strike—the general direction of the faults of this area. Horizontal slickensides on the sides of the " dykes " indicate lateral movement. There is a tendency to spheroidal structure in the main mass and typical spheroidal weathering is to be seen in vesicular basalt at an outcrop south of the quarry and 70 yds. W. 20° S. of the summit of the hill.

An outlier of the overlying grits and sandstones with a thin coal forms the top of the hill and a similar outlying patch of the grits and sandstones occurs to the south-east.

The top of the basalt, here finely vesicular, can be seen in a small quarry 500 yds. N.E. of Bardley Court and is overlain by calcareous sandstone. About 100 yds. east of the quarry highly vesicular basalt is exposed in the steep bank and here its top is a breccia of fragments of tachylyte which suggests the broken-up chilled surface of a lava flow. A highly vesicular top to the basalt is present again in the small inlier against the Kinlet Hall Fault about 350 yds. S.W. of Kinlet Hall and here also the overlying sandstones and grits include a thin coal near their base. This coal can be seen resting on dark clunch, and this in turn on basalt, alongside the shed about 600 yds. N.W. of the church.

The Dog-kennels about 300 yds. north of the Hall appear to be built on the igneous rock.

In the banks of the stream just north-west of the kennels there are exposures of a dark, coarse basalt and an old quarry in the bank above Dark Drive about 100 yds. north of the kennels shows about 15 ft. of spheroidal basalt. In the same bank, near the drive gate and 300 yds. N.N.E. of the kennels, a small exposure shows the rock in the form of a sharp fold which may possibly be interpreted as a blister in a cooling lava field. The Kinlet Hall Fault crosses the drive at the gate and throws the igneous rock against Coal Measures to the east.

The junction of the basalt with the overlying Coal Measures can be followed across the large field between the drive and High Wood. The bottom bed of the Coal Measures is here a dark shale, with rootlets and *Calamites*, upon which rests the thin coal. The igneous rock is exposed in tumps and shows joints of east-north-east trend.

---

[1] But see C. E. Marshall, *op. cit.*, p. 16.

The stream in the south-west corner of High Wood has cut through the Coal Measures to expose a small inlier of the basalt with the thin coal just above it. In an old quarry about 300 yds. to the south-west grit is brought against basalt by an east-and-west fault.

Northwards towards Mass House the basalt runs out eastwards in a spur against which the Coal Measures, with the small coal seam near their base, appear to rest.

East of the Kinlet Hall Fault there is an interesting section in an old quarry 650 yds. S.W. of Kinlet Church. It shows :—

		Ft.	In.
Massive but vesicular basalt	...	5	0
Coal smut on dark shale	...	1	0
Massive basalt	seen to	3	0

If the shale and coal smut were deposited on the surface of the lower basalt and subsequently covered by the upper vesicular basalt, then the igneous rock must be contemporaneous with the Coal Measures.

Of underground occurrences of the igneous rock in the neighbourhood of Kinlet the most interesting is that encountered in a boring, made in 1929, 250 yds. S.W. of Tiphouse Cottages (see Appendix II, p. 197, and Fig. 17) and just over two miles east of Knowle Hill. The presence of the basalt breccia bed from 376 ft. 9 in. to 383 ft. 10 in. appears to be strong evidence of contemporaneous erosion of some portion of the Kinlet basalt mass. It will be noticed that basalts occur in the boring below as well as above the Brooch Coal of Highley.

Basalt was met with in working the Brooch Coal near the Tiphouse Fault half a mile west of Netherton. Dr. Robertson reported that some distance east of the fault a bed of white trap 18 in. to 2 ft. thick appears in the roof, several feet above the coal worked and just below a thin top coal. A road driven through the fault found some 15 ft. of dolerite lying above the downthrown coal which it partly replaces or destroys. At a distance of about 200 yds. from the fault the coal ended abruptly against a vertical wall of basalt (E.15586).

Basalt occurs also in the workings of the Brooch Coal north of the Highley Pits near Hampton Loade.

R.W.P.

The following petrographical notes are contributed by Dr. J. Phemister :—

Kinlet.—The collection includes specimens from the outcrops and from borings in this neighbourhood. Olivine-basalts and theralitic dolerites occur both in outcrops and borings but the geological relation of the two types is not certain (*see below*, Claverley boring).

The normal fresh rock of the Kinlet outcrops is a fine to medium-grained grey dolerite speckled abundantly with small white or pink patches of alkali-feldspar and zeolites. Under the microscope it is seen to be composed of purplish augite, serpentinized olivine, long prisms of labradorite, skeletal plates of iron ore which are sometimes bordered by scraps of brown biotite, interstitial grains or stout prisms of alkali-feldspar, and interstitial analcite and other zeolites.

The pyroxene is a titaniferous variety, purplish in colour and pleochroic to yellowish grey, while the margins are sometimes coloured deep green. The purplish tint is often less intense in the centre of the crystals, and usually uniformly less intense throughout the grains which compose glomeroporphyritic aggregates (E.4882). Generally the pyroxene forms shapeless or prismatic plates ophitic to plagioclase and iron ore (E.17235). It occurs also as smaller grains in the base of the rock and these may be deep green when enclosed in alkali-feldspar (E.14496). Olivine forms idiomorphic crystals, generally of smaller size than the

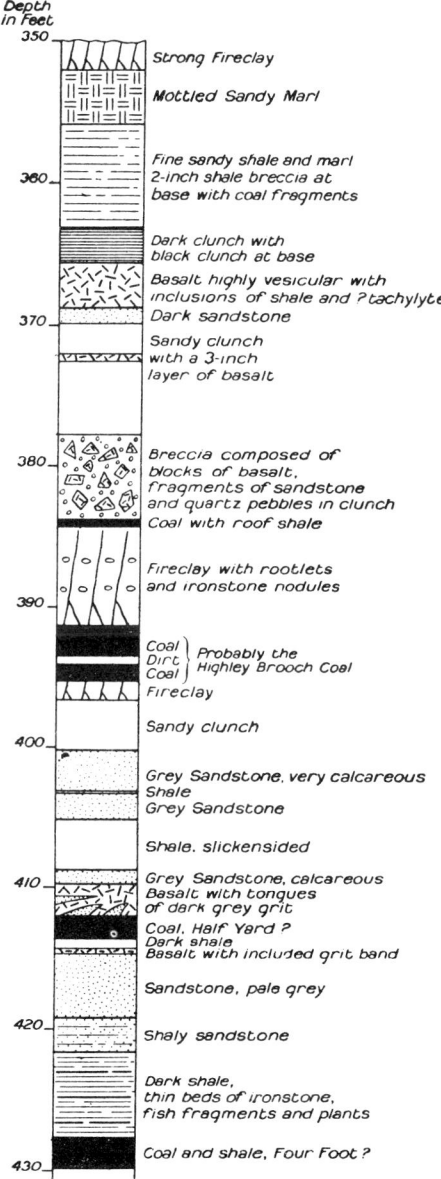

FIG. 17.—*Section of a portion of the Kinlet Boring,* 1929.

augite and plagioclase. They are distributed uniformly through the rock but occasionally compose groups of three or four individuals. Usually this mineral is converted to serpentine or a mica-like green mineral,[1] but in specimen E.17221 it locally appears as perfectly fresh large and small individuals subophitic to plagioclase and intergrown with augite. The plagioclase feldspar occurs as diversely arranged prisms of labradorite, containing about 60 per cent An (E.17221), which are usually pervaded by analcite and by chloritic material (E.17235). In some specimens these minerals may entirely replace the plagioclase (E.4882, 17217, 17220). Normal zoning to more sodic plagioclase is common. Alkali-feldspar may occur as a mantle to the plagioclase prisms, as interstitial grains, and as stout simply twinned prisms. The highest and lowest refractive indices are approximately 1.522 and 1.530 (E.17220), the sign is negative and the optic axial angle is small. These properties indicate a potash-oligoclase, but the typical cross-hatching of anorthoclase is not seen. The amount of alkali-feldspar varies very much, being only of accessory degree in specimen E.17235, but equivalent to the plagioclase in E.14496. This feldspar is idiomorphic against analcite and occasionally is seen to enclose stout rectangular and square sections of turbid, analcitized nepheline (E.14496). Analcite or another zeolite is always present interstitially and in general analcite is very abundant. While it seems clear that analcite may be present as a later pyrogenetic mineral, most of the rocks show evidence that it has been partly introduced by influx of sodic fluids. For example, it may occupy large spaces, less well defined than vesicles, at the edges of which the other minerals of the rock, except the alkali-feldspar, are partly dissolved away (E.4882). The distribution of analcite is, in general, patchy (E.17219). Two varieties are present, one brownish and turbid, the other clear, and the former presents idiomorphic polygonal or circular outlines to the latter and also to the other zeolite. The zeolites other than analcite have not been fully studied, but thomsonite in long, stout radiating prisms is present in E.17221. It is recognized by mean refractive index about 1.527, straight extinction, optic plane transverse to the elongation of the prisms, and double refraction of 0.012. Streakily analcitized idiomorphic crystals of nepheline are occasionally enclosed in analcite (E.17217, 17219). Apatite is a constant accessory mineral and occurs abundantly as slender prisms in the more alkaline patches of the rocks.

The collection from Kinlet includes many decomposed types. Augite is replaced by calcite, and plagioclase by analcite or by chloritic or serpentinous material. The titaniferous magnetite is sometimes thoroughly altered to leucoxene (E.17218).

From the Kinlet outcrops come also three specimens of decomposed olivine-basalt. Two of these (E.17224) have the groundmass greatly obscured by serpentinous material; the third, however, is a vesicular rock which contains much alkali-felsdpar as stout prisms and interstitial grains in the groundmass (E.17223).

The Kinlet rocks clearly belong to the theralite family. They resemble crinanite texturally but differ in the presence of alkali-feldspar. On the whole they are less rich in dark minerals and are of finer grain than true theralite, and they may be termed theralitic analcite-dolerites. They bear strong textural and mineralogical resemblance to many other basalts and dolerites of the Midlands, in particular to those of Rowley Regis and the Clee Hills.

Borings and workings near Kinlet.—From workings in the Highley—Brooch Coal comes a fine-grained variety of theralitic analcite-basalt (E.15586). It is microscopically similar to the rocks of the Kinlet outcrops but contains less alkali-feldspar and analcite, is finer in grain and has more perfect ophitic structure. Pale green pyroxene with intensely green borders, grains of aegirine pleochroic from deep emerald green to yellow-green, and scraps of barkevikitic hornblende occur among the feldspar laths of the base, and present idiomorphic faces to analcite.

---

[1] See also A. Scott in 'Geology of Stoke-upon-Trent' (*Mem. Geol. Surv.*), 1925, p. 90.

IGNEOUS ROCKS.

Igneous rocks met in a boring three quarters of a mile west of Kinlet Colliery and 250 yds. S.W. of Tiphouse Cottages prove to be thoroughly calcitized and serpentinized vesicular olivine-basalts. (14787, 15062-3, 15066).

Claverley Boring.—This boring pierced igneous rock between 1,937 and 1,960 ft., and the sliced specimens in the survey collection, from 1,940, 1,948 and 1,959 feet, are of the same type as the Kinlet theralitic analcite-dolerites. A specimen from 1,937 ft. 6 in., that is, at the top of the igneous body, is a thoroughly decomposed, vesicular, microporphyritic olivine-basalt. In the fresh rock from the centre purplish augite tends to form polysomatic groups with serpentinized olivine (E.4423). Laths of basic labradorite are ophitically related to the pyroxene. Alkali-feldspar, whose sign is negative and optic axial angle about 45°, forms interstitial grains between or mantles the plagioclase laths, and also builds stout prisms cemented by analcite. The latter, and another zeolite which is probably thomsonite, occur interstitially and also in large interspaces. The normal doleritic rock shows a passage in E.4422 to a more feldspathic facies in which long rods of iron ore, blades of chlorite, elongated calcitized pyroxene and lathy prisms of chloritized plagioclase are arranged in sheaves through a base of stout prisms of alkali-feldspar. This portion of the rock contains many vesicles which are filled, in order from periphery inwards, by minutely spherulitic serpentine, radiating zeolite, and coarsely granular calcite containing clusters of granules of ferriferous carbonate.

J.P.

CHAPTER VIII

GLACIAL AND POST-GLACIAL DEPOSITS

GLACIAL DEPOSITS

INTRODUCTION

The northern part of the present district was invaded by an ice-sheet mainly from the Irish Sea basin, but mingled probably with some ice from the mountains of North Wales. This ice-sheet reached an irregular line extending from the main watershed near Sedgley to the rising ground south of Morville (Fig. 20, p. 172).

South of this line the greater part of this district is free from deposits that are, or might be, of directly glacial origin, and even from glacial erratics except such as are contained in the gravels of the river terraces. In the south-east, however, rare erratic boulders, all apparently of North Welsh origin, and scattered patches of sand and gravel lying for the most part more than 300 ft. above O.D. must probably be attributed to an earlier ice-sheet. Some gravelly boulder clay lies on Romsley Hill in the extreme south-east of the district. Apart from this, no trace of an 'older' boulder clay was distinguished during the survey; but Prof. Wills mentions 'red boulder clay without Irish Sea material and with solifluxion contortions' at Wombourne.[1]

A period between the invasions of the two ice-sheets, during which the district was free from ice, appears to be indicated by the highest terrace of the River Stour. This, in point of age, seems to come between the high-level gravels referred to in the preceding paragraph and the deposits of the later ice-sheet, and in its gravels rocks of distant origin are very rare. A lower terrace of the Stour and Smestow Brook shows an evident connexion with the later glacial deposits, and can be regarded as a valley train washed down from the front of the ice-sheet, which had then left the district. It corresponds with the Main Terrace of the River Severn, the deposition of which, as shown by Prof. L. J. Wills,[2] was intimately related to the formation of the Ironbridge Gorge.

Tributaries of the Stour and Smestow Brook that flow from the areas not reached by the Irish Sea ice-sheet are bordered by terraces that grade to the Main Terrace and their gravels consist principally of local material. Near the Clent Hills and Enville Sheepwalks such gravels are made up largely of fragments derived from the Clent Breccia.

---

[1] Wills, L. J., 'The Pleistocene History of the West Midlands' (Presidential Address to Section C) *Rep. Brit. Assoc.*, 1937, p. 70 (see p. 77).

[2] 'The Development of the Severn Valley in the neighbourhood of Ironbridge and Bridgnorth,' *Quart. Journ. Geol. Soc.*, vol. lxxx, 1924, p. 274 (See pp. 295, 305-6).

GLACIAL DEPOSITS. 149

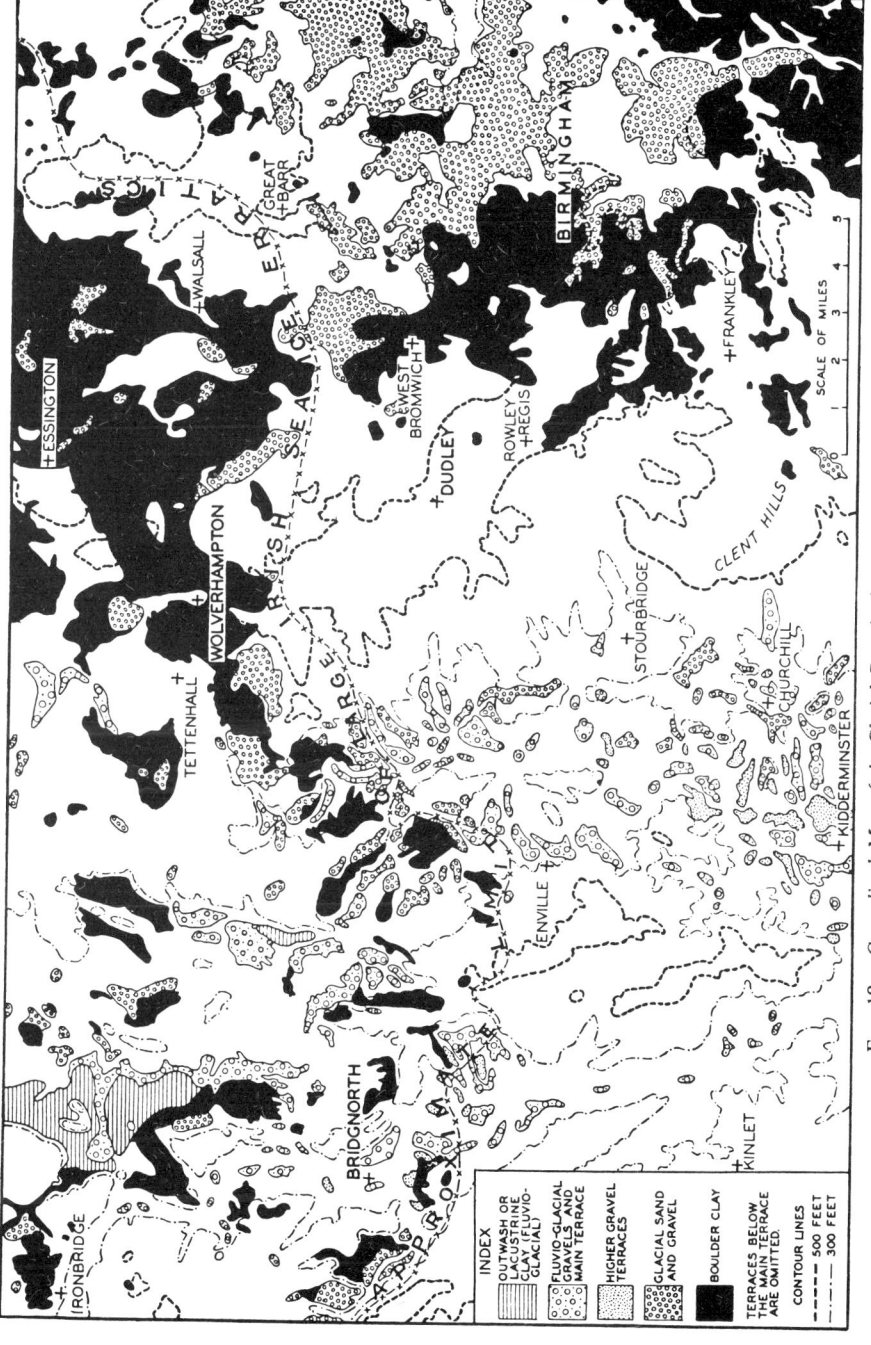

Fig. 18.—*Generalized Map of the Glacial Deposits between Bridgnorth and Birmingham.*

The earlier terraces are considerably dissected in consequence of the rejuvenation of the Severn and its tributaries, including the Stour. This rejuvenation was in part the result of the addition to the lower Severn[1] drainage of the waters of the upper Severn when the Ironbridge Gorge was cut through the previously existing watershed by the overflow from lakes impounded by ice to the west and north-west. It was, however, assisted by a general uplift of the area in relation to sea-level after the formation of the Main Terrace, and this led also to the dissection of later terraces, deposited when the ice lay far beyond the limits of the district or had melted away altogether.

OLDER GLACIAL GRAVELS

In the south-eastern part of the district areas of sand and gravel considered to belong to the "older drift" resolve themselves into two groups lying respectively east and west of the Bunter Pebble Beds escarpment and the Stapenhill Fault; but of those to the west it is doubtful whether some, at least, should be so classified (see below and p. 169).

Of those to the east the most considerable area lies south-west of Pedmore, where the deposits reach a height of about 430 ft. above O.D., with outlying patches capping knolls and spurs near Churchill. They appear to be the remnants of a dissected sheet of sand and gravel that lay in a shallow valley descending south-south-westward towards Churchill. The gravel pits in Norton Covert show highly false-bedded sands with lenses of moderately coarse gravel; erratic pebbles found include some of Scottish, Lake District and probably North Wales origin.

To the west of the Pebble Beds escarpment and the Stapenhill Fault patches of sand and gravel occur near Axborough Farm, west of Iverley House Farm, near Whittington Common and in Gibbet Wood, south-east of Stourton. A small area of gravel south of New Wood, north-west of Wollaston, may belong to the same group. Near Iverley House Farm the height of the base of these deposits is over 300 ft. Elsewhere they range between 270 and 300 ft. Near Whittington and in Gibbet Wood sand appears to predominate greatly over gravel. The sand of Gibbet Wood seems to terminate in a feature at a little below 300 ft. on Dunsley Bank, below and to the north-west of which is another mass of sand and gravel, perhaps not part of the same deposit (see p. 153), in which is the new exposure near the Stewponey Hotel referred to by Prof. L. J. Wills.[2] No stones foreign to the district were found west of the Pebble Beds escarpment, and the pebbles are overwhelmingly of "Bunter" type.

---

[1] By 'lower' Severn is meant that part of the river below the Ironbridge Gorge and south-east of the pre-Glacial watershed. Prof. L. J. Wills uses the term in a different sense. (*Rep. Brit. Assoc.*, 1937, p. 75; and *Quart. Journ. Geol. Soc.*, 1938, p. 162).

[2] 'The Pleistocene Development of the Severn from Bridgnorth to the Sea,' *Quart. Journ. Geol. Soc.*, vol. xciv, 1938, p. 161 (see p. 185).

Two isolated boulders, not necessarily connected with the gravels above described, were observed at Churchill. One, measuring 7 ft. by 3 ft. by 2 ft., lies by a path 500 yds. W.N.W. of the church. It consists of rhyolitic vitric tuff of a type which strongly suggests to Dr. David Williams a derivation from the Ordovician volcanic rocks of North Wales, though somewhat similar tuffs occur in the Borrowdale Volcanic Series. The other, a smaller boulder, lies in the village about 40 yds. N.N.W. of the church, and appears to consist of green rhyolite, also of North Welsh type. Over a considerable area to the east of Churchill no erratics were found. The nearest is a boulder of volcanic rock, possibly of Welsh origin, 530 yds. S.E. of Moor Hall. Farther east isolated boulders, mostly volcanic rock of "Welsh" type, become more common. On and near a sand and gravel ridge east of Madeley Heath (Sheets 182 and 183) large erratics include volcanic rocks, slates and grits, all apparently of North Welsh type, and cream-coloured quartzite, some specimens of which contained fossils, amongst which are *Spirifer bisulcatus* J. de C. Sowerby and *Camarotoechia pleurodon* (Phillips). For this rock Dr. J. Pringle suggests a source in the Cefn-y-Fedw Sandstone of North Wales. The gravels themselves have yielded to Prof. Wills a number of Scottish and Lake District rocks.

The work of Dr. M. E. Tomlinson[1] has shown that glacial deposits, including material of north-western origin, extend for some miles south-east of the area between Birmingham and Bromsgrove dealt with by earlier observers such as H. W. Crosskey,[2] F. W. Martin,[3] and W. J. Harrison.[4] The inclination would be to correlate the gravels of the Churchill area with the older "Western Drift" described by Dr. Tomlinson as covering the high ground about the Tame-Avon watershed and west of the Alne valley near Henley-in-Arden. But Prof. Wills's suggestion[5] that the West Midlands were invaded by two ice-sheets from Wales, in the earlier of which ice from the Irish Sea basin was not involved, perhaps makes a correlation of the Churchill gravels with the later "Western Drift" of the Blythe valley[6] more probable. The evidence in the present district can, by itself, hardly be expected to allow of the recognition of more than one advance of ice before the main Irish Sea glaciation, though the possibility cannot, of course, be excluded. For example, the gravels and the isolated boulders might be, respectively, the result of independent advances or oscillations.

THE KINGSWINFORD GRAVEL RIDGE

Beginning at Swindon a ridge of sand and gravel, broken in only two places where it is crossed by streams, extends some three miles to Belle-

---

[1] 'The Superficial Deposits of the Country North of Stratford-on-Avon,' *Quart. Journ. Geol. Soc.*, vol. xci, 1933, p. 423 (see p. 427-9).

[2] *Proc. Birm. Phil. Soc.*, vol. vi, 1887-9, p. 169.

[3] *Ibid.*, vol. vii, 1889-91, p. 85.

[4] *Proc. Geol. Assoc.*, vol. xv, 1897-8, p. 400.

[5] *Op. cit.* (1937), pp. 86, 87.

[6] Tomlinson, M. E., *op. cit.*, p. 452. Dr. Tómlinson attributes the later 'Western Drift' to a re-advance of 'Western' ice.

vue, Wordesley. The ridge has been interpreted as an "esker," in the sense of a deposit marking the course of a sub-glacial or englacial stream ;[1] but the complete lack of any closely associated and independent traces of the former presence of ice, such as boulder clay or erratics (other than those in the gravel itself), makes it difficult to accept this view. Its present form is probably in the main due to erosion, and the deposits may well have originally been more extensive.

From Swindon to Summerhill sand and gravel rest on a low ridge of Bunter sandstone, but sections near Wallheath (Survey Photograph 1961) show that, in places at least, the drift lies in a hollow excavated in the crest of this ridge, the base of the gravel sloping inwards. In one place, however, the drift was seen to be dipping away from the crest of the ridge. South of Summerhill the sand and gravel are banked against the dip slope of the Bunter Pebble Beds ; but here, again, the middle part of the deposit occupies, at least in places, a hollow in the solid rocks ; for, in a sandpit in Cot Lane, south of Mount Pleasant, the base was found to be about 25 ft. below the top of the Bunter Sandstone exposed in pits on the opposite (east) side of the road.[2] Near Belle Vue the drift forms an embankment-like ridge that ends abruptly southward.

Near Swindon the drift of the ridge consists largely of coarse to fine gravel, with some interbedded sand. Near Wallheath the deposit is largely sand, with lenses of gravel and beds of marly clay. Near Summerhill sand predominates. The Cot Lane sandpit (Survey Photographs 1959-60) showed in 1920 about 40 ft. of laminated, false-bedded sand with some pebbly lenses. The bedding of the deposits is in places contorted, a feature well seen in some of the pits near Wallheath. The pebbles are predominantly of Bunter type, though some of igneous rock were noted. Two small boulders of basalt were seen in one of the Wallheath pits. Prof. Boulton records sandstone, conglomerate, chert, cherty limestone, banded rhyolite, granite and andesite from various parts of the ridge. Pebbles of granite are, however, rare.

Sandy and pebbly drifts west of the cross-roads at Summerhill, on a spur near Ashwood Lodge and east and south of Ashwoodfield House may be parts of the same deposit.

In the nature and contents of its deposits the Kingswinford gravel ridge resembles the sands and gravels of the Churchill area (p. 150). Near Wallheath, however, the base of the Kingswinford deposits descends to 250 ft., or less, above O.D., not much above the level of the Main Terrace (p. 161), and this seems to allow of less erosion along the Smestow—Stour line of drainage since the deposition of the former than would, perhaps, be expected if (like the Churchill deposits) they belonged to the "older drift." Moreover, the level of the Kingswinford gravels near Wallheath relatively to that of some gravels west of Smestow Brook, near Hinksford (p. 153), suggests that the former are later, with a period of

---

[1] Boulton, W. S., 'An Esker near Kingswinford, South Staffordshire,' *Proc. Birm. Nat. Hist. & Phil. Soc.*, vol. xiv, 1916, p. 25 (see p. 33).

[2] Boulton, W. S., *op. cit.*, p. 29.

erosion between; if so, and if the Hinksford gravels represent the Kidderminster Terrace the Kingswinford gravels could not be relegated to the " older drift."

EARLIER RIVER TERRACES

Stour Basin

In the Stour valley a terrace of gravel, about 90 to 100 ft. above the present flood-plain, ranges in height from about 250 ft. near Stourton to 90 or 100 ft. above O.D. near the southern border of the district just north of Kidderminster, a considerable part of which is built upon it. This Kidderminster Terrace of Wills is believed to be later in date than the " older gravels " described on p. 150, for when they are found near one another the terrace is always at a lower level and there is in places a slope cut in the " solid " rock between them. On the other hand, the absence or rarity in the gravels of the Kidderminster Terrace of erratics that could have been derived from the deposits of the later ice-sheet[1] suggests that the former is the older, and this is confirmed by the fact that the Kidderminster Terrace is clearly older than the Main or fluvio-glacial Terrace, which was formed in intimate relation to the later ice-sheet.

Of certain scattered patches of sand and gravel near Stourton at a lower level than the sand of Gibbet Wood (p. 150) and separated from it by a feature, some may be the northernmost relics of the Kidderminster Terrace in the Stour valley itself. But in the pit near the Stewponey Hotel the base of the deposit has been proved, since the area was surveyed, to lie about 40 ft. below the level of this terrace as deduced from its gradient curve, and it is difficult to say in what category the sand and gravel in this pit should be placed.[2]

A series of gravel-patches to the west of Smestow Brook from Swindon to Hinksford seem to lie on the gradient-curve of the Kidderminster Terrace as extrapolated up the course of Smestow Brook. No far-travelled stones were found on the surface of these. There seem to be no connecting links between them and the deposits near Stourton, though there are traces of a possible drift deposit, too indefinite to map, in the northern part of Prestwood Park and the ground to the north-east. There is, however, room for doubt whether the Kidderminster Terrace actually extends up the Smestow Brook Valley.

Considerable areas of the Kidderminster Terrace lie to the west of the Stour south of Kinver, and about half a mile E.N.E. of Blakeshall some gravel at a higher level than these seems to constitute a fifth terrace, but

---

[1] A few ' Welsh ' erratics are recorded from the gravels of this terrace on Hartlebury Common and at Kidderminster; see L. J. Wills, ' The Geology and Soils of Hartlebury Common,' *Proc. Birm. Nat. Hist. & Phil. Soc.*, vol. xv, 1926, p. 95 (p. 97); and *idem.*, *op. cit.* (1938), p. 184.

[2] See L. J. Wills, *op. cit.* (1938), p. 185. On the map (Sheet 167) this deposit is shown as Fourth (i.e., Kidderminster) Terrace, the map having been engraved before the new exposure was seen in 1938.

may probably be regarded as merely an early phase of the fourth or Kidderminster Terrace.

The Kidderminster Terrace can be traced a short distance up tributaries of the Stour near Hurcott Wood to the east and Franche to the west, and up that which joins the Stour at Wolverley. Of some isolated patches of gravel near Kingsford, two to W.N.W. of High Hobro Farm, at a higher level than the rest, appear to belong to it. The gravels here consist of Bunter material with some angular fragments probably derived from the Clent Breccia.

Severn Valley

It is doubtful whether the Kidderminster Terrace extends up the valley of the Severn in the present district. Below Dudmaston there are no terraces above the level of the Main Terrace. From Dudmaston upstream higher terrace-like deposits do occur, but these are not necessarily contemporaneous with the Kidderminster Terrace of the Stour.[1] Prof. Wills, however, suggests that the Hoards Park-Eardington gravels, near Bridgnorth (below), may represent it.[2]

T.H.W.

Bridgnorth and Eardington.—At Knowlesands, one mile south of Bridgnorth, a thin gravel, largely unbedded, caps the hill at a height of 260 ft. above O.D. It contains a large proportion of fragments of nodular *Spirorbis* limestone with some perhaps of cornstone from the Old Red Sandstone (both of which could have been derived from the Tasley area) and does not appear to include any northern erratics. While its position within the main valley of the Severn suggests a high river-terrace, it is perhaps more likely that it was deposited during the retreat of the ice-front at some period between Stages I and II (see Fig. 20, p. 172). A small patch of gravel at about 275 ft. O.D. near Potseething Farm, about half a mile to the west, may be classified with this Knowlesands deposit.

From just north of Eardington an area of sand and gravel extends to the edge of the cliff overlooking the Severn at Cliff Coppice. Prof. L. J. Wills states that no material of northern origin could be detected, but *Spirorbis* limestone occurs. Although terrace-like, its material indicates derivation from the north-west and deposition during the retreat of the ice-front, as suggested above for the Knowlesands gravel.

Another problematical area of gravel occurs south of Hoards Park Farm and has been described by Prof. Wills[3] as "Hoards Park Gravel"; it lies about 250 ft. above O.D. Its terrace-like appearance and proximity

---

[1] To avoid the multiplication of symbols the terrace next above the Main Terrace in the Severn Valley has been given the " Fourth Terrace " symbol on the published geological map, but this must not be taken to imply correlation with the Fourth Terrace of the Stour.

[2] Wills, L. J., *op. cit.* (1938), pp. 179, 183, 218 ; cf. *idem.*, op. cit. (1937), p. 84.

[3] *Op. cit.* (1924), p. 293. For reference to another exposure of the Hoards Park gravel, overlain by red boulder clay with northern erratics, see L. J. Wills, *op. cit.* (1938), p. 183.

to the Severn suggest that it may be a high terrace of that river, but nevertheless it lies high on the side of the valley of Cantern Brook flowing from the north-west and may be connected with a retreat stage when the ice-front lay slightly to the north-east of the line of this valley and, to the east of the Severn, along the Pebble Beds outcrop at High Rock.

R.W.P.

Quatt.—Near Quatt, at the Holt and 300 yds. E.N.E. at Little Holt, patches of gravel lie well above the level of the fluvio-glacial gravels described on p. 163 and of the Main Terrace of the Severn. There are no sections, but the pebbles on the surface are of "Bunter" type with some "northern" erratics. Just east of the Bridgnorth road about 800 yds. N. by W. of Dudmaston Hall gravel occurs at a corresponding level, but here no northern erratics were seen. These gravels may correspond with the "Knowlesands Gravel" of Prof. L. J. Wills,[1] though, so far as can be judged, the composition is not the same.

Centred about 1,000 yds. N.W. of Dudmaston Hall, a deposit of gravel has its base just under the 200-ft. contour and lies a few feet above the Main Terrace gravels of the Severn. Pebbles consist overwhelmingly of rocks derived from the Wrekin—Longmynd area and the Welsh border, and include limestone of "*Spirorbis*" type. One of granite and a few of a porphyrite apparently of northern origin may not actually have come from the gravels on which they were lying. In level and composition these gravels appear to correspond with the "Eardington Sands and Gravels" of Prof. Wills.

BOULDER CLAY AND ASSOCIATED SAND AND GRAVEL

All the boulder clay in the district, with the exception of that on Romsley Hill mentioned on p. 148, appears to be due to the later ice-sheet in which ice from the Irish Sea basin seems to have predominated.

Wolverhampton, Penn and Wombourne.—Most of that part of the South Staffordshire Coalfield which lies within the present district is free from drift, but a patch of boulder clay has been mapped near Bilston, and the south-west edge of the drift-filled channel of Moxley, described in the Birmingham and Wolverhampton memoirs, crosses the extreme north-east corner of the district.

Boulder clay around Blakenhill is part of the spread covering most of the Wolverhampton area, which Dr. Robertson describes as chocolate-coloured sandy clay with small pebbles and occasional large boulders. Boulder clay also covers a considerable area north of Lower Penn, where it is a brown till with, in places, numerous boulders of granite and other far-travelled rocks.

Fringing the boulder clay near Penn Fields and Lower Penn, sand and gravel form in places hummocky ground and end somewhat indefinitely against the rise to Upper Penn and the Colton Hills. No glacial deposits

---

[1] *Op. cit.* (1924), pp. 292-3. See also above p. 154.

surmount the high ground of the Keuper Sandstone outlier of Lower Penn, but a boulder of granite lies at about 500 ft. O.D. 300 yds. N. by W. of Bearnett House, and several boulders of granite and other rocks occur at Upper Penn and to the west of Penn Common. The southernmost boulder seen in this neighbourhood is one of granite or syenite near the cross-roads 600 yds. N.E. of Wombourne Church.

Another area of sand and gravel occurs near Pool Hall and to the west of the Staffordshire and Worcestershire canal forms hummocky ground, with esker-like ridges near Langley Farm, 1,030 yds. S.E. of Pool Hall. To east of the canal it seems in most places to be thin. Dr. Robertson noted pebbles of flint, chalk and limestone on the surface near Furnace Grange, and obtained fragments from the Rhaetic and Lower Lias, flints with shell-casts, and fragments of Pleistocene shells, including *Turritella*, from a sand-pit just north of Langley Farm.

Between the Penn Keuper Sandstone outlier and Smestow Brook boulder clay forms hummocky ground on which erratics are numerous and include granites, rhyolites, sandstones of Keele and Triassic types and other rocks. Woodfield Grange marks the southern limit of boulder clay east of Smestow Brook. To east, south-east and south lie considerable areas of sand and gravel. Some of these, e.g., those to west of the Keuper Sandstone escarpment and south of Oreton and that around Clapgate, form ridges or hummocky ground. Two spreads, respectively north and south of Wombourne, have flattish surfaces with a general slope to west-south-west. The southern spread rises to a height of about 350 ft. near Foxhills. The bedding of these gravels is in places contorted, and the underlying Bunter sandstone disturbed and ploughed up. This is well seen in a pit opposite the lodge of the Foxhills, in another 340 yds. to the W.S.W. and in the large pits at Ounsdale. No far-travelled stones were found in these gravels, the pebbles being prevalently of "Bunter" type. It is possible, therefore, that these gravels belong to the "older drift"[1]; but, on the other hand, their position and direction of slope suggest their origin as an outwash from an ice-front resting just north of Upper Penn, and the fact that the melt waters from there would have crossed the Bunter Pebble Bed outcrop might explain the predominance of pebbles from that formation.

Trescott, Seisdon, Trysull, etc.—The concentration of erratic boulders near Trescott has been remarked by Murchison,[2] Mackintosh,[3] and others.[4] Dr. Robertson noted that grey granite with porphyritic feldspar, of "Criffel" type, was the most abundant, with another type of granite, non-porphyritic with much biotite. A pink granophyre (Ennerdale?) is also common. A count made at a roadside heap just north of the present district gave 80% of rocks certainly or probably of Scottish and Lake District origin, 14% of grey grits certainly or possibly of North

---

[1] See L. J. Wills, *op. cit.* (1938), pp. 184, 206.
[2] 'Silurian System,' 1839, pp. 535-6.
[3] *Quart. Journ. Geol. Soc.*, vol. xxxv, 1879, p. 437.
[4] On the distribution of boulders see F. W. Martin, *op. cit.* (with map).

Welsh origin and 6% of Keuper Sandstone.[1] The preponderance of northern over Welsh erratics is typical of the area generally and indeed of the whole of the northern part of the district.

On the north-east side of Smestow Brook a broad ridge of sand and gravel extends for about a mile north-westward from Trysull. From the south-east end to the Seisdon-Wolverhampton road the upper, very gravelly, part of this ridge is marked off by a distinct feature from the lower, more sandy, portion. A sand pit beside the road 80 yds. N.E. of Seisdon Mill showed (in 1930) gravel resting on a pockety surface of loamy sand, the top of which appeared to be disturbed. In another pit, 450 yds. W.N.W. of Trysull church, sand with lenses of coarse gravel with northern erratics rests upon well bedded sand in which the only stones found were soft, rather weathered-looking pieces of sandstone of Triassic type. It seems clear that there are here two deposits between the formation of which there was an interval with some erosion, and it is possible that they represent " newer " drift superimposed on " older " drift ; but, on the other hand, both deposits may belong to the same general phase, the upper, gravelly, one merely indicating an influx of more torrential water carrying coarser material and slightly eroding the previously deposited sand.

West and south of Seisdon and Trysull boulder clay ends off to north-east and east in a fairly distinct feature at about the 300-ft. level, below which the ground is largely covered by downwash or solifluxion drift. To the east, near Trysull and Feiashill, are kame-like mounds and ridges of pebbly sand. Erratic boulders are scattered over all this area as far south as the Wombourne to Bobbington Road ; but, east of the Pebble Beds escarpment, none was found farther south, except in, or on the level of, the fluvio-glacial terrace (p. 161).

The boulder clay to the west of Seisdon and Trysull ends somewhat indefinitely westward, and on the northward and eastward slopes of Abbot's Castle Hill there are only patches of thin remanié drift. Erratic boulders are, however, abundant and extend right up to the crest of the escarpment, with many on the scarp slope also. The great number of erratics on the Pebble Beds dip-slope here is in striking contrast to their apparent total absence (already alluded to) south of the Wombourne-Bobbington road, and the abruptness with which they cease at about the line of that road is remarkable. As elsewhere, boulders from south-west Scotland predominate, the "Criffel" type being the most numerous and generally providing the largest ; a volume of 20 to 25 cubic feet is common. Pink granite of " Eskdale " type and granophyre like that of Ennerdale is of frequent occurrence, but generally in smaller boulders. Volcanic rocks, either from North Wales or from the Borrowdale Volcanic Series, are also numerous.

Bobbington area.—To the south-west of Abbot's Castle Hill escarpment boulder clay occupies ground up to a little above the 300-ft. level. Between

---
[1] For details see T. Robertson in ' The Country between Wolverhampton and Oakengates ' (*Mem. Geol. Surv.*), 1928, p. 185.

it and the escarpment is a ridge of pebbly sand and gravel approximately parallel to the latter. The form and position of this ridge suggest that it was deposited between the edge of the ice-sheet and the escarpment. Amongst the pebbles found in the gravel are some of fossiliferous sandstone which Dr. J. Pringle suggests may have been derived from the Ordovician rocks of the Berwyn Hills, and one of "Eskdale" granite.

South and south-east of Halfpenny Green a somewhat featureless area of boulder clay terminates to the south, near Highgate Common, at a broad ridge of sand and gravel that is probably a terminal or recessional kame, its north-north-westward slope being, perhaps, an ice-contact slope. A low ridge of pebbly sand about a quarter of a mile north of Mere Farm may be a continuation of the kame. Knolls of pebbly sand on Highgate Common, fringing the boulder clay area to the east and north, may be parts of this or other kames.

Near Heathton and White Cross areas of clayey sand (mapped as boulder clay) or sand have rather flat surfaces which are at a somewhat lower level than the surrounding glacial deposits, and may be the relics of a lake resulting from some temporary obstruction of drainage by ice (p. 174). South of Morfe House Farm pebbly sand forms a pronounced east-to-west ridge—perhaps a terminal kame, for the ice cannot have extended much farther south. Few erratic boulders have been found south and south-east of a line joining this ridge with the kame-like mounds of Highgate Common, and these may have been carried by floating ice or flood-water beyond the position of the ice-front. Two of them, one of granophyre of Ennerdale type and the other of volcanic rock, lie in Enville village about 40 yds. east of the inn.

Farther west the southern limit of the ice was evidently controlled by the high ground about Four Ashes and Tuckhill. Erratic boulders, abundant near Beobridge, were found in decreasing numbers as far south as Tuckhill ; but, though traces of remanié drift occur, there are few areas of boulder clay sufficiently definite to map.

Rudge Heath, Claverley and Farmcote.—A somewhat clayey deposit at the foot of the Keuper Sandstone escarpment from Rudge Heath nearly to Ludstone may be in part boulder clay, but more probably is the clayey marginal portion of the fluvio-glacial deposits such as is seen in the Worfe Valley[1] and elsewhere (p. 162). Some boulder clay appears to be present near Hilton, in the roadside just east of which structureless sandy clay with pebbles may be seen overlying the Keuper Sandstone.

T.H.W.

Allscott, Barnsley and Mose.—Boulder clay is present between Hartlebury and Allscott, north of the lower Worfe valley, and contains boulders of granite and Uriconian rocks. South of the valley small patches of boulder clay occur at Bromley and near Burcote House. Farther south another considerable deposit is banked against the dip slope of the

---

[1] Wills, L. J. *op. cit.* (1924), p. 286 and T. Robertson in 'The Country between Wolverhampton and Oakengates' (*Mem. Geol. Surv.*), 1928, p. 194.

Pebble Beds south and west of Barnsley, and many boulders, including some of granite, occur on its surface. Farther south again at Mose a strong concentration of large boulders, mostly granite, in clay appears to be continuous to the north-east with a loamy clay with many granite boulders round Morfevalley.

Rhodes Farm and Morville.—A small deposit of unusual type, consisting mainly of local Old Red Sandstone debris, occurs near Rhodes Farm north of Tasley (Survey Photograph 6853). It has been described by Prof. Wills[1] as a morainic gravel and the section shows rude bedding, in alternate layers of sandstone slabs and coarse gravel containing calcareous pellets derived from the cornstones, dipping 20° S.W.

Clays and gravels cover considerable areas between Morville and Harpswood. They were apparently deposited in a lake which occupied the upper part of the Mor Brook valley at a time when outlet to the south-east was dammed by ice which lay across the valley approximately along the line Harpswood, Haughton, Round Hill and Upper Linley Brook (see Fig. 20, p. 172, Stage I). These deposits contain abundant northern erratics, probably brought by a marginal stream flowing across the Barrow col.

Prof. Wills[2] has postulated four sub-stages of the ice-front's retreat in this area and varying levels and extent of a lake to account for varying levels of the surface of the drifts ; thus round the village of Morville the maximum heights of the gravel spreads are about 370 ft. above O.D., the Bridgwalton sands and gravels rise to 300 ft. only and the clays of Morville Heath lie at a still lower level.

A small area of sand south-east of Cross Houses rises to about 370 ft., a slightly higher level than the Bridgwalton spread, and shows, in a section 300 yds. S.E. of the crossroads, 6 ft. of clean well washed reddish sand. Fine gravel, all apparently of local material, the even bedding of which suggests deposition in quiet waters, is exposed to a thickness of 6 ft. about 150 yds. west of Harpswood Bridge, and similar gravel is dug near Lye Farm a mile to the north-west.

The clays of Morville Heath are dominantly of red-brown colour and contain granites and possibly Welsh erratics. They are regarded by Prof. Wills as lake deposits.

Astley Abbots.—Patches of boulder clay occur between 300 and 400 ft. above O.D. west of Astley Abbots and near Colemoregreen to the north at a lower level.

Eardington.—A reddish loamy boulder clay with small northern erratics and granite boulders occupies the ground between Knowlesands and Eardington and extends to near Westwood, where it reaches a height of about 260 ft. above O.D. Immediately to the south the clay appears to be banked against rising ground which at one point reaches 351 ft.

---

[1] *Op. cit.* (1924), p. 287.

[2] *Op. cit.* (1924), p. 301.

A small tumultuous deposit of boulder clay containing massive blocks of sandstone, coal fragments and various pebbles, but apparently no granite boulders, can be seen overlying sandstone and marl of the Keele Group on the north side of Mor Brook valley just above Eardington Mill (Plate IX A). This drift may mark the most southerly extension of the ice-sheet in this area and may be related to the initiation of the gorge of Mor Brook at Stage I (Fig. 20, p. 172). A cliff of sandstone at 400 ft. O.D. on the southern side of the valley, about 400 yds. N. by E. of Uplands, is undercut to a depth of 10 ft. and otherwise strongly waterworn, showing what appear to be segments of potholes. It may be suggested that this cliff formed a part of the southern margin of a glacial channel whose northern margin was the ice-front at Stage I. No drift boulders were found on the ground to the south of Mor Brook, but one at about 240 ft., about a mile north of Hampton Loade, is recorded by Prof. Wills.

R.W.P.

FLUVIO-GLACIAL DEPOSITS AND MAIN TERRACE

The fluvio-glacial deposits can be interpreted as having been deposited along the valleys by waters derived from the melting ice-sheet, but not close to the ice-front, which had probably receded beyond the borders of the district. The torrents of melt-water seem to have filled the valleys, at least in their upper portions, depositing gravels more or less simultaneously across their whole width and even, in places, upon low divides between confluent valleys. In the higher reaches of the valleys the gravels, especially the lower beds, are coarse, containing even small boulders, and have a rude, tumultuous and occasionally contorted bedding. They contain a notable proportion of rocks foreign to the district, derived from the drifts of the Irish Sea ice-sheet. The surface of the gravels, while generally flat and terrace-like, is in places slightly hummocky. Farther down the valleys the deposits become less coarse and more evenly bedded, the terrace-feature more uniformly flat and regular, and the fluvio-glacial gravels thus pass insensibly into the Third or Main Terrace of the Severn and Stour.

All the deposits of this phase were not, however, due to waters from the melting ice-sheet. Streams flowing from areas that the later ice-sheet never reached are bordered by terraces that grade to the Main Terrace and are composed of gravels containing a large proportion of locally derived material such as breccia-fragments from the Clent Hills and Enville Sheepwalks, pieces of coal, sandstone, ironstone nodules and basalt from the South Staffordshire Coalfield, as well as much material from the Trias. In places such gravels contain fragments apparently derived from the "older drift" (p. 150). The materials were probably transported by waters derived from snow melting on the high ground in or near the present district, perhaps, as suggested by Prof. Wills, under the conditions of "taele" or frozen sub-soil. It seems probable that certain less definite deposits, on the higher ground, and not clearly related to the drainage system, such as the red marl wash that covers considerable

A.—Boulder Clay on Keele Beds near Eardington Mill.

A 6861

B.—View looking down valley of R. Severn; showing terrace features.

A 6885

areas near Hagley Wood and Hayley Green and some of the higher-level breccia gravels (p. 165), may have been, at least in part, formed at this same period by solifluxion.

Smestow Brook.—The fluvio-glacial gravels first appear as a low terrace, but little above the flood-plain alluvium of Smestow Brook, south-west of Trescott. They also form a low terrace, with a sharper fall towards the alluvium, from Seisdon to Trysull. East of Trysull, where Smestow Brook turns southward, the fluvio-glacial gravels form a broad spread about 25 ft. above the flood-plain alluvium, with solid rock appearing in places in the bank between them. Near Heath Mill the fluvio-glacial gravels extend across the spur between Smestow Brook and the tributary that comes in from the north-east, through a gap cut in the glacial gravels. A pit 400 yds. N.E. of Heath Mill shows highly false-bedded sand and pebbly sand with lenses of gravel, on coarse gravel with large stones and small boulders. Some of the latter are of Scottish granite and many are of red sandstone. Upper Mottled Sandstone is exposed in the bottom of the pit, the total thickness of the sand and gravel being 20 or 25 ft. The gravels extend up Wom Brook as far as Wombourne village. Large pits have been dug through these gravels to the Bunter sandstone at Giggety. Farther south the fluvio-glacial gravels form a broad expanse on the east side of Smestow Brook, but on the west there are only scattered patches, on the southern end of one of which stands Swindon church. Near Smestow several boulders of Scottish granite and other rocks lie on, or near the level of, the terrace, and since they are confined to this situation may be regarded as having been carried down by the waters that transported the terrace gravel, and not deposited directly from the ice (cf. p. 157 above).

Near Wallheath and Himley, on both sides of Holbeche Brook, are broad flats on the level of the fluvio-glacial terrace, upon which are remnants of gravel. Pebbles of Coal Measure sandstone and of basalt were found on the surface. Between Hinksford and Ashwood the fluvio-glacial terrace is broad on the east side of Smestow Brook. A gravel pit about half a mile south of Hinksford shows rather coarse gravel, the pebbles of which include some of black, unweathered-looking basalt.

From Ashwood to the junction of Smestow Brook with the Stour the fluvio-glacial terrace is represented by comparatively small patches. It can be traced up Spittle Brook to where it merges with the breccia-gravel terrace described below (p. 166).

River Stour.—When the thalweg of the Stour is plotted along with the gradient-curve of the base of the Main Terrace it is found that the point at which the two curves would meet corresponds with a position a short distance east of where the river crosses the Western Boundary Fault, thus accounting for the absence of any trace of the terrace within the coalfield. The terrace first appears at Amblecote, where, on the north side of the river, gravel and pebbly sand rest upon Upper Mottled Sandstone. In pits near Holloway End (Survey Photograph 2201) 8 ft. of the deposits were seen and about 200 yds. E.S.E. of Stourbridge gasworks a mineral railway cutting showed (in 1922) about 12 ft. of loamy sand and pebbly sand.

The pebbles included Bunter types, fragments of Clent breccia, Coal Measure sandstone and, at the junction with the Bunter Sandstone, a large slab of calcareous pellety rock of Keele or Enville type.

Just south of this was the section, no longer visible in 1922, in which mammalian remains were found by Prof. W. S. Boulton.[1] As noted by him the section was as follows :—

	Ft.
Red, tenaceous "india rubber" clay with occasional chips of rock, mostly small	3 to 10½
Sand, finely laminated and false-bedded, with layers of fine gravel and lenticular beds of red clay	2 to 5
Coarse gravel and sand with little or no signs of stratification, the first two feet or so at the base made up of very coarse gravel and large boulders	8 to 9

The pebbles in the gravel included, besides quartzite probably derived from the Bunter, ironstone nodules, lumps of coal, Coal Measure grit, sandstone and calcareous conglomerate (from the Keele and Enville beds), Carboniferous limestone (probably from the Enville conglomerates), dolerite (of Rowley type), banded hälle-flintas, rhyolitic ashes, felsitic grits and a large flat boulder of andesitic rock. This last, and perhaps some of the other pebbles of volcanic rock, would seem to have been derived from the "older drift" deposits.

The mammalian remains, some of which are preserved in the Public Reading Room, Stourbridge, consisted of the teeth and bones of mammoth (*Elephas primigenius*), hippopotamus (*H. major*), woolly rhinoceros (*R. tichorhinus*), horse (*Equus caballus*), bison (*B. priscus*), and ox or large deer. Most were found within 2 ft. of the base of the gravel. Hippopotamus, according to Prof. Wills, is represented by brown and highly glazed fragments of teeth, derived, in his opinion, from an earlier deposit, as may also be the bison.[2]

The occurrence of clay at the top of the deposits is interesting in view of the tendency observed elsewhere for the fluvio-glacial deposits to pass laterally or upwards into clayey material (pp. 158 and 163).

No further terrace deposits seem to occur in the Stour valley east of the Pebble Beds escarpment, but the dissected remnants of a terrace platform corresponding to the Main Terrace can be seen to west of Audnam and near Ashwood Farm. The Main Terrace gravels next appear just west of the escarpment, on the south side of the river at canal level between Bells Mill, 500 yds. S.W. of Ashwood Farm, and Stapenhill Farm (p. 127, footnote 2). A pit 200 yds. N.E. of the latter showed 12 ft. of sand, of which the lower part contains large pebbles which include Scottish rocks and others probably from North Wales, presumably brought down the Smestow Brook Valley.

From Prestwood to the southern border of the district the Main Terrace continues in a regular manner, appearing at short intervals on

---

[1] 'Mammalian Remains in the Glacial Gravels at Stourbridge,' *Proc. Birm. Nat Hist. & Phil. Soc.*, vol. xiv, 1917, p. 107.

[2] Wills, L. J., *op. cit.* (1938), p. 205.

both sides of the river. A few granite pebbles may be found almost everywhere upon it, in which respect it is in marked contrast to the Kidderminster Terrace. At Caunsall, on the north side of the river, the latter, the Main and two lower terraces can be seen in juxtaposition. It is clear that the deposits of the upper three rest upon rock-shelves, for traces of Bunter sandstone can be found in the "risers" between the terrace steps.

Along the stream that enters the Stour at Wolverley the Main Terrace is represented by gravel capping knolls near High Hobro Farm and Kingsford. The terrace can be followed up the Churchill Brook and near Churchill and Blakedown its gravels contain much breccia debris (p. 166).

Rudge Heath and Claverley.—A considerable expanse of sand and gravel extending from Rudge Heath[1] nearly to Upper Ludstone seems to belong to the fluvio-glacial group. The portion near Upper Ludstone has its counterpart on the opposite side of the brook there and these gravels extend southward to Danford Brook, north-east of Claverley. As already suggested (p. 158), somewhat clayey drift between these gravels and the Keuper Sandstone escarpment may represent marginal portions of the fluvio-glacial deposits thrown down in slack water. Near Aston and Draycott pits in smaller patches of the same gravels show that they consist of false-bedded sand with strings and lenses of fairly coarse gravel, with small boulders. Fluvio-glacial gravels also cover considerable areas near Claverley and Lower Beobridge.

T.H.W.

Dalicote and Worfe Valley.—A spread of fluvio-glacial gravel between 200 and 260 ft. O.D. occurs at Lea Farm and Dalicote and contains a fair number of small granite boulders. Many wind-etched pebbles were observed on the surface of a gravelly flat to the south-west of Dalicote.

The fluvio-glacial gravels of the Worfe Valley, lying above the 200-ft. contour, cover much ground to the west and north-west of Worfield. They contain granites, Llandovery sandstone and limestone, vein quartz and coal fragments. Smaller patches near Burcote House and east of Rindleford Mill cap the cliff of Pebble Beds through which the River Worfe here flows in a gorge. A similar gravel at 200 ft. O.D. north-east of Worfe Bridge connects these deposits with the Main Terrace of the Severn Valley.

R.W.P.

Wooton and Quatt.—Between Wooton, Quatt and Dudmaston Hall gravels form a dissected, flat-topped sheet falling gradually from about 250 ft. near the first-named place to about 200 ft. at the last. The few sections show moderately coarse gravel with many erratics, including northern types, and far-travelled stones are common on the surface. At and near Wooton and north of Comer Wood boulders of granite and other rocks lie on the surface of the gravel or at a corresponding level. Northern erratics seem commoner in this gravel sheet than on the Main

---

[1] On the Wolverhampton Sheet (153) the northward continuation of these deposits is coloured as "Glacial" (pink), but in the memoir (p. 193) Dr. Robertson suggests that they are fluvio-glacial flood plain deposits.

Terrace to the west. At Dudmaston Hall a rather sudden drop from the level of the base of this gravel sheet to that of the Main Terrace of the Severn (below) may indicate that the Wooton gravels belong to a somewhat earlier phase, perhaps to that of the Eardington Sands and Gravels (p. 154), for though their composition is different, this may be due to the material having come from a different direction. They are perhaps a downwash from the ice-front when it lay just to the north-east, behind the Triassic escarpments, through the gap at Morfevalley. Near Park Farm, south-west of Quatt, are traces of a superficial deposit which seems to be a remnant of the same gravel sheet. T.H.W.

River Severn.—North of Bridgnorth gravel of the Main Terrace occurs between 200 and about 240 ft. O.D. near Severn Hall and Stanley Hall. A pit at the latter showed 12 ft. of coarse gravel and sand on 2 ft. of finer sand. East of the Severn valley both sides of the mouth of the Worfe have patches of similar gravel at a corresponding height. At Bridgnorth, Panpudding Hill and small spurs to the north and south are capped by gravel at about 200 ft. O.D. Much larger areas occur on the east side of the valley and a 15-ft. section in one to the east of Low Town has been described by Prof. Wills.[1]

Near Eardington gravel of the Main Terrace at about 200 ft. O.D. ranges southward from the top of the river cliff opposite Danesford (Plate IX B) to the Mor Brook valley, but is not well exposed. A pit in an isolated portion, capping the cliff above Eardington Forge, shows some 12 ft. of coarse gravel resting on a pot-holed surface of Lower Mottled Sandstone. Four small remnants of the Main Terrace occur to south-east of Crateford, in one of which a pit shows very coarse gravel with pebbles of the same types as at Eardington.

In the Bridgnorth and Eardington area in general the pebbles of the Main Terrace gravels include Scottish and Lake District granites, rhyolites of Uriconian type, conglomerate and grit of Longmyndian types, Caradocian sandstone, Llandovery sandstone and limestone, and slates and grits of various types. The proportions of the various constituents vary somewhat from place to place. R.W.P.

The Main Terrace forms a considerable flat around Lodge Farm, near Dudmaston Hall (Plate IX B), its surface being about 90 ft. above that of the flood-plain alluvium. Erratics are plentiful and include granites of Scottish and Lake District types, but stones from the north-west, such as volcanic rocks of Uriconian type and Lower Palaeozoic sediments, seem to predominate. T.H.W.

Gravel of Main Terrace type occurs in patches on both sides of the small stream half a mile north of Hampton and would seem to have come through a depression trending south-eastwards from the Mor Brook valley near Eardington Mill. The gravel extends southwards to Hampton and a small patch just west of the station rests just below the 200 ft. contour. R.W.P.

---
[1] *Op. cit.* (1924), p. 295.

South of Hampton Loade small benches on the east side of the river in places appear to represent the Main Terrace feature, and on one of them, 450 yds. S.S.W. of Moor House, lies some moderately coarse gravel with northern and western erratics.

T.H.W.

About 300 yds. S.W. of The Heath, between Stanley and Severn Lodge, gravel lying somewhat below the 200 ft. contour appears to belong to the Main Terrace. A pit in it showed 15 ft. of well-sorted sand and gravel in thin seams and bands of purple and green clayey sand. The top 4 ft. consisted of coarser gravel.

R.W.P.

A small patch of Main Terrace gravel lies on the west side of the river about 670 yds. S.S.E. of Severn Lodge, and on the other side there are larger spreads at Upper Arley. Here, again, erratics of Welsh or Shropshire origin appear to predominate over those derived from the north.

The Main Terrace recurs near Arley Station (Survey Photograph 6906) to the south of which a pit shows about 25 ft. of coarse, well bedded gravel with lenses of sand ; the pebbles include many rocks of Welsh or Shropshire type, but granites, especially of "Eskdale" type, are numerous. The terrace is seen again, on the opposite side of the river, in the lower part of Eymore Wood, and a small patch on the west side, near Folly Point, is the last trace of it in the present district.

BRECCIA GRAVELS

To the west and south-west of the high ground formed by the Clent Hills and Wychbury Hill debris derived from the Clent Breccia, as well as from the Bunter Pebble Beds, is of common occurrence on the surface. Much of this debris is probably talus washed down during a long period up to and including the present time, during which conditions have not differed greatly from those now existing ; though the rate of transportation and accumulation has probably fluctuated with variations in annual rainfall. Certain more clearly defined deposits, consisting largely of breccia debris, would appear, however, to have been formed under conditions that have not since recurred.

One such deposit occurs at Pedmore, north-west of Wychbury Hill, where sandy to somewhat clayey material containing angular fragments from the Clent Breccia can be seen in places up to a thickness of 5 ft. On the surface, besides breccia fragments, "Bunter" pebbles also occur and one pebble of volcanic ash, possibly of Welsh origin, was seen. The base of the deposit slopes northward and may grade to the Main Terrace of the Stour at Amblecote, but there is more than a mile between them without any connecting link.

Another area of similar breccia gravel at Lower Clent appears to be related to a small valley opening from Clent Hill near Clent Grove (Plate VI A). The deposit slopes westward from a height of a little over 500 ft. at the mouth of this valley to about 450 ft. at Lower Clent. It consists of angular rock fragments and "Bunter" pebbles in a sandy matrix.

Near Field House the same kind of material forms a flat-topped expanse over a mile long and averaging a third of a mile broad. The level of the base falls from about 410 ft. O.D. at the eastern end to a little over 300 ft. at the western end. A small pit 330 yds. S.W. of Field House showed 2½ ft. of a breccia-like deposit with angular fragments derived from the Clent Breccia, up to 10 in. long, and "Bunter" pebbles. The eastern end of this deposit lies nearly opposite the mouth of the Clatterbach valley, between Clent and Walton Hills, and it is probably the remains of an alluvial fan related to that valley. The accumulation at Lower Clent may have been connected with it along a rather steep tributary. The Field House deposit clearly grades to the terrace-like gravels of Churchill and Blakedown, in which breccia fragments are again numerous, and these, in turn, grade to the Main Terrace of the Stour (p. 163). Thus the Field House breccia gravels, and perhaps also those near Lower Clent and Pedmore, were formed during the period of excessive run-off, caused by melting snow and perhaps frozen sub-soil (taele) conditions, of the fluvio-glacial phase that accompanied the melting of the ice-sheet farther north. It is possible, however, that the Lower Clent and Pedmore deposits belong to some post-Glacial pluvial phase.

Breccia gravels very similar to those already described lie to the east of the high ground near Enville. They are found in patches isolated by erosion near Compton, and form broad spreads near Enville Common and to the south of Highgate Common.

Near Compton the breccia gravels occur at two distinct levels, one about 25 ft. above the other. When the base of the higher group is plotted along with the thalweg of the stream that drains from Compton to the River Stour near the Hyde, it is found not to lie on a smooth gradient curve in that direction. If, instead, the base of the higher gravels is plotted along a line corresponding to a direction by way of Enville Common and Spittle Brook[1] to Smestow Brook (Fig. 19) it is found that it lies approximately on a smooth curve grading to the Main Terrace of Smestow Brook. The lower group of gravels, on the other hand, does appear to grade, though not quite regularly, to the Main Terrace of the Stour near the Hyde, by way of the present line of drainage (Fig. 19). It may therefore be inferred that the Compton area originally drained northward and then eastward by way of the Checkhill gap in the Pebble Beds escarpment, through which Spittle Brook now passes. The present drainage was established by capture during the fluvio-glacial phase, and the irregularity of the lower breccia gravels may be due to their having been deposited while the capture was taking place and before the new drainage was perfectly adjusted. Near Falcon the two groups of breccia gravels form a continuous spread, divided, however, by a distinct feature.

The two sets of breccia gravels do not appear to differ in composition. Near Compton the debris on the surface consists mainly of fragments from the Clent Breccia with some pebbles of chert derived from the

---

[1] Spittle Brook arises at the junction of Philley Brook with another stream coming down from Church Gorse. It is referred to by Prof. Wills (*op. cit.*, 1938, pp. 203, 205) as 'the Enville brook.'

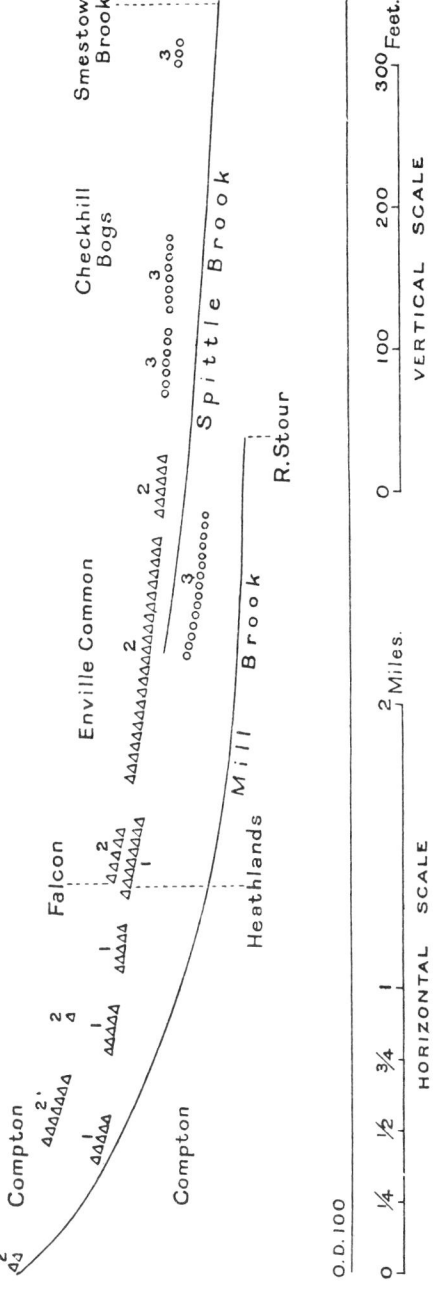

Fig. 19.—*Gradient profiles of Breccia-gravels near Enville, plotted with thalwegs of* (a) *Stream from Compton to the R. Stour near the Hyde.* (b) *Stream from near Blundies* (Enville) *to the junction of Smestow Brook and Spittle Brook.* 1. Lower Breccia-gravels of Compton. 2. Higher Breccia-gravels of Compton and Breccia-gravels of Enville Common, etc. 3. Main Terrace gravels of R. Stour, Smestow Brook and Spittle Brook.

conglomerates of the Bowhills Group and a few "Bunter" pebbles that presumably came from the direction of Kinver Edge. Near Enville the composition of the gravel, judging by surface debris and shallow sections, is similar. To the north of Spittle Brook one or two "northern" erratics lie on, or on the level of, the gravel terrace. Just north of Spittle Brook the gravels contain same angular debris, but farther north this dies out in favour of "Bunter" pebbles and the deposit becomes more sandy. At the south-west edge of Highgate Common the surface of the gravel becomes slightly hummocky and there is some doubt whether certain isolated patches should be assigned to this or to the kame gravels described on p. 158. Followed down Spittle Brook the breccia gravels merge with the regular terrace that grades to the Main Terrace of Smestow Brook (p. 161).

PEBBLE SPREADS

Though pebbles are usually common enough in the vicinity of the Bunter Pebble Beds certain areas where "Bunter" pebbles are especially abundant seem to call for remark. The best defined of these "pebble spreads"[1] occur between Compton and Kinver Edge, one north of Redcliff and another around Iron House. They form flat or nearly flat expanses upon which "Bunter" pebbles are noticeably more numerous than on the surrounding ground. There appears to be no definite deposit, with a matrix, the pebbles lying on, or just embedded in, the slightly weathered surface of the Lower Mottled Sandstone. The spreads lie at 300 ft. O.D. or a little over and have been left standing above the surrounding area by erosion, which has almost cut them off from the foot of the scarp of Kinver Edge. Another patch lies about 800 yds. south of Iron House and within 250 yds. of the crest of the Edge. It occupies a pronounced bench at the foot of the scarp, from which it also is rapidly being cut off by erosion. The level of these pebble spreads corresponds closely with that of the higher breccia gravels near Compton (p. 166), and the accumulation may perhaps date from the same period of active erosion and downwash. Possibly they are of the nature of "taele gravels," the interstitial material having been subsequently washed away.

Though the above-mentioned are the most striking of these pebble-covered surfaces, there are others elsewhere at about the same level. Two knolls rise on Enville racecourse to a little over 300 ft. O.D., and pebbles seem more numerous on them than on the surrounding lower ground. The smaller of the two, about half a mile west of the Pebble Beds escarpment, reaches 315 ft. (about 40 ft. above the surface of the breccia gravels near Enville Common) and its summit would approximately touch a plane sloping uniformly at about 1 in 50 from the crest of the escarpment to the edge of the breccia gravels. Another knoll with pebbles on it occurs at about the 300 ft. level south-west of Hampton Valley.

---

[1] The pebble spreads are not shown on the one-inch geological map, but the better defined ones near Kinver Edge are shown on the six-inch geological map, Staffordshire 70 S.E., a copy of which is deposited for reference in the Geological Survey Library, South Kensington, London.

Near the eastern escarpment of Pebble Beds, just north of Bunkers Hill, flat-topped spurs, rising to a little over 300 ft. and covered with pebbles, appear to be the dissected remnants of a bench projecting from the foot of the scarp feature and rather less than 100 ft. below its crest. Some of the deposits near Whittington Common mapped as "Glacial sand and gravel" and regarded as possibly "older drift" (p. 150) should perhaps be classed rather with these pebble spreads ; they lie at about the same level. A tendency for pebble-covered spurs to occur at about the 300-ft. level may be observed in part of the Ridgehill Wood escarpment, but here they are less distinct and constant in height.

GLACIAL HISTORY

The meagre evidence provided by the older drift deposits renders any deductions concerning the earlier Glacial history of the district, or the form of the pre-Glacial surface, somewhat speculative.

As Prof. Wills has shown,[1] the relation of the drift deposits to the Ironbridge Gorge seems to make it clear that the watershed across the site of that gorge, dividing the drainage to the Dee estuary from that to the Bristol Channel, persisted up to the time of the last invasion of the district by ice and it must therefore have formed part of the pre-Glacial topography. A study of the contours suggests the probability that the pre-Glacial headwaters of the lower Severn are represented approximately by Mor Brook, and that a subsidiary watershed extended from the high ground near Upper Farmcote (about 3 miles east of Bridgnorth) across the present course of the Severn near Hoards Park and thence north-westward to meet the Dee—Bristol Channel watershed near Barrow. The whole stretch of the Severn from Ironbridge to Bridgnorth would appear to be the result of adjustments to glacial conditions, though not necessarily all of the same date. The pre-Glacial drainage of the Worfe valley area and that part of the Coalbrookdale Coalfield plateau which lay to south-east of the old watershed probably flowed south-eastward, between the Abbot's Castle Hill escarpment and the high ground of Gatacre and Four Ashes, to the Stour basin.[2] It may be presumed that this drainage passed through the Checkhill gap, in view of the evidence (p. 166) that this gap took the drainage from the Compton area up to the time of the formation of the Main Terrace.

The present divide between streams flowing to the Worfe and those to Smestow Brook is at a low saddle at Mere Farm, about two and a half miles south-east of Bobbington. This, which has little or no drift upon it, is at a height of about 270 ft. above O.D., but since it may have functioned as an overflow channel, and suffered erosion, during the later glacial episode (p. 174), its level when it carried the Worfe drainage must be

---

[1] *Op. cit.* (1924), pp. 279-283, 298, 299.

[2] *Cf.* F. W. Harmer, *Quart. Journ. Geol. Soc.*, vol. lxiii, 1907, p. 479, who appears to have assumed that this drainage persisted up to the time of formation of the Ironbridge Gorge, i.e., up to the Main Irish Sea glaciation. See also L. J. Wills, *op. cit.* (1937), p. 75 and Fig. 1, p. 74.

supposed to have been, relatively, higher. Some of the (newer) drift of the Worfe valley descends well below the present level of this saddle and must, therefore, have been deposited after any drainage from the Worfe area that may originally have flowed that way had been diverted to the lower Severn. That is, the diversion must have taken place before, or during an early phase of, the main Irish Sea glaciation. In view of the height of the Mere Farm saddle drainage from the Worfe valley could not have flowed that way unless the height of the whole of that valley was over 300 ft. (in relation to present sea level), and much of it must have been over 350 ft. Since some of the glacial deposits in the valley lie well below 250 ft., it would seem that the amount of erosion that occurred between the supposed diversion of drainage and the deposition of the glacial deposits is so great as to render it unlikely that the former took place during an early phase of the later glaciation. Since there would seem to be no sufficient cause of such a diversion[1] during the interglacial episode represented by the Kidderminster Terrace it seems reasonable to connect it with the earlier glaciation, during which the drifts near Churchill were deposited.

The absence of any recognizable "older drift" deposits in the Worfe valley and the Bobbington area precludes any direct reconstruction of the surface upon which the older ice-sheet rested; but it may be surmised that if it entered the district by the way of the Worfe valley, it may have impounded water against the subsidiary watershed alluded to above (p. 169). Such impounded water may have escaped over a low point in the Pebble Beds escarpment near the present mouth of the Worfe, where, perhaps, the watershed had already been notched by a tributary of the pre-Glacial lower Severn, and thus initiated the present westward course of the Worfe from near Worfield to the Severn. This, in Prof. Wills's view, was already the direction of drainage of the Worfe, Mad Brook and other streams from the Coalbrookdale Coalfield plateau (including the small one presumed to have flowed down from the neighbourhood of Ironbridge) at the time of the invasion by the Irish Sea ice-sheet.

If the older, "Welsh," ice-sheet entered the district down the Worfe valley, it may have been by the Claverley and Bobbington route that it reached the Churchill area. Although, as already remarked, no "older drift" has been recognized near Bobbington, it is, of course, not impossible that some of the erratic boulders, especially those of Welsh origin, within the area covered by the Irish Sea ice-sheet may have been brought by the older ice-sheet. Prof. Wills has suggested that the Churchill gravels were deposited in temporary lakes held up by ice against the higher ground; but it is not easy to see how an ice-sheet coming from the north or north-west could have thus obstructed the natural drainage of the area towards the south.

In its comparative freedom from glacial deposits outside the area of the Irish Sea ice-sheet the present district exhibits a contrast to the Birmingham district, very striking when the two geological maps (167 and

---

[1] Simple capture appears improbable in view of the apparent relative unimportance of the lower Severn before the formation of the Ironbridge Gorge.

168) are compared. The difference is no doubt due in part to the much greater amount of erosion that has taken place in the Severn basin than on the other side of the main watershed. But the survival of relics of the "older drift" in and near the Stour valley, an old drainage line where erosion must have been intense and long continued, suggests that the absence of such relics elsewhere cannot be attributed entirely to denudation, but must indicate that much of the belt between the Rowley and Dudley Hills and the Clee Hills, south of the limit of the Irish Sea ice-sheet, escaped glaciation by the older ice-sheet also.[1] The curious fact that the erratic boulders increase in number from Churchill eastward and not northward or north-westward may have a significance in this connexion.

The regularity of the Kidderminster Terrace and its virtual freedom from foreign material appear to indicate a period of normal fluviatile conditions[2] after the earlier glacial episode. The extent of this terrace in the Stour valley suggests that this constituted the principal line of drainage up to the time of the last invasion of the district by ice, the Severn being merely a tributary stream which, unless the correlation of the Hoards Park—Eardington gravels with the Kidderminster Terrace (p. 154) be accepted, may not have attained a mature stage of development.

The last, or Irish Sea, ice-sheet entered the district over the Ironbridge watershed (p. 169), down the Worfe valley and, probably somewhat later, over the main watershed to the west of Wolverhampton. Since it extended some two and a half miles farther south to the west of the main Bunter Pebble Beds escarpment than it did in the Smestow Brook valley, the Worfe valley was probably the route of the principal ice-stream. At its maximum probably nearly the whole of the district within the limits indicated approximately by the distribution of large Irish Sea ice erratics (Fig. 18, p. 149) was covered continuously by ice.

The ice-front appears to have surmounted the main watershed near Sedgley Park and probably lay just north of the crest of the Colton Hills and the high ground of Upper Penn (Fig. 20), whence the boulders to the west of Penn Common (p. 156) were perhaps washed out. At or near its maximum the ice-front probably lay to the west of the Keuper escarpment near Oreton and the ridge of gravel south of that place may be the remains of a kame deposited between the ice and the scarp feature. Farther southwest some of the mounds and ridges of sand and gravel between Smestow and Trysull may indicate local stands of the ice at or near the maximum. At least the absence of any glacial drift or erratics, except those in the Main Terrace gravels, farther south seems to enforce the conclusion that the ice-sheet terminated north of Smestow. The abundance of boulders on the dip-slope of Abbot's Castle Hill suggests that the ice-front lay for a prolonged period in this neighbourhood, and more continuous and well defined terminal deposits hereabouts might therefore have been expected.

---

[1] It must be emphasized that these observations are intended to apply only to the present district and its near neighbourhood, including, as it does, a considerable area of high ground and lying near the major watersheds, present and pre-Glacial. *Cf.* L. J. Wills, *op. cit.* (1938), p. 226.

[2] Prof. Wills assigns this to the Acheulian interglacial period; *op. cit.* (1938), table, p. 232.

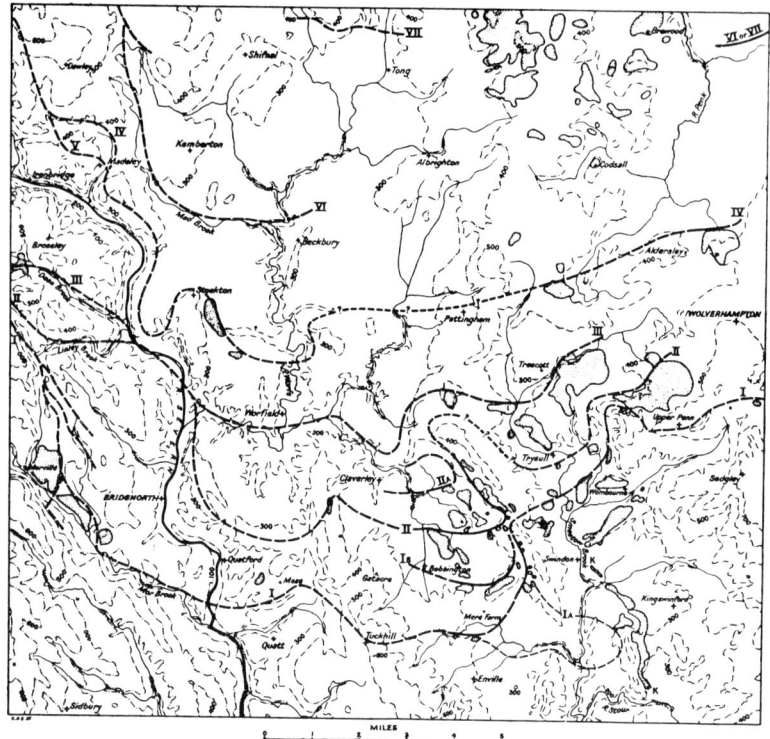

Fig. 20.—*Sketch-map showing Retreat Stages of the Irish Sea Ice-sheet near Bridgnorth and Wolverhampton.* Stages I-VII in the Severn and Worfe Valleys after L. J. Wills (*Quart. Journ·Geol. Soc.*, vol. lxxx, 1924, Pl. XII). Stage VI or VII east of Brewood after E. E. L. Dixon ('Wolverhampton Memoir,' 1928, Fig. 9, p. 177, where it is numbered IV). Stippled areas indicate glacial sands and gravels (Kames, esker-chains, outwash, etc.). K-K, the Kingswinford Gravel-ridge.

It is likely, however, that the depression in the Pebble Beds escarpment north of Blackhills Plantation, where the Enville Fault crosses it, acted for a time as an overflow for water from the ice-front to the west (p. 158), and this flow may have destroyed such terminal deposits.

At its maximum the ice-front seems to have skirted the ridge of Blackhills Plantation and lain to the west of and parallel to the Pebble Beds escarpment as far south as Highgate Common. The mounds of sand and gravel at the eastern edge of the boulder clay area (p. 158) and those on the Common may mark its limits. From Highgate Common the ice-front appears to have extended south-westward to a point about a mile north of Enville, where the gravel ridge near Morfe House Farm (p. 158) may perhaps be regarded as a terminal kame.

The possibility may, however, here be considered that, at its maximum, the ice-sheet thrust a lobe through the Checkhill gap and across the Smestow Brook valley to impinge upon the Ridgehill Wood escarpment (Fig. 20, I A). On the assumption that the deposits of the Kingswinford gravel ridge belong to the "newer drift" this hypothesis would seem to offer an explanation of their origin and at the same time to account for the somewhat anomalous course of the Stour through the Bells Mill gap (p. 162) if, as seems probable, its original course was north-north-westward from where it leaves the coalfield and then round the north end of the Ridgehill Wood escarpment into the Smestow Brook valley. The supposed ice-lobe would have impounded a lake in the neighbourhood of Swindon and Wallheath and farther north, in which the Kingswinford sands and gravels might have been deposited. The lake might have risen until it overflowed at a low point in the escarpment on the site of the Bells Mill gap. The overflow could not have begun at a level of less than 350 ft. in relation to present sea-level, for the drift near Mount Pleasant and Bellevue reaches this height ; but the eastern edge of the drift is at a height of about 320 ft. on the saddle near Summerhill, so that 30 ft. would appear to be the minimum amount of erosion required to divert the Stour permanently to its present course. Under the conditions envisaged an amount of erosion of this order could have taken place very rapidly, so that the hypothetical ice-lobe may have been very short-lived, which would perhaps account for the want of any direct evidence of its presence in the area it is supposed to have covered. It must, however, be admitted that this hypothesis does not account for the rarity of northern erratics in the gravels of the Kingswinford ridge, and it is possible that a sequence of events similar to that above suggested took place in connexion with the earlier "Welsh," ice-sheet ; though, as already mentioned (p. 152),[1] there are difficulties in relegating the Kingswinford gravel to the "older drift."

From near Morfe House Farm for some distance westward no definite deposits mark the termination of the ice-sheet, the limits of which can only be gauged approximately by the cessation of erratic boulders, the most southerly of which were found at Tuckhill. The numerous boulders near Wooton probably indicate the proximity of the ice-front at its maximum, while Prof. Wills[2] has suggested that the concentration of boulders at Mose may be terminal in origin. Beyond the Severn the ice-front at its maximum appears to have followed approximately the line of Mor Brook and to be marked by some of the gravels at Morville (p. 159).[3]

Prof. Wills has suggested a number of stages in the retreat of the ice-sheet from the neighbourhood of the Severn and Worfe valleys.[4] These cannot be connected with any certainty with stages farther east in the present district, but some tentative correlations may be suggested (Fig. 20, p. 172).

At Stage II of Prof. Wills's scheme the ice-sheet lay along Linley Brook (Sheet 153) to its mouth and just east of the course of the Severn from near the mouth of the Worfe southward (Fig. 20, II). The Bridgnorth reach of the Severn thus functioned as a marginal drainage channel and erosion proceeded vigorously, but, as Prof. Wills remarks, the deposition

---

[1] *Cf.* L. J. Wills, *op. cit.* (1937), pp. 88, 90.
[2] *Op. cit.* (1924), p. 300.
[3] Wills, L. J., *op. cit.* (1924), pp. 289, 301.
[4] *Op. cit.* (1924), Fig. 7, p. 302 and Pl. XXII.

of the Main Terrace gravels may have begun farther downstream. It seems likely that the ice-front was continued along the dip-slope of the Pebble Beds north of Quatford. It was possibly at this stage that the Wooton and Quatt gravels (p. 163) were deposited as outwash from the ice through the little valley near Morfevalley. The ridge of gravel near Wounsdale perhaps marks a point on the ice-front at this stage. Its further continuation eastward may be indicated by the kame-like ridges near Leaton Hall and Highgate Farm (p. 158) or, more probably perhaps, by the gravel mounds east of Halfpenny Green and the ridge parallel to the Abbot's Castle Hill escarpment (p. 158 and Fig. 20, II). At this stage it would seem that the ice must have prevented drainage from the high ground to the south from escaping in the Claverley direction, and a lake may have been formed that overflowed by the Mere Farm saddle. It was perhaps under these conditions that the flat-surfaced sands near Heathton and White Cross (p. 158) were deposited. The further continuation of the ice-front may have lain along the dip-slope of the Bunter Pebble Beds and is perhaps indicated by some of the gravels near Trysull. The large, somewhat hummocky spread of sand and gravel near Penn Fields (p. 155) would seem to represent outwash at some stage subsequent to the maximum, but whether it corresponds with the stages farther west above described is necessarily doubtful.

At Stage III postulated by Prof. Wills the ice-sheet still covered the site of the Ironbridge Gorge, the excavation of which could therefore not have begun. In the present district the suggested position of the ice-front is along the line of the Worfe to about its junction with Claverley Brook. An ice-stand that may correspond with this stage seems to be suggested by the gravels near Shipley and south-west of Trescott ; also in a general way by the gravel ridge of Trysull and Seisdon (p. 157) and possibly by the spread of sand and gravel near Pool Hall (p. 156).

At Prof. Wills's Stage IV the ice-front appears to have lain wholly outside the present district, though it seems to have reached to within a little over a mile north of Worfield. The ice-sheet had then begun to split into two lobes, laying bare the watershed at Ironbridge, and excavation of the gorge at Ironbridge was begun.

It was perhaps at about this stage (IV) that the ice-front lay across the Aldersley gap in the main watershed near Wolverhampton. With a further retirement of the ice-front this gap would have acted as the overflow channel of a lake impounded at the head of the Penk valley,[1] and it was probably as the result of material carried through the Tettenhall gap by this overflow that the deposition of the fluvio-glacial gravels in the upper part of the Smestow Brook valley began ; though the formation of the Main Terrace farther down Smestow Brook and along the Stour may have begun earlier.

Subsequent recession of the ice-sheet increased the gap between the two lobes and laid bare an increasing area of the Worfe valley in which

---

[1] Dixon, E. E. L., H. Dewey and T. Robertson in ' The Country between Wolverhampton and Oakengates ' (*Mem. Geol. Surv.*), 1928, pp. 179, 189, 190.

accumulation of the fluvio-glacial deposits then began. Erosion caused by the access of water through the Ironbridge Gorge, aided by concurrent uplift of the area, dissected the Main Terrace gravels and the fluvio-glacial deposits that grade to them.

POST-GLACIAL DEPOSITS

LATER RIVER TERRACES

Two or three gravel terraces later than the Main Terrace can be recognised in the valleys of the Stour and Severn. That next below the Main Terrace (the second terrace) may, in general, be identified as the Upper Danesford or Worcester Terrace of Prof. Wills[1]; but near Bridgnorth the somewhat younger Lower Danesford Terrace has been given the same symbol on the geological map. The first terrace may probably be correlated with the Power House (Power Station) Terrace of Prof. Wills[2] in the Stour valley, in the Severn valley as far up as Highley and possibly near Bridgnorth also.

In the case of the Upper Danesford Terrace, at least, the distinction between Glacial and post-Glacial is somewhat arbitrary; for, though there was certainly no ice in the immediate neighbourhood of the present district when its gravels were deposited, there is some evidence that ice of the Welsh Re-advance (or "Little Welsh Glaciation") lay in the Severn Valley above Shrewsbury at that time.[3]

Smestow Brook and River Stour.—The second terrace first appears in the Smestow Brook valley just east of Trysull, where its surface is from 4 to 10 ft. above the flood-plain alluvium. From here to the junction of Smestow Brook with the Stour it is found at intervals with its lower edge in contact with the alluvium. Neither this nor any lower terrace is recognisable along the Stour down to its junction with Smestow Brook. At Caunsall the second terrace forms a well-marked bench between 20 and 30 ft. above the flood-plain alluvium, north-east of the smithy, separated by a rock-step from the larger one of the Main Terrace. Small patches of the second terrace occur near Cookley and Wolverley.

A lower terrace seems to begin near the Hyde, Kinver, and occurs at Caunsall, where most of the village is built upon it, with traces near Cookley, Wolverley and Franche. T.H.W.

Worfe Valley.—Along the lower Worfe and its tributary the Hilton Brook there are terraces at two levels below that of the fluvio-glacial gravels which grade to the Main Terrace of the Severn.

---

[1] *Op. cit.* (1937), p. 84 and *op. cit.* (1938), p. 206.
[2] *Proc. Birm. Nat. Hist. & Phil. Soc.*, vol. xv, 1926, p. 99; and *op. cit.* (1938) pp. 213, 214, 215.
[3] See T. H. Whitehead in 'The Shrewsbury District' (*Mem. Geol. Surv.*), 1938, p. 209, and L. J. Wills, *op. cit.* (1937), p. 94.

The higher of these, or second terrace, may be as much as 40 ft. above the alluvium of the flood plain, while the lower seldom attains a height of more than 10 ft. above that plain.

The largest area of the second terrace lies between Hilton and Wyken, where a small section at the inn shows 5 ft. of well bedded sand and gravel containing quartzites and granites. The lower or first terrace covers considerable areas near Worfield and westwards to Burcote Mill, where the river enters the steep-sided cut across the outcrop of the Pebble Beds. It is also present west of Rindleford and again from Worfe Bridge to the mouth of the Worfe.

Severn Valley.—Two terraces (combined on the geological map as second terrace) below the Main or Third terrace in the Severn Valley in the neighbourhood of Bridgnorth may be identified with the Upper and Lower Danesford Terraces of Prof. Wills ; but in places a still lower terrace is present. This last occurs on the right bank above Severn Hall, where it rises to about 10 ft. above flood level, as an island in the alluvium east of Stanley Hall and as the flat about 6 ft. above flood level upon which the Low Town of Bridgnorth is built on the left bank of the river.

A terrace present all along the right bank between Severn Hall and Bridgnorth rises in places to as much as 25 ft. above the alluvial flat and may be assigned to the Lower Danesford Terrace.

At Danesford a terrace, mapped as the Lower Danesford Terrace by Prof. Wills, extends from the Priory nearly to Quatford and at the highest part, about a quarter of a mile south of the Priory, reaches a height of 38 ft. above the alluvium. The Upper Danesford Terrace of Prof. Wills has not been recognised at Danesford.

Opposite Quatford a flat at 6 ft. above the alluvium extends for three quarters of a mile. Above it a terrace at 60 ft. above the alluvium appears to be the Upper Danesford Terrace. This terrace also occupies the depression to the west of the patch of Main Terrace gravel above the Eardington Forge.

R.W.P.

On the east side of the river the Upper Danesford Terrace appears, with its surface about 60 ft. above the flood-plain alluvium, south-west of Lodge Farm, Dudmaston. There is then no trace of it on the east bank for more than two miles ; but two patches of gravel at Hay House, a mile south of Hampton Loade, and two more at Eymore Farm are at the appropriate level.

A lower terrace appears between Hay House and Potters Loade Ferry (west of Hallclose), near which it shows signs of division into two benches. The terrace occurs again on the east bank from a point nearly opposite the mouth of Borle Brook to Upper Arley. Here again it is in two benches separated by an abrupt drop of about 10 ft. Traces of the same terrace may be found south-east of Arley Station, on the west bank, and near Eymore Farm and Folly Point on the east bank of the river. These may correspond with the Power Station Terrace ; but possibly only the lower bench does so, the upper representing the Lower Danesford Terrace (p. 175).

ALLUVIUM

Smestow Brook and River Stour.—Smestow Brook is bordered by a narrow alluvial flat throughout its course in the present district, and this extends through the Tettenhall—Aldersley gap (Sheet 153).

Alluvium begins to appear along the Stour near Cradley. It narrows somewhat through the Bells Mill gap, but extends uninterruptedly along the whole course of the river. The alluvium both of the Stour and of Smestow Brook consists mainly of silt or sand.

T.H.W.

River Worfe.—The alluvium of the Worfe is narrow above Worfield but widens considerably at the junction with Stratford Brook at Wyken. It narrows again to less than 50 yds., where the river cuts through the Pebble Beds below Burcote.

River Severn.—The flat of the River Severn expands below Severn Hall and is about 300 yds. wide at the junction with the Worfe. It contracts at Bridgnorth to about 120 yds. but expands again about half a mile downstream to nearly 400 yds. (Survey Photograph 6847) and maintains a considerable width as far as the junction with Mor Brook except for a narrowing to 200 yds. at Quatford Ferry.

R.W.P.

Near Dudmaston Hall, where the Severn leaves the Trias outcrop, the alluvial flat narrows considerably, and to the southern border of the district is rarely more than 100 yds. or so broad. The alluvium nevertheless attains a fair thickness in places, for 10 ft. can often be seen at the riverward edge when the water-level is low. Besides silt and sand, lenses of fairly coarse gravel are often present. Near Upper Arley about 10 ft. of silty sand overlying more than a foot of coarse gravel were seen resting on Coal Measures in the summer of 1932.

T.H.W.

CHAPTER IX

MINERAL PRODUCTS, WATER SUPPLY AND AGRICULTURE

COAL

Some account of the qualities and methods of working of the coal seams of South Staffordshire will be found in an earlier memoir.[1]

The characters of the coals of the Highley and Kinlet groups of the Wyre Forest Coalfield are suggested by the names originally given to these two divisions, namely the Sulphur Coal Group and the Sweet Coal Group.

The coals of the Upper Measures (Highley Group) vary considerably in quality; some are fairly free from sulphur, others are very sulphureous but burn well, while some contain much earthy matter and are not free-burning. These coals were at one time extensively mined in shallow pits and open workings round Tasley, in the Deuxhill outlier, along the Chelmarsh ridge, at Stanley near Highley, at Bank Farm south of Severn Lodge, at Shatterford and in Eymore Wood. The principal market appears to have been the hop-drying industry, for the purpose of which the sulphur is considered to be beneficial. The coal was first converted into coke (called charcoal).

None of the seams in the Upper Measures is now worked within the district, although to the south in the Mamble and Bayton area they are still actively mined.

The coals of the Middle Measures (Kinlet Group) command a ready sale over a large area. There are four principal seams (see p. 41) but only one, the Highley-Brooch Coal, seems to have been worked extensively.

The sweet coals at one time worked at Baveney had a considerable reputation among local blacksmiths, like the Smith coal seam of the Clee Hill Coalfield.

IRONSTONE

The ironstones of the South Staffordshire Coalfield will be found mentioned in their relative positions amongst the other measures in the table on p. 35 (see also Plate X). Those below the Thick Coal are in general superior to those above; but they have all almost ceased to be worked.[2]

---

[1] 'The Geology of the Southern Part of the South Staffordshire Coalfield' (*Mem. Geol. Surv.*), 1927, pp. 189-192.

[2] For further information see 'The Geology of the Southern Part of the South Staffordshire Coalfield' (*Mem. Geol. Surv.*), 1927, p. 192; and 'Special Reports on the Mineral Resources of Great Britain' (*Mem. Geol. Surv.*), vol. xiii (Pre-Carboniferous and Carboniferous Bedded Ores of England and Wales), 1920, pp. 59-62.

In the Wyre Forest Coalfield there are no ironstones of value in the beds of the Highley Group ; but in the Kinlet Group ironstone has been worked in the past near Chorley (see p. 49) and ironstones are also associated with the coals of Billingsley, where at one time a company was formed to work them and blast furnaces were erected. The products were conveyed by railway along the course of Borle Brook to the Severn. Cakes of ironstone weighing from 60 to 100 lbs. were commonly mined and some of as much as 5 cwt.

Ironworks also existed at Eardington (Survey. Photographs 6862-3) and at Hampton Loade but these were supplied with pig iron from Coalbrookdale. Their position was due to the proximity of supplies of charcoal, to available water power and to the convenience, at that time, of barge traffic on the Severn. At Eardington a tunnel was cut to convey water for power from the valley of Mor Brook at the upper forge through the intervening ridge to the main ironworks on the Severn (Survey Photographs 6859-60).

Fireclays and Pottery Clays

The present district includes the area of the finest fireclays of South Staffordshire (Survey Photographs 1981-6), extending northward from Stourbridge to Oak Farm (Shut End) and Gornal and eastward to the neighbourhood of Overend (Cradley) and Tansley Hill, south-east of Dudley. The fireclays are in greatest number and, in general, of the highest quality in the vicinity of Stourbridge and Brettel Lane. The best-known seam, the Stourbridge Old Mine Fireclay, is a workable measure over most of the southern part of the ground mentioned above, and is recognizable in a debased condition in the Shut End area. It occurs beneath the Sulphur or Stinking Coal and, as mentioned on p. 38, a good fireclay has been found in this position near Tansley Hill. A variable number of workable fireclays is found below the horizon of the New Mine Coal and others occur in places in association with the Heathen Coals (see Table, p. 35 and Plate X). Refractory and semi-refractory clays suitable for the manufacture of sanitary ware, drain pipes, and pottery are found among the measures above the Thick Coal and also, outside the area mentioned above, among those below that coal.[1]

Fireclays apparently in the upper part of the Kinlet Group were formerly worked at Shatterford,[2] in trenches parallel to and to the west of the basalt quarries, for coarse pottery, bricks and tiles.

Fireclays were also worked by the side of the Severn at the foot of Seckley Wood (p. 54), where firebricks and common red bricks were made.[3]

---

[1] For further information see *op. cit.* (*Mem. Geol. Surv.*), 1927, p. 193 ; and ' Special Reports on the Mineral Resources of Great Britain ' (*Mem. Geol. Surv.*), vol. xiv (Refractory Materials : Fireclays, Resources and Geology), 1920, chap. xi, and vol. xxviii (Refractory Materials : Fireclays : Analyses and Physical Tests), 1924.

[2] In G. E. Roberts, ' The Geologist ', 1861, p. 421, and Daniel Jones ' The Forest of Wyre Coalfield ', *Mining Journal*, circa 1872.

[3] Jones, D., *op. cit.*

Brick and Tile Clays

The Etruria Marl furnishes the material for the well known blue engineering bricks at Oldhill, Cradley, and also near Brierley Hill, Kingswinford and Shut End. Red bricks also are made from these clays by firing at a lower temperature and without subjecting the clay to a reducing atmosphere.

Weathered Silurian shale has been used for brick-making near Sedgley, and also Middle Coal Measures clay to some extent.

The use of the clays at Shatterford and Seckley Wood for brick-making has already been mentioned (p. 179). Clay associated with the Bank Farm Coal was formerly worked for brick-making near Bank Farm.

At Billingsley, about 400 yds. S.E. of the church, the mottled clays in the lower part of the Kinlet Group have been dug for brick and tile making and bricks have also been made from red clays near the top of the group between Southall House and Bradley.

Clays in the Highley Group have been used for bricks and tiles north and south-east of Tasley, near Highley Colliery and close to Kinlet.

The clays of the Keele Group are generally too calcareous for brick-making, but they have been utilized near Astley Abbots and on a small scale at several other places. They are being worked actively at the Bridgnorth Brickworks, Knowlesands (Survey Photograph 6879), and have also been dug 250 yds. north of Highley Station. Red clay in the Clent Group, above the higher breccia belt (p. 102), was formerly dug for making bricks for the Enville Estate in a field north-east of Church Gorse. Some old pits near " Brickyard Cottages," 350 yds. S.E. of Newhouse Farm, must have been dug in clay interbedded with the Clent Breccia.

Lime and Cement

Both the Sedgley or Aymestry Limestone and the Dudley or Wenlock Limestone were extensively worked, the former for lime-burning and the latter for lime and also for use as a flux for ironsmelting. The burning of the Wenlock Limestone on the Wren's Nest ceased only a few years ago. It was obtained by a shaft, and any further supplies would have to be sought by that means, as, except for pillars supporting the roof of the old workings, there is none left that could be wrought open-cast. The area underlain by the Sedgley Limestone is now largely built upon.

The cornstones of the *Psammosteus* Limestones group in the Trimpley inlier have in the past been considerably worked for lime-burning, and the remains of kilns can still be found (p. 25).

The *Spirorbis* limestones in the Keele Beds of the western part of the district have been burnt for lime to some extent. The pits near Witnells End (p. 86) were probably opened for that purpose, and the old workings in Hallclose Coppice (p. 87) certainly were, for the remains of a kiln are still visible. They have also been quarried for metalling bye-roads and farm tracks.

MINERAL PRODUCTS. 181

The *Spirorbis* limestones of the Highley Group seem to have been but little worked south of Tasley. T. C. Cantrill[1] found traces of the Bank Farm Limestone having been burnt for lime at Bank Farm, but in spite of its continuity and comparative purity it does not seem to have been worked elsewhere near Upper Arley, or near Winwoods.

Highly calcareous beds in the Keele Group have been burnt for lime in one or two places and the remains of a kiln for this purpose exist about three quarters of a mile north-north-west of Alveley.

SAND AND GRAVEL

Sand suitable for moulding and for pig-iron beds is obtained from the Upper Mottled Sandstone at Amblecote, Stourbridge, Wombourne and Kingswinford.

Silica sand for use in the fireclay industry is got from the Gornal Sandstone (p. 37).[2] According to the " List of Quarries,"[3] silica sand is also obtained at Iverley, near Stourbridge, and at Wollaston, presumably from the Upper Mottled Sandstone.

The Bunter Pebble Beds have been dug in places for gravel, for example, near Smestow, on Calcot Hill south-east of Clent and to a small extent near Kinver.

The glacial and fluvio-glacial deposits yield sharp sand suitable for builders' purposes, and also gravel, the former being dug at Trysull (Wombourne), between Penn and Wolverhampton, and in the Kingswinford gravel-ridge near Wallheath, while fluvio-glacial deposits are worked near Heath Mill, Wombourne, at Gigetty, at Draycott, Aston and near Littlegain. At the last mentioned place the gravel is crushed, the finer product for tar spraying and rough-casting, the coarser, mixed with fine, for concrete. The presence of fragments of soft Triassic sandstone in some of these gravels is a disadvantage.

The terrace gravels of the River Severn have been worked at many points in the neighbourhood of Bridgnorth ; at Eardington the coarse gravel of the Main Terrace is crushed for use as a top dressing for roads.

The terrace gravels of the River Stour have been worked to some extent, but not on a large scale in the present district.

BUILDING STONE

Although suitable stone for building is to be found in the district, very little, if any, is quarried at the present time. The Wenlock Limestone was used in building Dudley Castle, while the flaggy and more massive beds of the Old Red Sandstone have provided material for walls and for farm and other buildings in the western area of the sheet.

---

[1] ' A Contribution to the Geology of the Wyre Forest Coalfield ', 1895, p. 22.
[2] See ' Special Reports on the Mineral Resources of Great Britain ' (*Mem. Geol. Surv.*), vol. vi (Refractory materials : Ganister and Silica rock), ed. 2, 1920, p. 98.
[3] Mines Department, 1930.

The Gornal Sandstone has been used for buildings and walls near its outcrop. Some of the sandstones of the Halesowen Group have been locally used in the past, and similar sandstones in the Highley Group have been quarried near Upper Arley; for Arley Castle, amongst other buildings. Of sandstones in the Keele and Enville Groups the best is that from the Keele Group of the Alveley and Upper Arley area, which is capable of yielding large blocks. Stone from the quarries near the Butts and Hextons Farm was formerly transported by water, as is testified by the remains of a wharf, with quarried blocks still lying on it, beside the Severn. The Alveley stone was quarried for building as late as 1930. Claverley church, although built in the main of local Keuper sandstone, has buttresses of Alveley stone that are a later addition.

The Lower Keuper Sandstone on the whole furnishes the best building stone and has been extensively used for churches and other buildings, but is not now quarried.

ROAD METAL

The only sources of good road metal within the district are the comparatively small outcrops of basalt and dolerite near Dudley and at Pensnett, Shatterford and near Kinlet. The more easily accessible parts of these are largely worked out, though in the Kinlet area there are some indications of massive basalt which may prove valuable. There are, however, ample supplies of excellent material a short distance to the east of the district in the Rowley Regis area, and to the west in the Clee Hills.

WATER SUPPLY[1]

The district is in general well supplied with water from underground sources. Most of the eastern portion, up to a line indicated approximately by the valleys of Smestow Brook and the River Stour, is served by public undertakings. The Elan Aqueduct of the Birmingham Waterworks enters the district near Pound Green in the south-west and leaves it near Uffmoor Wood in the south-east, and supplies for the villages of Shatterford and Upper Arley are taken by a pipe-line from it at a point near the Kidderminster and Bridgnorth road west of Wolverley.

Silurian and Old Red Sandstone.—The Silurian inliers of the South Staffordshire Coalfield are within an area furnished with public water-supplies and are, besides, too small to be of importance as sources of underground water. Supplies are obtained, however, from a borehole in the Downtonian rocks at the Lye (47). Water from the lower part of the Downtonian rocks is frequently saline. A saline spring issues at Saltwells, near Netherton, where there was formerly a spa,[2] and salt water has been encountered in several colliery sinkings.

---

[1] The numbers in brackets refer to Appendix I, p. 186.

[2] See L. Richardson, Wells and Springs of Worcestershire ' (*Mem. Geol. Surv.*), 1930, p. 108.

WATER SUPPLY.

In the Trimpley inlier the sandstones yield sufficient supplies for farms and houses from wells, and good springs issue in places at the junctions of sandstones and cornstones with marls, especially at the base of the *Psammosteus* Limestones Group. Springs also occur in places along the lines of faults.

The village of Shatterford formerly depended mainly upon local springs, of which one of the most important issues on the north side of the Kidderminster road 380 yds. E.S.E. of Bellman's Cross Inn. Though apparently just within the outcrop of the Coal Measures, this spring very likely comes from the Downtonian rocks, here faulted against the Coal Measures.

The Downtonian and Old Red Sandstone rocks of the western margin of the sheet yield, in general, satisfactory supplies from wells (9) for the use of farms and cottages, and good springs are of fairly frequent occurrence.

Coal Measures.—As already mentioned, the South Staffordshire Coalfield is served by public undertakings, but some industrial concerns have private boreholes (31, 38, 46), and a few outlying houses and cottages still depend upon wells.

Sandstones in the Coal Measures are capable of yielding small amounts, but are generally unreliable for large quantities of water. Large volumes of mine-water, suitable at least for some industrial purposes, exist in many parts of the coalfield.

The Etruria Marl may be regarded as a non-water-bearing formation. The sandstones of the Halesowen Group yield water in sufficient quantity for small local supplies both in wells and from springs where they rest upon the clays. The same may be said of the similar sandstones in the Highley Group. The village of Upper Arley formerly depended mainly upon springs coming from these rocks.[1]

Sandstones in the Keele and Enville Groups yield supplies sufficient for local purposes in rural areas. They also give rise to strong springs in the neighbourhood of faults. The town of Bridgnorth is in part supplied by a powerful spring, apparently arising from a minor fault line in these rocks, at the conduit in a small valley about a third of a mile west of the station. Highley gets its supply from a spring which appears to arise from Keele Beds on the line of the Tiphouse Fault where it crosses Londonderry Coppice half a mile south of Hampton Loade Station.

Trias.—In this district, as elsewhere, the Bunter sandstones are the most important natural reservoir of underground water. From boreholes in the Bunter Pebble Beds and Lower Mottled Sandstone at Hinksford (18), Ashwood (20, 21, 22), Prestwood (23) and Kinver (24, 25, 26, 27), the South Staffordshire Waterworks Company supplies Dudley, Tipton, Cradley, Brierley Hill, Sedgley (in part) and Coseley (in part), and furnishes supplies in bulk to Bilston, Seisdon and Stourbridge, besides supplying a large area outside the district.

---

[1] See L. Richardson, *op. cit.*, p. 90.

The Stourbridge and District Water Board, from a borehole in all three divisions of the Bunter at Amblecote (42), supplies Stourbridge, Lye and Wollescote, Wordsley, Hagley and Clent, and, in bulk, Churchill and Broom. There are, besides, in Stourbridge private boreholes drawing water from the Upper Mottled Sandstone (33, 34, 35, 36, 37).

Wolverhampton Corporation Waterworks draws part of its water from boreholes in all three divisions of the Bunter at Dimmingsdale near Lower Penn (10, 11, 12) and in the Upper Mottled Sandstone at Hilton near Worfield (3, 4) ; and furnishes part of the supplies of Bilston, Coseley, Sedgley and Wombourne and that of Lower Penn.

Bilston Corporation Waterworks obtains its water from two boreholes in the Upper Mottled Sandstone and Bunter Pebble Beds at the Bratch, Wombourne (15, 16), and supplies, besides Bilston, Trysull, Swindon and parts of Coseley and Seisdon.

The supply for Bridgnorth referred to above (p. 183) has recently been augmented by boreholes in the Bunter Pebble Beds and Lower Mottled Sandstone at Rindleford (5).

Supplies for villages, farms and dwellings are obtained from the Bunter sandstones in many other places in the district. The Keuper Sandstone also yields supplies in some localities.

Superficial Deposits.—Small cottage supplies are obtained in places from wells in the Glacial sands and gravels.

Agriculture

The several geological formations that crop out within the district have resulted in a considerable variation of soil texture and conditions and this has to some extent influenced the type of husbandry to be found in different parts ; but a factor of perhaps greater importance has been the relative proximity to the markets afforded by the neighbouring industrial areas.

Thus the light soils, on the Triassic rocks, which everywhere show the greatest proportion of plough-land are, in the strip that borders the South Staffordshire Coalfield and in the south-eastern part of the district near Kidderminster, largely given over to market gardening and the growing of vegetables, especially potatoes and peas.

Elsewhere dairy or mixed farming prevails. The clays and marls of the Halesowen, Highley, Keele and Enville Groups are almost everywhere under permanent pasture, such arable land as is maintained on these groups being usually confined to places where an outcrop of sandstone affords a lighter and more easily worked soil. The same is to a considerable extent true of the areas underlain by the pre-Carboniferous rocks. The greater part of the Downtonian and Old Red Sandstone area in the west is under permanent pasture and is essentially a dairy-farming district[1].

---

[1] The observations in this paragraph apply, in the main, to pre-war conditions.

Considerable areas of the Bunter outcrops are abandoned to heath or waste woodland and the Clent Breccia at its higher levels affords little but rough, ill watered pasture.

The forest area on the Coal Measures of the south-west corner of the district consists of oak coppice, cut periodically for oak-bark (for tanning), charcoal burning and small timber. The Forestry Commission has in recent years planted areas near Upper Arley and Pound Green with coniferous trees, chiefly larch and spruce. Cherry orchards are a feature of clearings in the forest.

T.H.W., R.W.P.

APPENDIX I—TABLE OF WELLS AND BOREHOLES FOR WATER, SHEET 167 (DUDLEY)
(as in 1939)

Note.—A dash (—) indicates lack of information; f^5 signifies Lower Keuper Sandstone; f^3 Upper Mottled Sandstone; f^2 Pebble Beds; f^1 Lower Mottled Sandstone.

Name	Six-inch map	Approx. height of surface above O.D. in ft.	Depth in ft.	Standing water-level in ft. below surface	Yield: gallons per hour (H) or day (D)	Strata passed through: depth in ft. to base of each formation except lowest	Remarks
1. Wyken, north-east of Bridgnorth	Salop 59 N.W.	140	150	33	720 H	Drift 5, f^3 150	Water struck at 84 ft. and 132 ft. Pumping at 720 gal. per hour depresses water level to 36 ft.
2. Wolverhampton Waterworks, Worfield, 600 yds. N. of Hilton Bridge	Salop 59 N.W.	156	860	3	20,000 H (test)	Drift 12, f^5 200, f^3 648	Trial bore
3. Wolverhampton Waterworks, Worfield, 50 yds. N.E. of No. 2	Salop 59 N.W.	160	705	10		Drift 13, f^5 76½, f^3 616	No. 1 bore — Keuper Sandstone apparently brought in by faulting.
4. Wolverhampton Waterworks, Worfield, 30 ft. S.E. of No. 3	Salop 59 N.W.	163½	705	13		Drift 12⅔, f^5 94, f^3 598½	No. 2 bore
5. Bridgnorth Waterworks, 100 yds. S.W. of Rindleford Mill	Salop 59 N.W.	127	300	17	25,000 H (test)	Terrace gravel 36, f^2 143, f^1 300	Shaft 36 ft. Two bores 5½ ft. apart from same shaft; 2nd bore to 250 ft.
6. Claverley. 135 yds. S. × 245 yds. E. of Claverley Church	Salop 59 N.E.	180	124	overflows (40 gal. per hour)	500 H	Drift 10, f^3 124	

	Name	Sheet	Depth			Yield	Strata	Remarks
7.	Sandford Hall, Claverley	Salop 59 S.E.	200+	305	57	1570 H	Drift 184, f³ 305	
8.	King's Barn, Farmcote, east of Bridgnorth	Salop 59 S.E.	300+	200	94	630 H (test)	Drift 4, f⁵ 200	
9.	Rectory, Glazeley, south-south-west of Bridgnorth	Salop 66 N.E.	390	108	20	750 H	Old Red Sandstone 108	Pumping at 750 gal. per hour lowers water level to 90 ft.
10.	Dimmingsdale Waterworks, Lower Penn. 165 yds. S. and 800 yds. W. of Inn at Lower Penn.	Staffs. 61 S.E.	313	694	—	3,000 H for 50½ hrs. (test)	Drift 82, f³ 496, f² 694	Trial bore
11.	Dimmingsdale Waterworks, Lower Penn. 750 yds. W. of Inn at Lower Penn	Staffs. 61 S.E.	324	1024	30—40	3,000,000 D	Drift 6, f³ 487, f² 844½, f¹ 1,024	No. 1 bore
12.	Dimmingsdale Waterworks, Lower Penn. 750 yds. W. of Inn at Lower Penn	Staffs. 61 S.E.	325	1025			Drift 6, f³ 494, f² 851, f¹ 1,025	No. 2 bore
13.	Wolverhampton Waterworks, Goldthorn Hill	Staffs. 62 S.W.	590	630	—		Drift 16½, Enville Beds 630	Abandoned
14.	Midland Counties Dairy Co. Ltd., Cobden Lane Wolverhampton	Staffs. 62 S.W.	500?	200	—		Drift 20—30, Enville Beds 200	Abandoned
15.	Bilston Waterworks, by Canal Bridge at the Bratch, Wombourn	Staffs. 66 N.E.	258	150	40	1,514,044 D	f³ 150	Yield measured after water had been kept down for 67 days for purpose of putting in tubing

Name	Six-inch map	Approx. height of surface above O.D. in ft.	Depth in ft.	Standing water-level in ft. below surface	Yield: gallons per hour (H) or day (D)	Strata passed through: depth in ft. to base of each formation except lowest	Remarks
16. Bilston Waterworks, by Canal Bridge at the Bratch, Wombourn	Staffs. 66 N.E.	250	641	17	—	f^3 316, f^2 641	Well to 140 ft. Later bore than No. 15 above
17. Earl of Dudley Sand and Gravel Mines, north-west of Swindon	Staffs. 66 N.E.	300+	300	65½	21,500 H	f^2 72, f^1 300	Pumping water level 77 ft.
18. South Staffordshire Waterworks. Hinksford Pumping Station; beside Inn at Hinksford	Staffs. 66 S.E.	203	279	12½	1,500,000 D (test)	Drift 5½, f^2 270, f^1 279	Well to 65 ft.
19. Heathbrook, near Kingswinford	Staffs. 67 S.W.	228	80	—	1,300 H	f^3 80	
20. South Staffordshire Waterworks. Ashwood Pumping Station	Staffs. 70 N.E.	205	607½	19		Drift 6½, f^3 106, f^2 374, f^1 607½	Well to 30 ft. South Staffordshire Waterworks have 6 boreholes in all at Ashwood Pumping Station
21. South Staffordshire Waterworks. Ashwood Pumping Station	Staffs. 70 N.E.	205	612	27	2,885,363 D	Similar to No. 20 above	Well to 30 ft.
22. South Staffordshire Waterworks. Ashwood Pumping Station	Staffs. 70 N.E.	205	429	—	—	f^3 99, f^2 307, f^1 429	Date 1892

APPENDIX I : WELLS AND BOREHOLES.   189

23. South Staffordshire Waterworks, Prestwood. 400 yds. S. and 800 yds. W. of Prestwood House	Staffs. 70 N.E.	176	626	—	—	f² 374, f¹ 626	
24. South Staffordshire Waterworks, Mill Lane, Kinver	Staffs. 70 S.E.	155	151	5	—	Alluvium 31, f¹ 151	
25. South Staffordshire Waterworks, Kinver. 220 yds. S. and 60 yds. W. of Lock at Kinver	Staffs. 70 S.E.	152	756	4	—	Alluvium 24, f¹ 756	Trial bore : later than No. 24
26. South Staffordshire Waterworks, Kinver, about 150 yds. N. by E. of No. 25	Staffs. 70 S.E.	151	760	Overflows	3,500 H (artesian flow)	Alluvium 14½, f¹ 760	No. 1 bore
27. South Staffordshire Waterworks, Kinver, about 150 yds. N. by E. of No. 25	Staffs. 70 S.E.	151	704	Overflows	3,500 H (artesian flow)	Alluvium 14, f¹ 704	No. 2 bore
28. Stourbridge Waterworks, The Tack, Wordsley. 140 yds. S. of Ashwood Farm	Staffs. 71 N.W.	225	200	27	—	f² 59½, f¹ 200	No. 1 bore. Well to 10 ft.
29. Stourbridge Waterworks, The Tack, Wordsley. 140 yds. S. of Ashwood Farm	Staffs. 71 N.W.	225	220	29¼	50,000 D	f² 60, f¹ 220	No. 2 bore. Well to 10 ft.
30. Stourbridge Union Workhouse : on Kingswinford - Stourbridge road	Staffs. 71 N.W.	300	316	—	—	f³ 111, f² 257½, separated by fault from Clent Beds 316	Old well to 111 ft.

Name	Six-inch map	Approx. height of surface above O.D. in ft.	Depth in ft.	Standing water-level in ft. below surface	Yield: gallons per hour (H) or day (D)	Strata passed through: depth in ft. to base of each formation except lowest	Remarks
31. Messrs. Marsh and Baxter's Bacon Factory, Hall St., Brierley Hill	Staffs. 71 N.W.	500	550	312	—	Etruria Marl 165+, Middle Coal Measures 550	Boring probably in or near a fault. Another bore entered broken ground with rush of water at 330 ft. and was abandoned
32. G.W.R. Goods Station, Stourbridge	Staffs. 71 S.W.	230	907	14	15,400 H	f^3 208, f^2 590, f^1 845, Clent Beds 907	Water level falls to 16 ft. after pumping
33. Messrs. Jones and Attwell, Stourbridge	Staffs. 71 S.W.	—	145	17	—	f^3 145	
34. Corporation Baths, Stourbridge	Staffs. 71 S.W.	—	255	26	5,400 H	f^3 255	Well to 10 ft.
35. North Worcester Brewery, Duke St., Stourbridge	Staffs. 71 S.W.	—	255	41	8,000 H	f^3 255	Well to 39 ft. Premises now owned by A. Harris
36. Messrs. Mark Palfrey & Co., Forward Works, Giles Hill, Stourbridge	Staffs. 71 S.W.	—	200	45	—	f^3 200	
37. Messrs. W. J. Turney & Co., Ltd., Mill St., Stourbridge	Staffs. 71 S.W.	—	210	6	9-10,000 H	f^3 210	Water level has risen about 6 ft.

APPENDIX I : WELLS AND BOREHOLES.

38. Messrs. Harper & Moores, Stourbridge	Staffs. 71 S.W.	—	48	—	—	Middle Coal Measures 48	
39. Messrs. Isaac Nash and Sons, Ltd., Wollaston Mills, Coalbournbrook	Staffs. 71 S.W.	255	203	—	15,000 D	Drift 2, f^3 and f^2 203	
40. Stourbridge Gas Works, 480 yds. S. and 125 yds. E. of Hospital at Amblecote	Staffs. 71 S.W.	231	215	23	6,000 H	f^3 215	Old bore. Well to 24 ft.
41. Stourbridge Gas Works, 480 yds. S. and 125 yds. E. of Hospital at Amblecote	Staffs. 71 S.W.	245	301	30	12,000 H	f^3 and f^2 301	New bore
42. Stourbridge Waterworks, Coalbournbrook, 650 yds. W.N.W. of Hospital at Amblecote	Staffs. 71 S.W.	218	501	16	1,500,000 D	f^3 $42\frac{1}{2}$, f^2 $409\frac{1}{2}$, f^1 501	
43. Stourbridge Waterworks, Millmeadow, Amblecote	Staffs. 71 S.W.	228	$201\frac{1}{2}$	$13\frac{1}{2}$	500,000 D	f^3 $79\frac{1}{2}$, f^2 $201\frac{1}{2}$	
44. Stourbridge Waterworks; Glebe Lane, Gigmill, Stourbridge, 290 yds. E. and 140 yds. N. of Glebe Farm	Staffs, 71 S.W.	250+	403	34	500,000 D	Drift (or downwash) 10, f^1 31, f^2 (or Clent Beds faulted in) 403	Bore never used. After a day's pumping water level fell to $81\frac{1}{2}$ ft. below surface
45. Messrs. Thos. Plant and Co., Steam Brewery, St. John's St., Netherton, near Dudley	Staffs. 71 N.E.	500+	300	210	Nil	Middle Coal Measures 300	Bore never used. Premises now belong to Ansells Brewery Ltd.

Name	Six-inch map	Approx. height of surface above O.D. in ft.	Depth in ft.	Standing water-level in ft. below surface	Yield: gallons per hour (H) or day (D)	Strata passed through: depth in ft. to base of each formation except lowest	Remarks
46. Messrs. King Bros., Fireclay and Brick Works, Cradley	Staffs. 71 S.E.	—	387	—	—	Etruria Marl 289, Middle Coal Measures 387	
47. Messrs. Eveson Bros. Works, Providence St., Lye	Staffs. 71 S.E.	300—310	314	—	1,000 H	Coal Measures 36, Downtonian 314	Well to 36 ft.
48. Messrs. James and Philip Pounds Galvanizing Works, Jackson St., Lye	Staffs. 71 S.E.	—	93	76	2,000 D	Middle Coal Measures 93?	No permanent lowering in the rest level of the water has taken place
49. Wolverley Waterworks, Cookley	Staffs. 74 N.E.	132	170	overflows 11½ ft. above surface	110,000 D	f^2 170	
50. Lea Castle, Wolverley	Staffs. 74 N.E.	200+	180	160	—	Drift 5-10, f^3 and f^2 180	
51. Winwood Heath, adjoining Romsley Sanatorium, near Great Farley Wood.	Worcs. 9 S.E.	850	231	198	15 H	f^2 129½, Clent Beds 231	"Romsley Hill" in 'Wells and Springs of Worcestershire,' p. 77.

APPENDIX II

DETAILS OF CERTAIN BOREHOLES

A.—SMESTOW, NEAR WOMBOURNE

Boring for coal 580 yds. N.W. of Heath House, Smestow, Swindon parish. Six-inch map Staffordshire 66 N.E. Lat. 52° 31′ 59″, Long. 2° 12′ 47″. Height above O.D. about 250 ft. Date 1912. By percussion to 1200 ft. (or lower). Section compiled and abridged by T. H. Whitehead from a section from Mr. W. B. Cleverley, communicated by Mr. W. W. King, a section communicated by Mr. J. Bloomer, and notes by W. Gibson from part of the cores on the site. Grouping of strata and notes by T. H. Whitehead (see also above pp. 39 and 132). Notes by A. J. Butler on the Silurian fossils in the Geological Department of Birmingham University.

	Thickness Ft. In.		Depth Ft. In.	
Fluvio-glacial Gravel—				
Soil, sand and gravel	5	0	5	0
Upper Mottled Sandstone—				
Sandstone, light red	53	0	58	0
Bunter Pebble Beds—				
Sandstone, red, rather coarse, with some layers of red marl and sand	279	0	337	0
Lower Mottled Sandstone (and Clent Group?)—				
Sandstone, red, finer grain, with some clays	589	0	926	0
Clent Group—				
Sandstone, red, coarser grain, with layers of red marl and clay	269	0	1195	0
Clent and Bowhills Groups—				
Sandstone, dark red, with dark red marls	72	0	1267	0
Bowhills Group—				
Sandstone, dark red, and dark red marl, spotted	199	0	1466	0
Bowhills and Keele Groups—				
Marl, very hard, red, with some pebbles	27	0	1493	0
Keele Group—				
Marl, red, grey and green, with layers of sandstone	47	0	1540	0
Marl, red, with small nodules	10	0	1550	0
Marl, purple, with some red clays	15	0	1565	0
Sandstone, red, with marls, gritty and spotted	21	0	1586	0
Marl, red, grey streaks and nodules, gritty	22	0	1608	0
Sandstone, green, very gritty	1	0	1609	0
Sandstone, red, with marl	21	0	1630	0

	Thickness Ft. In.	Depth Ft. In.
Marl, red and mottled, with sandstone	53 0	1683 0
Marl, red, sandy, with nodules and yellow bands...	5 0	1688 0
Marl, red and purple mottled, with sandstone	58 0	1746 0
Sandstone with nodules (*Note* 1)	4 0	1750 0
Marl, red and purple mottled, sandy	150 0	1900 0
Espley rock	1 0	1901 0
Marl, mottled	3 0	1904 0
Spirorbis **Limestone** ; black (seen by W. Gibson. Note 2)	?1 0	1905 0
Marl, red, mottled, sandy	30 0	1935 0
Sandstone, red and grey	10 0	1945 0
Marl, red, mottled, and sandstone	10 0	1955 0
Sandstone, red and grey	5 0	1960 0
Marl, red and grey, sandy, and sandstone	40 0	2000 0
Espley rock	11 0	2011 0
Marl, red and grey, and grey gritstone	4 0	2015 0
Binds, red, grey and purple, mottled, sandy	90 0	2105 0
Rock, red, sandy	9 0	2114 0
Binds, red and purple mottled, sandy, with grey gritstone	65 0	2179 0
Shale, grey (? ironstone nodules)	1 0	2180 0
Binds, grey and mottled	35 0	2215 0
Limestone ; grey, with ostracods (seen by W. Gibson)	?1 0	2216 0
Clod	19 0	2235 0
Binds, red and mottled	9 0	2244 0
Halesowen Group—		
Clod, dark	2 0	2246 0
Shale, black	1 0	2247 0
Coal, with pins of ironstone (*Note* 3)	1 0	2248 0
Fireclay	1 0	2249 0
Clod and binds, grey, red, and purple mottled	61 0	2310 0
Conglomerate, sandy	5 0	2315 0
Binds, red and grey, and grey shale	30 0	2345 0
Conglomerate, mottled, and grey rock	20 0	2365 0
Binds, dark grey	5 0	2370 0
Coal (sediment)	6	2370 6
Clod	1 0	2371 6
Coal (sediment) (*Note* 4)	6	2372 0
Clod, bind and clunch, grey	33 0	2405 0
Sandstone, grey, very fine grained	10 0	2415 0
Sandstone, mottled	20 0	2435 0
Binds, red mottled, sandy	10 0	2445 0
Sandstone, grey with coal markings	35 0	2480 0
Clod, grey and yellow, and grey clunch	10 0	2490 0
Rock, grey	15 0	2505 0
Fireclay	5 0	2510 0

APPENDIX II : BOREHOLE DETAILS. 195

	Thickness Ft. In.		Depth Ft. In.	
Clod and binds, grey	32	0	2542	0
Coal (sediment) (*Note* 5)		6	2542	6
Fireclay and dark grey clod	22	6	2565	0
Sandstone, grey, fine	5	0	2570	0
Clod	10	0	2580	0
Rock, grey	10	0	2590	0
Clod, with ironstone	5	0	2595	0
Sandstone, grey, with mica, fine to coarse (*Note* 6)	65	0	2660	0
Etruria Marl—				
Fireclay and clod	1	6	2661	6
Coal (sediment) (*Note* 7)		6	2662	0
Binds, mottled (Fine grained greenish grit-espley rock, 2 ft. with base at 2663 seen by W. Gibson)	3	0	2665	0
Binds, mottled and sandstone	25	0	2690	0
Rock, grey	10	0	2700	0
Clod and mottled binds	10	0	2710	0
Fault probable hereabouts (*Note* 8)				
' Middle ' Coal Measures—				
Clod and fireclay, dark	10	0	2720	0
Black bat	5	0	2725	0
Coal	7	0	2732	0
Fireclay with coal markings	3	0	2735	0
Fault hereabouts ?				
Binds, dark, with fossils (dark grey shale with *Lingula*, 1 ft., with base at 2743, seen by W. Gibson (*Note* 9)	15	0	2750	0
Fireclay and dark clod	20	0	2770	0
Rock and fireclay	20	0	2790	0
Rock, dark grey, very fine	5	0	2795	0
Rock, dark brown, coal markings	5	0	2800	0
Clod, dark	5	0	2805	0
Grit, very coarse (seen by W. Gibson)	9	0	2814	0
Silurian (*Wenlock*)—				
" Silurian shales and thin limestones, *Atrypa reticularis* in abundance, examined by Prof. Lapworth " (W. Gibson) (*Note* 10)	30	0	2844	0

Note 1. Possibly these nodules were of *Spirorbis* limestone, in which case their position would be comparable with that of the *Spirorbis* limestone at 775 ft. in the Claverley borehole, p. 85 (W. Gibson, *Trans. Inst. Min. Eng.*, vol. xlv, 1913, p. 36) and that at 414 ft. 6 in. in No. 1 shaft of Baggeridge Colliery (' Southern Part of South Staffordshire Coalfield,' *Mem. Geol. Surv.*, 1927, p. 133 and Vert. Sect. *Geol. Surv.*, Sheet 94, No. 1).

Note 2. The position of this *Spirorbis* limestone appears to correspond with that of the *Spirorbis* limestone nodules at 972 ft. in the Claverley borehole. p. 85 (W. Gibson, *loc. cit.*).

Note 3. *Cf.* a coal at 822 ft. 8 in. in No. 1 Shaft of Baggeridge Colliery and the coal of Uffmoor Wood, p. 63 (See Vert. Sect. *jam. cit.*).

Note 4. *Cf.* a coal at 968 ft. 6 in. in No. 1 Shaft of Baggeridge Colliery and the Wassel Grove Coal, p. 63 (' Southern Part of South Staffordshire Coalfield,' 1927, p. 118).

Note 5. *Cf.* a coal at 1211 ft. 9 in. in No. 1 Shaft of Baggeridge Colliery and the Halesowen Coal, p. 63 (' Southern Part of South Staffordshire Coalfield,' 1927, p. 118).

Note 6. These thick sandstones are evidently the Basal Sandstone of the Halesowen Group, (p. 63).

Note 7. *Cf.* a coal at 1259 ft. 9 in. in No. 1 Shaft of Baggeridge Colliery (' Southern Part of South Staffordshire Coalfield,' 1927, p. 86).

Note 8. This is presumed to be the Enville Fault (p. 132).

Note 9. In the abstract section compiled from his notes W. Gibson puts the 7-ft. coal below the shale with *Lingula* ; but he did not see the coal and gives no depth for it, and as he gives the depth of the shale as 2743 ft. it was probably below the coal. It may represent the Pennystone-Whitestone Marine Band of the South Staffordshire Coalfield, which is represented in Baggeridge Colliery by a *Lingula* band (p. 38), and the *Lingula* band at 2056 ft. 6 in. in the Claverley borehole (p. 40). If this is so, and if the 7-ft. coal represents the Thick Coal, the two horizons are closer together than is normal, possibly because the westward downthrow mentioned on p. 132 passes through the borehole between them.

Note 10. Samples of the cores from the Silurian beds are preserved in the Geological Museum of the University of Birmingham. By kind permission of Prof. L. J. Wills they have been examined by Mr. A. J. Butler, who reports as follows :—

The cores contain the following fossils :

Plectodonta transversalis (Davidson), var. *lata* (O. T. Jones). 2814 ft.

Leptaena rhomboidalis (Wilckens). 2818 ft., 2830 ft., 2840 ft.

Strophonella sp. 2814 ft.

? *Meristina sp.* 2844 ft.

? *Orbiculoidea sp.* 2840 ft.

Syringopora cf. *serpens* (Linnaeus). 2844 ft.

? Annelid jaw. 2822 ft.

The presence of *Plectodonta transversalis* var. *lata* shows that the cores are of Wenlock age ; a more exact determination of their position within the Wenlock Series is not possible from the fossils. Their lithology, however, indicates that they come from the Wenlock Shale, and not from the Wenlock Limestone. Most of the pieces of core consist of a rather fine-grained impure grey limestone, with flecks of pyrites. Such limestones are frequent as thin bands and nodules in the grey calcareous and micaceous mudstones which form the top 130 ft. of the Wenlock Shale as developed in the Birmingham area. The mudstones themselves are very like the specimen from 2840 ft.

It is probable that the cores come from this upper part of the Wenlock Shale. Although a similar lithology characterizes the beds which form the lowest 140 ft. of the Wenlock Shale immediately above the Barr Limestone, correlation with this horizon seems less likely on structural grounds, since the total thickness of the division in the country to the east is nearly 700 ft. (see A. J. Butler, *Geol. Mag.*, 1937, p. 248).

Dr. J. Pringle has seen the specimens and agrees with these conclusions.

APPENDIX II : BOREHOLE DETAILS.
B.—KINLET

Boring for coal 6 furlongs W.10 S. of Kinlet Colliery. Six-inch map, Shropshire 73 N.E. Lat. 52° 25′ 55″, long. 2° 24′ 10″. Height above O.D. about 200 ft. Date 1929. Communicated by the Highley Mining Co. Ltd. Description by R. W. Pocock from cores (see also p. 51).

	Thickness Ft. In.		Depth Ft. In.	
Boulder clay...	4	0	4	0
'Middle' Coal Measures (*Kinlet Group*)—				
Sandstone, grey	3	6	7	6
Fireclay	5	6	13	0
Sandstone, greenish-grey ...	2	0	15	0
Marl, mottled green, red, and yellow ...	28	0	43	0
Fireclay and shale, marly...	1	3	44	3
Grit, greenish-grey, in part mottled, coarse, slightly calcareous, small brownish concretions	18	3	62	6
Marl, mottled	9	6	72	0
Clunch, sandy and grey grit	6	0	78	0
Marl, mottled, sandy and micaceous	12	0	90	0
Sandstone, greenish-grey, coarse, gritty, with fragments of quartz, etc., up to ½ inch at base ...	3	6	93	6
Marl, red, and sandy clunch	4	0	97	6
Sandstone, greenish-grey, in part red mottled, medium-grained, micaceous	6	10	104	4
Marl, mottled, sandy, with soft red slickensided marl at top	2	6	106	10
Sandstone, mottled...	2	6	109	4
Sandstone, greenish-grey, fine to medium-grained, micaceous	2	8	112	0
Marl, mottled, sandy, with sandstone bands up to 5 in.	4	3	116	3
Sandstone, greenish-grey, in part mottled, fine-grained, micaceous	1	0	117	3
Sandstone, greenish-grey, fine to coarse, lower part calcareous, very coarse at base with green and red fragments and pink quartz	7	0	124	3
Sandstone, greenish, marly, micaceous ...	2	8	126	11
Sandstone, grey, with green pebbles up to 1 in. near base ...	7	10	134	9
Grit, grey, coarse, with pebbles up to 1 in. of quartz, limestone, and igneous rock	3	8	138	5
Fireclay, grey, and dark marly clunch	5	0	143	5
Marl, mottled red, green and yellow, micaceous ...	64	3	207	8
Sandstone, grey, shaly, with plant stems ...	5	0	212	8
Sandstone, greenish, fine to coarse, with plant stems, slightly calcareous	14	3	226	11
Marl, dark mottled, with rootlets...	5	0	231	11
Sandstone, shaly	4	6	236	5
Marine Band, dark shale, with *Lingula, Orbiculoidea, Productus,* obscure Lamellibranchs, etc.	5	9	242	2
Sandstone, greenish-grey, fine-grained, marly ...	1	8	243	10

	Thickness Ft. In.	Depth Ft. In.
Marl, with rootlets and lycopod megaspores	3 0	246 10
Shale, sandy	7 6	254 4
Shale, dark, sandy	1 10	256 2
Shale, mottled, fine-grained, *Neuropteris*, etc.	8 9	264 11
Sandstone, pale greenish-grey, fine to medium with a few marl lenses	8 6	273 5
Marl, red mottled, and sandy shale	18 7	292 0
Sandstone, greenish-grey, fine to medium	2 6	294 6
Shale, dark, with *Neuropteris, Cordaites, Lepidodendron, Sphenophyllum*, etc.	6	295 0
Clunch and sandy shale with plants	14 0	309 0
Coal	2	309 2
Fireclay, passing down into sandy clunch	3 6	312 8
Sandstone, mottled, marly	5 0	317 8
Sandstone, grey, fine to coarse, with plants	11 4	329 0
Marl, mottled, in part sandy	11 6	340 6
Shale, sandy, with plants	5 6	346 0
Shale, dark, *Neuropteris, Annularia*, etc.	6	346 6
Fireclay, strong, with *Stigmaria*	5 6	352 0
Marl, mottled, sandy	3 10	355 10
Shale, fine sandy, and marl, 2-in. shale breccia at base with coal fragments	7 4	363 2
Clunch, dark, black at base	2 3	365 5
Basalt, highly vesicular, greenish-grey, with veins and patches of clunchy shale breccia and fragments of ? tachylyte	3 2	368 7
Sandstone, dark, with green fragments and coaly films	1 3	369 10
Clunch, grey to dark grey or black, fine-grained, sandy, with a 3-in. layer of **Basalt** about 2-ft. down in contact with hard concretionary rock, possibly clunch indurated by contact; thin layers of quartz grit	6 11	376 9
Breccia of fine-grained sandy clunch with large included blocks of **Basalt** surrounded by quartz grit, angular fragments of sandstone, pebbles of quartz, shale and coal	7 1	383 10
Coal, with coaly roof shale	5	384 3
Fireclay, with rootlets and ironstone nodules	7 0	391 3
Coal	2 1	393 4
Dirt — Probably the **Brooch Coal** of Highley	7	393 11
Coal	1 2	395 1
Fireclay	1 5	396 6
Clunch, sandy	3 8	400 2
Sandstone, grey, medium-grained, very calcareous	2 10	403 0
Shale	2	403 2
Sandstone, grey, fine to medium-grained, coaly	1 10	405 0
Shale, slickensided	3 8	408 8
Sandstone, grey, medium to coarse, calcareous	1 0	409 8

APPENDIX II : BOREHOLE DETAILS.

	Thickness Ft. In.		Depth Ft. In.	
Basalt, greenish, with tongues of dark grey grit ...	2	4	412	0
Coal, unaltered, probably the **Half-yard** of Highley	1	7	413	7
Shale, dark, sandy and bright coaly films		7	414	2
Basalt, black shale film at top, thin coarse sandy layer in middle		5	414	7
Sandstone, pale grey to white, with carbonaceous streaks and patches	4	7	419	2
Sandstone, shaly, with carbonaceous layers, 6 in. of dark shale on 6 in. sandstone at base	2	5	421	7
Shale, dark, thin layers of ironstone, fish fragments, traces of *Carbonicola* ?, *Neuropteris*, etc.... ...	5	11	427	6
Coal, and shale, ground up, probably the **Four-foot** of Highley	2	5	429	11
Fireclay	1	8	431	7
Clunch, with rootlets, hard and part sandy with ironstone nodules, *Cordaites*, etc.	11	2	442	9
Shale, black, coaly, with rootlets and megaspores	2	0	444	9
Fireclay, grey	1	0	445	9
Clunch, dark shaly		2	445	11
Coal, probably the **Two-foot** of Highley (not seen by R.W.P.)		4	446	3
Clunch, sandy, ferruginous pellets in lower part ...	5	0	451	3
Marl, mottled purplish red and buff...	5	8	456	11
Clunch, hard, sandy, with plants, *Sigillaria*, etc. ...	1	10	458	9
Clunch, sandy, *Stigmaria* and indet. plant stems ...	1	2	459	11
Shale, black, coaly		4	460	3
Clunch and fireclay	3	7	463	10
Marl, mottled, and sandy clunch	6	6	470	4
Marl, red		8	471	0
Marl, mottled, sandy, with ferruginous pellets ...	3	0	474	0
Clunch, mottled, marly	1	8	475	8
Sandstone, marly, fine-grained	1	9	477	5
Sandstone, greenish-grey and mottled, fine to medium, with thin grit bed	5	6	482	11
Conglomerate, fine	1	3	484	2
Sandstone, shaly, large plant stems		3	484	5
Sandstone, greenish-grey, medium-grained ...	1	6	485	11
Shale, dark greenish-grey, slickensided		8	486	7
Shale, mottled, sandy, part marly	2	3	488	10
Sandstone, green, medium to coarse, faulted against fine shaly sandstone and a 3 in. shale with *Neuropteris*, *Calamites* and *Cordaites*	1	2	490	0
Shale, sandy		3	490	3
Marl, mottled, sandy		8	490	11
Marl, hard, sandy and micaceous	3	2	494	1
Shale, sandy, micaceous, with rootlets	2	8	496	9
Sandstone, greenish-grey	1	5	498	2
Sandstone, grey, shaly, coaly plant stems	1	9	499	11

	Thickness		Depth	
	Ft.	In.	Ft.	In.
Grit, coarse, false-bedded, ferruginous concretions, green fragments and plant stems	7	1	507	0
Shale, dark grey, sandy, indet. plants		4	507	4
Conglomerate, fine, with coaly plant stems	3	1	510	5
Coal (not seen by R.W.P.)		5	510	10
Clunch, light grey, sandy, ironstone nodules and *Stigmaria*	2	7	513	5
Sandstone, greenish, with few rootlets	3	0	516	5
Shale, mottled dark red and green, sandy	1	9	518	2
Shale, dark, with ironstone nodules, *Calamites* and *Neuropteris*	1	6	519	8
Broken ground		11	520	7
Marl, hard, greenish-grey, sandy, with green specks	2	6	523	1
Sandstone, grey to dark grey, shaly	5	9	528	10
Shale, dark grey, with *Calamites*, *Cordaites*, *Neuropteris*, large seeds and ironstone nodules	2	9	531	7
Coal		1	531	8
Fireclay, light coloured		8	532	4
Bat, dark, much slickensided		3	532	7
Clunch, sandy	2	0	534	7
Marl, mottled green and red, sandy	3	5	538	0
Marl, mottled	2	4	540	4
Clunch, in part sandy	4	0	544	4
Clunch, with ironstone nodules up to 5 in.	6	8	551	0
Marl, green and mottled	4	3	555	3
Marl, green, hard, sandy and gritty, with 6 in. ironstone at base	7	6	562	9
Sandstone, greenish	1	11	564	8
Marl, mottled, hard, in part sandy and micaceous	1	0	565	8
Marl, mainly reddish, mottled, soft	7	9	573	5
Sandstone, green to grey, medium to coarse, with coaly films	6	9	580	2
Conglomerate, fine		8	580	10
Marl, mottled		3	581	1
Sandstone, greenish, marly	8	6	589	7
Marl		4	589	11
Shale, grey, sandy, with few ironstone nodules	3	11	593	10
Shale, fine sandy, with ironstone nodules, *Cordaites*, *Neuropteris*, *Calamites*, *Pecopteris* and seeds	4	4	598	2
Shale, dark greenish-grey, with ironstone nodules and indet. plants	1	8	599	10
Coal, and bat		3	600	1
Fireclay, much slickensided		3	600	4
Broken ground, sandy clunch, ironstone nodules	1	5	601	9
Fireclay	2	2	603	11
Clunch, grey, fine to sandy, ironstone nodules	3	0	606	11
Marl, dark, with ironstone, in part sandy	3	8	610	7
Marl, mottled	9	0	619	7

	Thickness Ft. In.	Depth Ft. In.
Sandstone, pale greenish-grey, fine to medium ...	2 8	622 3
Marl, slightly mottled	1 7	623 10
Marl, dark greenish grey, sandy	3 6	627 4
Shale, greenish, fine-grained sandy, *Cordaites*, *Alethopteris*, seeds, with 4 in. ironstone band at base, horizontal slickensides, calcite	4 8	632 0
Clunch, greenish, sandy	1 4	633 4
Marl, greenish mottled, hard sandy, grey pellets ...	4 0	637 4
Marl, red with green streaks (like bole)	1 8	639 0
Marl, mottled, hard shaly, part fine-grained, sandy	6 0	645 0
Marl, mottled, shaly and sandy	3 0	648 0
Marl, mottled, part sandy and micaceous	21 4	669 4

APPENDIX III

LIST OF GEOLOGICAL SURVEY PHOTOGRAPHS IN SHEET 167

(TAKEN BY MR. J. RHODES)

Copies of these photographs are deposited for public reference in the Library of the Geological Survey, Exhibition Road, South Kensington, London. Prints and lantern slides can be supplied at a fixed tariff.

All numbers belong to Series A.

No.

1936.—Quarry, Tansley Hill, near Dudley. Basalt intruded into Old Hill Marls with espleys.

1952.—Brewin's Bridge, near Netherton. Coal Measure sandstone with conglomerate at base resting on Downtonian sandstones. (Plate IIIA).

1953.—Ditto. Temeside Beds on Downton Castle Sandstone.

1954.—Ditto. Temeside Beds on Downton Castle Sandstone truncated by a fault. Basalt intrusion in fault plane.

1955.—Doulton's Clay Hole, 100 yds. south of Brewin's Bridge. Section in Middle Coal Measures.

1956.—Ditto. Thick Coal with Ten Foot Measures above.

1957.—Canal Basin, west of Brewin's Bridge. Turn-table for emptying trucks of clay, from Doulton's Clay Hole, into barges.

1958.—The Hayes, Lye. Basal Coal Measure conglomerate resting unconformably upon and overstepping the Downton Castle Sandstone.

1959.—Sand Pit, Cot Lane, Kingswinford. Section in Glacial Sand.

1960.—Ditto. Glacial Sand, contorted.

1961.—Sand Pit, Wall Heath, Kingswinford. Glacial Sand and Gravel resting on Upper Mottled Sandstone.

1962.—Wren's Nest Hill, Dudley, south end. Fold in Wenlock Limestone.

1963.—Ditto. East side. Entrance to caverns in Wenlock Limestone.

1964.—Ditto. Caverns cut in Wenlock Limestone.

1965.—Ditto. Another view.

1966.—Ditto. Quarry, west side. Shaft head and dip slope of the Middle Limestone shales.

1967.—Ditto. Dip slope of the Wenlock Limestone Nodular Beds.

1968.—Upper Gornal, north-west of Dudley, Ruiton Sandpit. Stone crusher for crushing the Gornal Sandstone.

1969.—Ditto. Section in Gornal Sandstone, near base of Middle Coal Measures.

1970.—Lower Gornal, north-west of Dudley. Basalt showing spheroidal weathering.

1971.—Ditto. Another view.

1972.—Hurst Hill, near Sedgley. Wenlock Limestone, section in the Nodular Beds.

1973.—Beacon Hill, north of Dudley. Eastward scarp of Aymestry Limestone on Lower Ludlow shales.

APPENDIX III : LIST OF PHOTOGRAPHS. 203

1974.—Beacon Hill Quarry, north of Sedgley. Aymestry Limestone.
1975.—Escarpment east of Sedgley Hall. Westward scarp of the Aymestry Limestone on Lower Ludlow shales.
1976.—Ditto. Another view.
1977.—Baggeridge Wood, Sedgley. Old quarry in calcareous conglomerates of the Enville Beds.
1978.—Ditto. Block of calcareous conglomerate of Enville Beds.
1979.—Sandyfields Farm, Sedgley. Calcareous conglomerate of the Enville Beds.
1980.—Ditto. Nearer view.
1981.—Brick and Fireclay works, Blower's Green, Dudley. Brick Kilns (old type).
1982.—Ditto. Another view.
1983.—Ditto. Fire-house. Furnaces for gas production.
1984.—Ditto. Brick Kiln (new type), gas heated.
1985.—Ditto. Another view.
1986.—Ditto. Store shed, with stacks of glazed bricks and tiles.
2029.—Witley Colliery, Halesowen. Fossil tree-trunk in Halesowen Sandstone.
2030.—Clent Hills from Witley Colliery. Type of scenery produced by the Clent Breccia.
2194.—Road-cutting near Foresters Arms, Stourbridge. Lower Mottled Sandstone, false-bedded.
2195.—Ditto. Irregular junction of Bunter Pebble Beds and Lower Mottled Sandstone.
2196.—Ridge Sand-pit, Foresters Arms, Stourbridge. Lower Mottled Sandstone, showing large scale false-bedding.
2197.—Ditto. Pebbly lens, forming irregular base of Bunter Pebble Beds, resting on Lower Mottled Sandstone, false-bedded.
2198.—Ditto. Pockets of gravel at the junction of Bunter Pebble Beds with Lower Mottled Sandstone. (Plate VB).
2199.—Wollaston, Stourbridge. View from north end of New Wood, showing the escarpment of Bunter Pebble Beds breached by the River Stour.
2200.—Holloway End, Stourbridge. Sand pit in Upper Mottled Sandstone with a capping of pebbly drift.
2201.—Ditto. Pebbly drift with lenses of sand on Upper Mottled Sandstone.
2202.—View of Hodge Hill, Wychbury Hill and Clent Hills.
2203.—Hodge Hill, south of Wollescote. Escarpment of the highest sandstone of the Halesowen Group.
2204-5.—Wychbury Hill, south-east of Pedmore. North-east side showing steep slope of Clent Breccia.
2206.—Wollescote Dingle, south-west of Wollescote. Waterfall over hard sandstone bed in the Halesowen Sandstone.
2207.—Ditto. False-bedded sandstones in Halesowen Group.
2208.—Oldnall Colliery, Wollescote. Halesowen Sandstone in tramline cutting.
2209.—Ditto. Halesowen Sandstone, near base of the group.
2210.—Clent Hills, Clent. Topography of the Breccia outcrop with dry valley.

2211.—Ditto. Clent Breccia with dry valley in foreground. Looking over the Trias country to the west. (Plate VIA).

2212.—Clent Hills. Section in Clent Breccia.

2213.—Ditto. Another view.

2214.—Valley between Clent and Walton Hills.

2215-16.—Panoramic view of Clent Hills from Walton Hill.

2217.—Romsley Hill, from Walton Hill, Clent. Escarpment of the Clent Breccia.

2218.—Castle Hill, Dudley. Pericline of Wenlock Limestone. Rowley Basalt capping hills in distance.

2219.—Wren's Nest Hill, Dudley. Pericline of Wenlock Limestone. (Plate IIB).

6837.—Holly Bush Road, Bridgnorth. Lower Mottled Sandstone, false-bedded.

6838.—Ditto. (Plate VIIB).

6839.—Ditto.

6840.—Queen's Parlor, Bridgnorth. Base of Bunter Pebble Beds on Lower Mottled Sandstone.

6841.—Ditto. (Plate VIIA).

6842.—View across the Severn Valley at Bridgnorth, looking west.

6843.—Hermitage Hill, Bridgnorth. Base of Bunter Pebble Beds.

6844.—River Severn at Bridgnorth, looking upstream. (Plate I).

6845.—Ditto.

6846.—View across the River Severn at Bridgnorth, from High Town.

6847.—Valley of River Severn below Bridgnorth from Castle Hill.

6848.—Road-cutting, 1 mile north-east of Claverley. Basement beds of Lower Keuper Sandstone.

6849.—Abbot's Castle Hill. Escarpment of Bunter Pebble Beds.

6850.—Ditto. Another view.

6851.—View from Abbot's Castle Hill, Clee Hills in distance.

6852.—Abbot's Castle Hill. Basement beds of Bunter Pebble Beds on Lower Mottled Sandstone.

6853.—Gravel Pit south of Rhodes Farm, Astley Abbots. Glacial flood gravel.

6854.—View of Camp Hill, Quatford. Lower Mottled Sandstone.

6855.—Ditto.

6856.—Road-cutting south of Quatford. Lower Mottled Sandstone, false-bedded.

6857.—Ditto.

6858.—Old dam on Mor Brook, 300 yds. north-east of Astbury Hall, Eardington.

6859.—Valley of Mor Brook at Eardington Upper Forge.

6860.—Ditto.

6861.—Boulder Clay on Keele Beds at Eardington Mill, Valley of Mor Brook. (Plate IXA).

6862.—Eardington Iron Works on River Severn. River Cliff of Lower Mottled Sandstone.

6863.—Ditto.

6864.—Ditto. Pillar of Lower Mottled Sandstone showing current bedding.

APPENDIX III : LIST OF PHOTOGRAPHS.

6865.—View from near Morville. South end of Meadowley Hill, escarpment of *Psammosteus* Limestones on right and the northern end of the Deuxhill outlier of Coal Measures.

6866.—View looking south from near Morville to the outcrops of Coal Measures at ' The Hill ' and the northern end of the Deuxhill outlier.

6867.—The escarpment of Meadowley Hill from Bridgwalton Farm. Outcrop of the *Psammosteus* Limestones and overlying sandstones.

6868.—Escarpment of Meadowley Hill. (Plate IIA).

6869.—Valley of River Severn near Quatford.

6870.—Quarry north-west of Alveley Church, Keele Beds, Alveley grindstone and building-stone beds.

6871.—Ditto. (Plate VA).

6872.—Ditto.

6873.—Road-cutting near Ounsdale Bridge, Wombourn. Upper Mottled Sandstone showing false-bedding.

6874.—Road-cutting, Worfe Bridge on Bridgnorth-Shifnal road. Base of Bunter Pebble Beds on Lower Mottled Sandstone.

6875.—View of southern end of Deuxhill outlier. Coal Measures on Old Red Sandstone.

6876.—Old quarry, near Hawkswood Farm, Billingsley. Massive yellow sandstone, (? Brownstones).

6877-78.—Chelmarsh Ridge from Cape of Good Hope, Billingsley. Keele Beds on Highley Beds.

6879.—Bridgnorth Brickworks, Knowlsands. Keele Beds.

6880.—Cutting on new road near Kinlet Colliery. Sandstone on coal and fireclay, Highley Beds.

6881.—Old quarry south-east of Highley Colliery. Massive grey sandstone in Highley Beds.

6882.—Bunter Pebble Beds in river cliff east of Worfe Bridge.

6883.—Valley of River Worfe above Worfe Bridge.

6884.—Road-cutting in Bunter Pebble Beds, 1 mile east by south of Bridgnorth.

6885.—View looking down the Severn Valley showing Main Terrace and lower terraces. (Plate IXB).

6886.—Ditto.

6889.—View looking east from Trimpley Anticline. (Plate VIB).

6890.—Man Wood, Trimpley, and feature of *Psammosteus* Limestone.

6891.—Quarry at Bird's Green, Alveley. Enville Beds, Bowhills Group, calcareous conglomerate.

6892.—Ditto.

6893.—Holy Austin Rock, Kinver Edge. Crag of Lower Mottled Sandstone with rock house.

6894.—General view of Trias outcrop from Kinver Edge.

6895.—General view of Lower Mottled Sandstone outcrop from Kinver Edge.

6896.—Kinver Edge. Escarpment of Bunter Pebble Beds on Lower Mottled Sandstone.

6897.—Kinver Edge and Blakeshall Common.

6898.—Kinver Edge from Greyfield's Court, Kinver.

6899.—Distant view of South Staffordshire Coalfield from near Stourton Field, Kinver.

6900.—View over Ridge Hill towards the South Staffordshire Coalfield from near Stourton Field, Kinver.
6901.—View across valley of River Stour to Clent Hills from near Stourton Field, Kinver.
6902.—Kinver Edge from Enville Sheepwalks.
6903.—View across valley of River Stour to Clent Hills, from Enville Sheepwalks. (Plate IIIB).
6904.—View over the Meaton Upper Coal Measures outlier to Clee Hill and Titterstone.
6905.—Seckley Cliff, west bank of River Severn below Arley.
6906.—View across the Severn Valley near Arley.
6907.—View across River Severn to Eymore Wood and Shatterford.
6908.—View from near Baveneywood towards Clee Hill and Titterstone.
6909.—The western margin of Wyre Forest Coalfield at Baveneywood from near Wall Town.
6910.—View from Shutley, Stottesdon, towards Brown Clee Hill and the eastern end of the Clee Hill Carboniferous syncline.
6911.—Walton Quarry, south of Stottesdon, Upper Old Red Sandstone.
6913.—View from Bardley Court over Birchen Park, Kinlet, to Wyre Forest.
6914.—Raggits Quarry, Knowle Hill, Kinlet. Kinlet Basalt.
6915.—Ditto. (Plate VIIIB).
6916.—Mass House Quarry, Kinlet Park. Spheroidal weathering in basalt. (Plate VIIIA).
6917.—Scenery of Bunter Pebble Beds outcrop near Hill House, Wolverley.
6918.—Bunter Pebble Beds country near Bury Farm, Wolverley.
6919.—Ditto.

INDEX

Abbot's Castle Hill, 106, 110–1, 122, 129, 157, 169 ; boulders on dip slope of, 171 ; ridge parallel to, 174.
Acanthodian spines, 23.
Acheulian, 171
Aeolian origin, of Lower Mottled Sandstone, 106.
Agnathous Vertebrates, Relationships of, 31.
Agriculture, 178, 184–5.
Albynes (Hobbins) borehole, 66–8, 75 ; limestone in, 76 ; Silurian in, 20.
Aldersley gap, 174.
Alethopteris, 201.
ALLPORT, S., 134.
Allsbarn, 13.
Allscott, 158.
Alluvium, 2, 177.
Alne valley, 151.
Alveley 1, 82–4, 86, 88–9, 101, 129, 181–2 ; grindstone beds and building stones, 84–5, 129 ; shaft, 133 ; stone, 182.
—— grindstone beds, plant remains in, 88.
Amblecote, 84, 96, 101, 127, 161, 165, 181 ; borehole, 127, 184.
Ammanford Sheet, 7.
Anglaspis, 22, 24.
Annelid jaw, 196.
—— tubes, Claverley, 40.
Annularia, 198.
—— *stellata*, 52, 74.
Anthracoceras cf. *aegiranum*, 54
Anthracomya sp., 70.
Anthraconauta cf. *phillipsii*, 70.
Anthracopupa britannica, 83.
Anticline, 6, 11–2, 121 ; Enville and Bobbington, 124 ; Linley, 120 ; Longmynd-Wrekin, 118 ; Netherton, 123 ; Sedgley and Dudley, 120, 123 ; Trimpley, 124 ; Trimpley, Enville and Bobbington, 105.
ARBER, E. A. N., 33, 39, 59, 64, 66–7, 70–1, 94–5.
Arley, 61, 69, 79, 82, 86.
—— Castle, 79, 87, 182.
—— Mill, 79, 88.
—— Park, 79, 87 ; Fault, 54, 76, 87, 129.
—— Station, 54, 76, 165, 176.
—— Wood, 12, 56, 80, 138–9.
Ashwood, 161 ; boring, 183.
—— Farm, 111, 162.
—— Lodge, 152.

Ashwoodfield House, 152.
Astbury Hall, 90, 110.
Asterophyllites equisetiformis, 54.
Asterotheca oreopteridia, 74.
Astley, 85, 99, 104 ; outlier of breccia, 94.
—— Abbots, 84, 89, 159, 180 ; Albynes borehole near, 66.
Astley Court Limestone, in Warwickshire Coalfield, 99.
Aston, 102, 163, 181.
Atrypa reticularis, 14–5, 195.
Auchenaspis band, of Ledbury, 17.
—— Sandstone, of Ledbury district, 22.
Audnam, 110–1, 162.
Austcliff, 113.
Aviculopecten cf. *gentilis*, 42.
Avonian Series, S. Staffs., 92.
Axborough Farm, 150.
Aymestry Limestone, 4, 9, 10, 15, 17, 20 ; of Sedgley, 14 ; of Shropshire, 9.

Baggeridge, 85 ; Halesowen beds of, 65.
—— Colliery, 10, 38–9, 61, 82, 127, 133 ; Halesowen group in shafts of, 63 ; *Spirorbis* limestone at, 85.
—— Wood, 84, 96–7.
Bagginswood, 44–5, 50, 73 ; ganister at, 44.
Ballstone, of Wenlock Edge, 14.
Bank, The, 103.
Bank Farm, 52, 66–7, 69, 70, 76, 178, 181.
—————— Coal, 61, 66–9, 71–3, 75–81, 180 ; sandstone above, 74, 79.
—————— *Spirorbis* Limestone, 67, 69, 71–2, 76, 79, 80 ; burnt for lime, 181 ; fauna of, 69.
Bannering Cottage, 25.
Bannut Tree Farm, 53, 77.
Barchan dunes, 106.
Bardley Court, 24, 50, 143.
Barium sulphate, 110.
Barnsley, 115, 158–9.
Barnt Green rock, 92.
BARROIS, C., 7.
Barrow col, 159.
Barrow Hill, Pensnett, 137.
Basalt, 4, 51, 131, 144, 182, 198–9 ; Claverley, 40 ; Clee Hills, 134 ; Highley, 142 ; Kinlet, 50, 142 ; in Kinlet boring, 51 ; in Shatterford

Pit, 55 ; outcrop near Harcott Pit, 50 ; vesicular, 51.
Basalt breccia, in Kinlet boring, 51.
—— quarries, Witnells End, 56.
Basement Beds, of Lower Keuper Sandstone, 109.
Bastard cornstone, at base of Coal Measures, 75.
Bat (carbonaceous shale), 36.
Batemans Dingle, 79.
Batty Coal, 47.
Baveney, 178.
Baveneywood, 44–5, 48, 50 ; old workings, 51.
Baxter's Monument, 110.
Baynham's Cottage, 87–8.
Bayton, 33, 66, 69, 70, 78–9 ; area, 178.
Beacon Hill, 15.
Bearmore (Cradley Heath), 60.
Bearnett House, 116, 156 ; level of Thick Coal near, 131.
Bellevue, Wordesley, 151–2, 173.
Bellerophontid, 51.
Bellman's Cross, 79, 86,
——————— Inn, 55–6, 79.
Bells Mill, 162 ; gap, 173, 177.
Bench Coal, 78.
Beobridge, 98, 158.
Billingsley, 44–5, 48, 50, 69, 73 ; blast furnaces, 179 ; brick and tile clays, 180 ; Church, 48 ; Colliery, 44, 49, 68, 70, 73, 131 ; Engine Pit, 73 ; Fault, 130–1 ; Hall Farm, 48, 73 ; ironstones, 179 ; section of coals, 47.
Bilston, 1, 10, 36 ; boulder clay near, 155 ; Corporation Waterworks, 114, 184, 187 ; water supply, 183–4.
Bind, The, 91.
—— Farm, ? Keele Beds at, 73 ; red clay at, 73.
—— Seam, 70, 72–5 ; of Eardington, 75 ; in Highley Shaft, 66 ; Main Sulphur of Cantrill, 72 ; Sulphur Coal, 68.
Bine Farm, 117.
Binnal, 89.
Birch Farms, 26, 29, 52, 71 ; section near, 27.
—— Hill Limestone, 12.
—— Wood, 13, 24–5 ; outlier of Coal Measures, 128.
Birchen Park, 51 ; coal in, 50.
Birchwood, 25, 128.
Birds Green, 99.
Birmingham, 151 ; district, 170 ; Map, 7 ; Syncline, 118, 124 ; Waterworks, 182 ; Waterworks trench, 29.
BISAT, W. S., 38.
Bison priscus, 162.
Black Brook, 20.

Black Country, 1.
—— Mountains, 24.
Blackband Group, 58.
Blackgroves Copse, 52–3.
Blackheath, 61.
Blackhills Plantation, 111, 128, 132, 172.
Blackstone Ironstone, 35.
Bladders Bank, 103.
Blakedown, 163, 166.
Blakenhall, 126.
Blakenhill, 155.
Blakeshall, 113, 153 ; Common, 110.
Bleeding rock, 80.
BLOOMER, J., 132, 193.
Bloomery (old furnace), 49.
Blue Flats Ironstone, 35.
Blythe valley, 151.
Bobbington, 92–3, 96, 101–2, 122, 157, 169–70.
Bodenham Farm, 25.
Bogs Farm, 63.
Bole, 201.
Bone beds, in Temeside Beds, 9, 20.
BONNEY, T. G., 93, 108.
Boreholes : Amblecote, 127 ; Borle Mill, 44, 50, 71 ; Hampton, in Keele Beds north of, 91, 101 ; Highgate Common, 103, 124 ; Rays Farm, 72 ; Schoolhouse Lane, 53 ; Tack, Wordesley, 111, 127 ; White Cross Schools, 102 ; Wordesley, 111, 127.
Borings, need for, 133.
Borle Brook, 2, 23, 48, 51, 70–75, 91, 130–1, 134, 176.
——————, mineral railway, 179.
Borle Mill, Boring, 44, 50, 71.
Borrowdale Volcanic Series, 151, 157.
Bothriolepis, 24.
Bottom Coal, 35–7 ; Brown Clee, 47.
——————— Holers, 35–6.
Boulder clay, 2, 6, 155 ; ' older ', 148.
Boulders : near Churchill, 151 ; in Enville village, 158 ; erratic, 156–7 ; granite, 158 ; of Scottish granite, 161 ; of Uriconian rocks, 158.
BOULTON, W. S., 33–4, 61, 82, 93–6, 135, 152, 162.
Bowhills (Bowells or Bowels), 95, 98–100, 104, 129 ; outlier of breccia, 94.
—— Beds, 121.
—— Dingle, 85, 99.
—— Group, 3, 94–101, 103–4, 107, 124, 129.
—— (Middle Permian), 98.
Bradley, 51, 77, 180.
Bragginslye Covert, 142.
Bratch, Wombourne, boreholes at, 184 ; Pumping Station, 114.
Breccia gravels, 2, 165–8.
—— or Clent Group, 3, 95, 97, 101–3 ;

lithology, 92 ; unconformity at base of, 93–4.
Brettel Lane, 10, 36, 97, 127, 179 ; Station, 17.
Brewin's Bridge, Netherton, 10, 17, 19; 136 ; canal section, 18.
Brick clay, 76, 102, 180 ; at Bank Farm, 180 ; in Highley Group, 180 ; in Kinlet Group, 180.
—— and tile clays of Coalport Group, 69.
—— Clay or Espley Group, 62.
Bricks, 179 ; common red, made at Seckley Wood, 179 ; from Middle Coal Measures clay, 180 ; from Silurian shale, 180 ; Staffordshire blue, 59.
Brick-making, 48.
Brickyard Cottages, 180.
Bridgnorth, 1–2, 20, 39, 44, 69, 81, 106, 110, 113, 120, 124, 134, 148, 154, 164, 175, 177 ; Brick Works, 90, 110, 180 ; High Town, 110 ; Station, 90 ; water supply, 183–4 ; Waterworks, 186.
Bridgwalton, 22 ; sands and gravels, 159.
Brierley Hill, 1, 10, 36, 62, 180 ; Trough, 123 ; water supply, 183.
Bristol Channel, drainage to, 169.
Brittle's Farm, 80, 86.
Broad Lanes, 84–5, 98.
—— Oak, 98.
Brock Hall, 74, 91, 130.
———— Coal, 65, 69, 71–4, 78–80.
———— (or Brockholes) Coal, 66.
———— (Sulphur Coal), 68.
———— Fault, 49, 50, 72–3, 90–1, 97, 100–1, 127, 129–31.
Brockmoor, 61–3 ; Fault, 123.
Bromley, 10, 36, 158.
—— Farm, 79, 87.
Bromsgrove, 151.
Brooch Coal, 35–7, 49, 59, 68 ; group at Shatterford Pit, 55 ; of Highley, 51, 57, 144, 198 ; of S. Staffs, 47.
—— Ironstone, 35.
Brooksmeeting Bridge, 49.
Broom, 109, 115 ; water supply, 184.
Broseley, 66–7, 69 ; limestone, 67, 70 ; Old Willey Furnace, 49.
Brown Clee, ganister at, 44 ; quartz in *Psammosteus* Limestones of, 26 ; section of coals at, 47.
———— Hill, 45.
Brown lime, 9.
Brownstones, 24.
BUCKLAND, W., 93.
Buckpool, 10.
Budleigh Salterton Pebble Bed, 108.

Building sand, 181.
—— Stone, 71, 109, 181–2 ; of Lower Keuper Sandstone, 182 ; quarries, 182 ; wharf on Severn, 182.
Bulwardine, 39.
Bumble Hole, 136.
Bunker's Hill, 106, 108, 111, 128,˙169.
Bunter Pebble Beds, 104, 106, 108–9, 111–12, 127–8, 152 ; basement bed of, 106 ; debris from, 165 ; escarpment, 150 ; gravel from, 181 ; origin of, 107.
—— Sandstone, 152 ; heavy minerals of, 107–8.
—— Series, 3, 6, 57, 105, 123, 127.
Burcote, 113, 177 ; House, 113, 158, 167 ; Mill, 113, 176.
Burf Castle, 113.
Bush Wood, 50.
Bushbury Fault, 127.
BUTLER, A. J., 8, 14, 193, 196.
Butter Cross, 130.
Butterton dyke, 134.
Buttonbridge, 45, 52–3, 67, 77, 130.
Button Oak, 77.
Butts quarry, tracks of vertebrates and arthropods in, 83.
Butts, The, 87, 182.
Bythocypris of. *siliqua*, 18.

———

Calamites, 143, 199, 200.
—— *sp.*, 55–6.
Calcareous Conglomerate Group, 3, 33, 92–7, 99 ; lithology, 91 ; overstepped by Keele Group, 83.
Calcareous tufa, 71, 88, 116 ; in Cantern Brook, 90 ; in Londonderry Coppice, 91 ; near Hempton Mill, 91 ; in Spadeley Rough, 90.
Calcot Hill, 181.
Caldecote rock, 92.
Caledonian folding, 121.
Caledonoid, 121 ; direction, 118.
Camarotoechia nucula, 9, 15, 18.
—— *pleurodon*, 151.
Cambrian Quartzite, 107.
Caninia, 99.
Cannock Chase, 41, 123.
Cantern Brook, 89, 155 ; fault at waterfall in, 90.
CANTRILL, T. C., 39, 40, 44–5, 54, 57, 65–7, 70–1, 76–80, 82, 86–8, 102–3, 121, 124, 129, 138, 181.
Cape of Good Hope inn, 49, 73, 131.
Carbonicola ?, 199.
Carboniferous, 32–4, 58.
—— Limestone, 24 ; pebbles, 100, 112–3 ; pebbles in Enville Conglomerates, 92.

Carbonita fabulina, 70.
—— *rankiniana*, 70.
—— *secans*, 70.
Careless Green, 20.
Castle, Gatacre, 98.
—— Hill, 8–9, 25, 56, 81, 103, 123 ; coal pits at, 79 ; pericline of, 14.
Cat brain, 109, 115–6.
Catherton Common, ganister at, 44.
Catholic Lane, 15.
Catsley, 51.
Caunsall, 163, 175.
Cefn-y-Fedw Sandstone, 151.
Cement, 180.
Cephalaspidae, The, 31.
Cephalaspis fletti, 11.
—— *sp.*, 22.
—— Sandstone, 7, 11, 23, 29, 31 ; sandstone-cornstone, 13, 26, 29.
Chalcot Furnace (Wricton Forge ?), Cleobury North, 49.
Charcoal burning, 185.
Charles Marine Band, S. Staffs, 40–1, 51, 59.
Charnoid, 121 ; direction, 118.
Checkhill Farm, 112 ; gap, 166, 169, 172.
Chelmarsh, 89, 90 ; Church, 90 ; Hall, 90 ; Ridge, 73–4, 178 ; Smithy, 90.
Cherry Orchard Farm, 98.
—— orchards, 185.
Cheshire or Prees Syncline, 118.
Chesterton, 115–6.
Chonetes flags, 9.
—— *minimus*, 15.
—— *sp.*, 51.
—— *striatellus*, 9, 15, 18.
Chonetoidea (Aegirina) cf. *grayi*, 15.
Chorley, 44–5, 48–9, 131 ; Covert, 49 ; Hall, 24 ; ironstone, 179 ; shafts at, 49.
Church Gorse, 102, 166, 180.
Churchill, 114, 150, 152, 163, 166, 170–1 ; boulders near, 151 ; Brook, 163 ; gravels, 151, 170 ; water supply, 184.
Cinder Hill, 14.
Clapgate, 156.
Clatterbach valley, 166.
Claverley, 1, 40, 43, 45, 65, 102, 114–6, 158, 163, 170, 186.
—— boring, 39, 41, 44–5, 51, 61, 64–6, 82, 120, 124, 133–4, 144 ; Basal Sandstone in, 65 ; Bowhills Group in, 98 ; Clent Group in, 102 ; Etruria Marl in, 62 ; fossil plants of, 84 ; igneous rocks in, 134, 147 ; Keele Goup in, 85 ; marine band in, 42 ; Silurian in, 8 ; variegated marls in, 40.

Claverley Brook, 174.
Clay-ironstone, 32.
—— pit, Messrs. Doulton's, 18.
Clee Hill, 32, 45, 51, 120, 126 ; basalt, age of, 135 ; ganister at, 44 ; Syncline, 120, 122.
—— Hills, 2, 8, 11, 126, 146, 171 ; basalt of, 48 ; road metal of, 182.
Cleiothyridina royssii, 99.
Clent, 95, 106–7, 111, 115, 122, 127 ; water supply, 184.
—— Beds, 96, 105, 111, 121–2, 127–8 ; age of, 96 ; of Boulton, 95.
—— Breccia, 33, 83, 92, 96, 101, 107, 111, 148 ; debris from, 165–6.
—— breccias, analagous to piedmont deposits, 93.
—— Group, 3, 102–3 ; of Arber, 95 ; brick clay in, 180.
—— Grove, 165.
—— Hill, 165–6.
—— Hills, 2, 6, 33, 92, 96–7, 101, 106, 111, 148, 165 ; breccia fragments from, 160.
Cleobury Mortimer, 44, 126.
CLEVERLEY, W. B., 193.
Cliff Coppice, 154.
—— Wood, 54.
Coal, 178 ; ' peacock ', 80 ; ' sulphur ', 66.
Coalbournbrook, 10, 17.
Coalbrookdale, 43, 118, 120, 126 ; Coalfield, 42, 69, 121, 126 ; Coalfield plateau, 169–70 ; pig iron from, 179.
Coalfield, ' Old,' 37.
Coalfields, extensions of, 118.
Coal Measures, 3, 24, 29, 56–7, 74, 124, 129 ; basal, 20 ; base of, at Schoolhouse Lane, 53 ; basement conglomerate, 19 ; below basalt, 56 ; faulted outlier of, 57 ; floor of, 10 ; Lower, 32 ; lowest beds of, 49 ; Middle, 4, 32 ; Middle, S. Staffs., 3 ; Middle, Wyre Forest, 3 ; ' Middle ' (Yorkian), 120 ; outliers, 44 ; outliers on Old Red Sandstone, 48 ; Productive, 32 ; southward attenuation of, 58 ; Upper, 6, 32, 58, 120 ; Upper, S. Staffs., 3 ; Upper, Wyre Forest, 3 ; water from sandstones in, 183.
Coal Pit Slang, 71.
Coalport, 66.
—— Beds of Coalbrookdale, 121.
—— Group, 65, 69, 121.
Coal Seams ; aggregate thickness of workable, 32 ; deterioration of, 132 ; in Etruria Marl Group, 61 ; of Highley Group, 178 ; of Kinlet Group, 45, 178 ; of S. Staffs., 36 ; ' sulphur ' in Highley Group, 66.
Coats Farm, 23.

Coke (charcoal), for hop-drying, 178.
Coldridge Wood, 12, 24–5, 56, 139.
Coldwell Copse, 52.
Colemoregreen, 159.
Colton Hills, 111, 155, 171.
Comer Wood, 163.
Common Heath, Coppice of, 49.
Compton, 66, 79, 80, 84, 86–7, 103, 128, 166, 168–9 ; shaft and boring, 57 ; sinking, 57, 80.
—— Court Farm, 80, 97.
Conglomerate, at base of Lower Mottled Sandstone, Eardington, 106.
Conglomeratic base, of Trias, 90.
Conodont, 40.
Cookley, 108, 112–4, 175.
Coopers Bank, 137.
Copper staining, 90.
Corbyn's Hall, 10 ; Fault, 123.
Cordaicarpus cordai, 56.
Cordaites, 198–201.
—— *sp.*, 53.
Chorley Beds, of Boulton, 94.
—— Conglomerates, of Vernon, 94.
—— Group, of Eastwood, 94.
Cornbrook, section of coal, 47.
Cornstone, conglomeratic, of Glazeley, 23.
Cornstones, 81.
Corporation Baths, Stourbridge, 190.
Coseley, 10, 36 ; water supply, 183–4.
Cot Lane, 152.
Coton, 85.
—— Farm, 98, 104.
—— Hall, 99, 104 ; outlier of breccia, 94.
Cotwall End, 15, 17, 126.
Covert Lane Bridge, 74.
Cox Green, 97.
Cox, L. R., 83.
Cradley, 1, 36–7, 61–3, 177, 180 ; water supply, 183.
Crateford, 164.
Criddon Bridge, 22.
Crinanite, 146.
Crinoid ; columnals, Kinlet Colliery, 42 ; fragments, 99.
Croft, 22.
Crog-balls, 14.
CROOKALL, R., 39, 52, 57, 71.
CROSFIELD, M.C., 14.
Cross Houses, 159.
——, The, 117.
CROSSKEY, H. W., 151.
Crosslanehead, 89.
Crunnell Brook, 23, 48, 74.
Cypris, 75.
Cyrtia exporrecta, 15.

*D*adoxylon *kayi*, 64.
Dalicote, 115, 117, 163.
Dalmanites sp., 14.
—— *vulgaris*, 15.
Danesford, 164, 176.
Danford Brook, 163.
Daniels Mill, 90.
Dark Drive, Kinlet, 143.
DAY, T. C., 136.
Dayia navicula, 9, 20.
—————— Beds, 9.
Dee estuary, drainage to, 169.
Deep Pit, Shatterford, 55, 140.
Deepfields, 10.
Deer, remains of, 162.
Dennis Park, 17.
Deuxhill, 23, 44, 48, 73, 75, 126 ; limestone of, 70.
—— Farm, 131.
—— Fault, 23, 131.
—— outlier, 74, 77, 131, 178 ; eastern boundary fault of, 75.
Devonian fossils, in Bunter pebbles, 108.
DEWAR, W., 24, 56–7, 71.
DEWEY, H., 174.
Diamond Ironstone, 35.
Didymaspis grindrodi, 17, 24.
Dimmingsdale boreholes, 184.
——————— Waterworks, 187.
Diplodus, 83.
Ditton Series, 4, 7, 8, 11, 23–4.
Dittonian, 7, 12 ; of King, 8.
DIX, E., 95.
DIXON, E. E. L., 32, 39, 54, 57, 118, 174.
Dodds Green, 99.
Dog kennels, Kinlet, 143 ; thin coal near, 50.
Doggetts Batch, 87–9.
Dolerite, 4, 182 ; Claverley, 40 ; intrusion, 20.
Dowles Valley, 44 ; borings in, 45.
Down Mill Bridge, 75.
Downton Castle Sandstone, 4, 7, 9–11, 17, 19–20, 137.
Downtonian, 7 ; base of, 11 ; beds, 127 ; of King, 8 ; rocks, 7, 10, 55–6 ; Trimpley, 12, 120 ; upper limit, 7.
Drakelow, 110, 112.
Drain pipes, 179.
Drainage channel, marginal, 173.
Draycott, 163, 181.
Drift, 2.
DUBOIS, G., 7.
Dudhill, 98.
Dudley, 1, 4, 10, 14, 36–7, 179, 182 ; water supply, 183.
—— Hills, 171.
—— (Wenlock) Limestone, 4, 8, 14 ; burnt for lime, 180.
Dudley-Sedgley ridge, 10.

Dudley Wood, 36.
Dudley Woodside, partings absent in Thick Coal at, 36.
Dudmaston, 154.
—— Hall, 100, 110, 155, 163–4, 177.
Dunsley, 112.
—— Bank, 150.
—— Hall, 112.
Dunval, 89.

Eardington, 89–90, 110, 134, 154, 159, 164 ; conglomerate at base of Lower Mottled Sandstone, 106 ; ironworks at, 179 ; Main Terrace gravel worked, 181 ; sands and gravels, 155, 164 ; tunnel at, 179.
—— Boring, 74.
—— Deep Pit, 44, 66–8, 74, 90.
—— Forge, 164, 176.
—— Mill, 74, 160, 164.
Earl of Dudley Sand and Gravel Mines, 188.
Earlswood, 127.
Earnwood, 71.
Eastham Sandstone, 13.
Eastham's Farm, 57.
EASTWOOD, T., 40, 59, 62, 94, 99, 111, 122.
Ebleys, coal worked in the, 49.
Echinoid spine, 39.
Elan Aqueduct, 54, 77, 182.
—— pipe-trench, 76.
Elephas primigenius (mammoth), 162.
ELLES, G. L., 9, 19–20.
Ellowes Hall, 17.
—— Park, 10, 17, 123.
En-glacial stream, 152.
Entomostraca, in limestone, Smestow boring, 85.
Enville, 1, 96–7, 101–2, 110, 166, 168, 172 ; knolls on racecourse, 169.
—— Beds, 1, 3, 32–4, 57, 84, 87, 91, 96–7, 101–2, 126–7 ; distribution of, 96 ; nomenclature and classification of, 94.
—— Brook, 166.
—— Common, 166, 168.
—— Estate, bricks for, 180.
—— Fault, 80, 97, 103, 112, 122, 124, 126, 128, 132–4, 172.
—— Group, 83 ; water from, 183.
—— Hall, 101.
—— Series, of Arber, 94–5.
—— Sheepwalks, 2, 6, 94, 97, 101, 122, 128, 148 ; breccia-fragments from, 160.
Equus caballus, 162.
Erratic pebbles, Lake District, North Welsh and Scottish, 150.

Erratics, distribution of, 171.
Escarpment, of Nash End sandstone, 88.
Escarpments, of Bunter Pebble Beds, 105, 110 ; of Lower Keuper Sandstone, 105.
Esker, 152.
Espley ; grits, 51, 54–5, 136 ; rocks, 32, 45, 47 ; rough rock, 61 ; sandstone, 49–53, 55–6.
Etruria Marl, 3, 33, 47, 59, 121, 123, 126–7, 136 ; basalt in, 137 ; blue bricks from, 180 ; espleys of, 64 ; in Coseley Syncline, 123 ; lowest part of, 6 ; on basalt, 135 ; red bricks from, 180 ; sections of, 60.
—— Marl facies, 41–2, 45.
—— Marl Group, 32–3, 41–2, 69 ; lower limit, 58 ; of North Staffordshire, 58 ; thickness of, 61.
—— (Old Hill) Marl Group, 58, 62.
Ettingshall, 37.
Eudon Burnell, 74.
—— George, 23, 74–5, 131.
—— Mill, 131.
Eupecopteris volkmanni, 57.
—— *sp.*, 74.
Eurypterid Sandstones, 13, 26.
Eurypterus, 19–20.
Eve Hill, 14.
EVESON BROS., Lye, 192.
Extensions of the Coalfields, 131–4.
Eymore Farm, Upper Arley, 42, 54–5, 78, 176.
—— Lane, 29.
—— Wood, 25, 29, 54–6, 77, 79, 138, 141, 165, 178.

Falcon, 166.
Farlow, 11, 24.
—— Sandstones, 8.
—— Series 4, 7.
Farmcote, 116, 158.
—— Hall, 117.
Fastings Coppice, 52.
Faults, 118.
Feiashill, 157.
FELLOWS, J. M., 78.
Fenestella cf. *membranacea*, 99.
Ferries, 1.
Field House, 166.
Fillets, 85.
Firebricks, at Seckley Wood, 179.
Fireclay ; New Mine, of Stourbridge, 39 ; Stourbridge Old Mine, 179 ;
—— Balls Ironstone, 35.
—— Coal, 35–6, 39, 138.
Fireclays, 37, 39, 179 ; S. Staffs., 32, 36 ; Tansley Hill, 38.

Fish, beds, 17, 24 ; fragments, Claverley, 40 ; remains, 76.
Five Foot Coal, 47, 49, 50, 53 ; of Highley, 51.
FLEET, W. F., 92–3, 107.
FLETT, J. S., 136.
Fluvio-glacial deposits, 160–5.
—— Gravels, 2.
Flying Reed, 36.
Folds, 118.
—— and Faults, 118–31.
Folly Point, 55, 128, 165, 178.
Footprints ; in Calcareous Conglomerate Group, 95 ; in Keele Group, 83, 87.
Forest of Wyre, 1.
—————— Coalfield, 32 ; floor of, 4.
Forestry Commission, 185.
Fossil wood, in Halesowen Group, 64.
Fossils ; of Highley Group, 69 ; of Keele Group, 83.
Four Ashes, 94, 97–8, 101, 158, 169.
Four Foot Coal, 41, 45, 47, 49, 50, 53 ; of Highley, 51, 199.
Fourth Terrace, 154.
Foxhills, 11, 156.
Foxholes, 74.
Foxyards, 10.
Franche, 113, 154, 175.
Furnace Grange, 156.

Gags Hill, 110, 113.
Ganister, 41, 44, 48–50 ; sandstone, 37 ; rock, 126.
Gastropods, 83.
Gatacre, 97–8, 101, 169 ; Park, 98.
Geological Sequence, 2.
Getting Rock Ironstone, 35.
Gibbet Wood, 116, 150, 153.
GIBSON, W., 39, 62, 65, 82, 85, 98, 102, 132, 193, 196.
Giggety, 161, 181.
Gilbert's Cross, 101.
Gills Rough, 57.
Gin Mine, of N. Staffs., 38.
Glacial deposits, 2, 6, 148–60 ; map of, 149 ; older, 150.
——History, 169–75.
—— lake, 158 ; in Mor Brook valley, 159.
Glazeley, 11, 23.
—— Bridge, 23, 73–4.
—— Church, 66, 131.
—— rectory, 187.
Goldthorn Hill, 97.
Goniatite Fauna, 38.
Goniophora cymbaeformis, 15.
Gornal, 36–7, 123, 179.

Gornal Sandstone, 181 ; used for building, 182.
Gornalwood, 36.
Gorten's Dingle, 88–9.
—— Mill, 88.
Grace Mary Colliery, basalt in No. 2 Shaft, 135.
—————— pit, 136.
Gradient profiles, of Breccia gravels, 167.
Grains Ironstone, 35.
Grammysia cf. *triangulata*, 20.
Granite, 'Criffel' type, 156–7 ; 'Eskdale' type, 157–8.
Granophyre, Ennerdale, 156–7.
Gravel, 181.
Graveyard, Gornal Wood, 137.
Great Coal, 45.
—— Farley Wood, 111.
Green Hall, 91.
—— House, 99.
Greyfields Court, 81, 103, 128.
Grindstone quarry, Alveley, 88.
Gubbin and Balls Ironstone, 35.
—— Ironstone, 35, 37, 137.
Guildings bone-bed, 29.
—— Farm, 29.
—— stream section, 29.
Gunhill Wood, 55.
G.W.R. Goods Station, Stourbridge, 190.
Gypidula galeata, 17.

Habberley Valley, 113.
HADEN, H. J., 38.
Hadleys, 87, 129.
Hagley, 1, 10, 83, 106, 111, 115, 123, 127 ; water supply, 184.
—— church, 83.
—— Hall, 1.
—— Hill, 84.
—— Park, 83, 93, 101, 122.
—— Station, 115.
—— Wood, 63, 161.
Halesowen, 61–3, 121.
—— Beds, 63–4 ; economic products of, 64.
—— Coal, 62–3.
—— Coal Group, 65.
—— Group, 3, 32–3, 45, 61, 64, 69, 77, 81, 132 ; Basal Sandstone of, 62–3, 65 ; base of at Smestow, 132 ; beds above Third Sandstone, 65 ; economic products of, 64 ; fossils of, 64 ; fossil wood in, 64 ; lithology of, 62 ; plant remains in, 64 ; sandstones at base of, 66 ; sandstones for building, 182 ; sandstones in upper part of, 63 ; Second Sandstone of,

H

62–3, 65 ; sequence, 63 ; in Smestow boring, 65 ; S. Staffs, 62, 65 ; thickness of, 64 ; Third Sandstone of, 63, 65 ; water from, 183.
Halesowen Sandstone Group, in Claverley Boring, 65.
—— Sandstones, 70.
—— type sandstone, 52.
Halfpenny Green, 158, 174.
Half Yard Coal, 41, 47, 49, 50, 53 ; of Highley, 51, 199.
Hall Barn, 13.
———————— Cornstone, 23 ; group, 31.
———————— Cornstones, 13, 29, 31.
Hall Close, 129, 176.
Hallclose, 87–9.
—— Coppice, 86–7, 180.
Hall Orchard, 23.
Hall's Barn, 31.
Halls Farm, 13, 31.
Ham Dingle, 63.
Hampton, 1, 67, 164.
—— Loade, 89, 129–30, 144, 160, 165, 176 ; ferry, 87, 130 ; ironworks at, 179 ; Station, 91.
—— Valley, 112, 168.
Hamstead, 95 ; Permian fossils at, 34.
—— Group, 94.
—— Quarries, 83.
Handsworth Station, 82.
Harcott, 45 ; section of coals, 47.
—— Pit, 50.
Harcourt Pit, 50, 131.
HARDAKER, W. H., 34, 83, 95.
HARMER, F. W., 169.
HARPER AND MOORES, Stourbridge, 191.
Harpsford, 67 ; limestone, 67.
—— Colliery, 74.
—— (Harpswood) Pit, 66, 68.
—— Shaft, 75.
Harpswood, 22, 74, 131, 159.
—— Bridge, 159.
HARRISON, W. J., 151.
Hartlebury, 115, 158.
—— Common, 153.
Hartsgreen, 85–6.
Haughton, 22, 159.
Hawkswood, 23.
—— Estate, boring on, 49.
Hay House, 176.
Hayes, near Lye, 8–10, 20 ; section at, 21.
Hayley Green, 63, 161.
Hazelwells, 89, 130.
Heath, The, 165.
—— Mill, 161, 181.
Heathbrook, 188.
Heathen Coals, 35–6, 179.
Heathton, 158, 174.
Heightington, 126.

Helium Method Time-scale, 135.
Hemicyclaspis (*Cephalaspis*) *murchisoni*, 17.
—— cf. *murchisoni*, 19.
Henley-in-Arden, 151.
Hercynian movements, 96.
Hermitage, The, 110, 113.
—— Farm, 113.
Herring Coal, 35.
Hexton's Farm, 87–8, 182.
High Green, 49–50, 73, 142.
—— Hobro Farm, 154, 163.
—— Rock, 110, 155.
—— Wood, 51, 143–4.
Highfields, 36.
Highgate Common, 106, 110, 128, 158, 166, 168, 172 ; borehole at, 103, 124 ; kame-like mounds of, 158.
———————— (or Forest) borehole, 133.
Highgate Farm, 103 ; kame-like ridges near, 174.
Highlands, The, 23.
Highley, 1, 6, 32, 41–2, 45, 49, 53, 57, 61, 71, 79–80, 84, 89, 175, 178 ; Halesowen beds of, 65 ; igneous rocks, 134 ; water supply, 91, 183.
—— Beds, 50.
—— Boring, 42, 44–5, 51.
—— Brooch Coal, 41–2, 44, 47, 49–50, 52–3, 57, 133, 142, 178 ; depth to, at Highley, 53 ; gob of, 53 ; Shatterford, 55.
—— Brooch Seam, southward deterioration of, 53.
—— Church, 91.
—— Colliery, 45, 72, 180 ; record of Shaft, 53.
—— Group, 3, 33, 42, 44, 48, 50–1, 53, 55, 57, 65, 67, 69, 71, 73–5, 77, 79–81, 86, 91, 129–30, 179 ; coals of, 178 ; Eardington Pit, 90 ; economic products, 71 ; Forest of Wyre, 65 ; lithology, 65 ; outlier of, 77 ; overstep by, 57 ; overstep of, 68 ; sandstones for building, 182 ; sandstones near base of, 66 ; sequence, 68 ; sulphur coals of, 6 ; thick sandstone near base of, 73, 76 ; unconformity at base, 68 ; upper limit, 69 ; water from, 183 ; west of Bridgnorth, 90.
—— Kinlet area, 41, 43 ; deterioration of coals southward from, 45 ; Fault, 52–3, 71–2, 91, 130.
—— MINING CO., 197.
—— Pit, 69.
Highley Shaft, 55, 68, 72 ; depth of, 53 ; hard light rock in, 67 ; section of coals, 47.
—— Station, 53, 72, 91, 129, 180.

Hightrees, 79.
—— Farm, 80, 86, 138.
Hill, The, 22–3, 74, 131.
—— End, 111, 129.
—— Farm, 31, 129.
Hillfield House, 86, 88.
Hillhouse Farm, 86.
Hilton, 115–117, 158, 176 ; boreholes, 184.
—— Brook, 116–17, 175.
Himley, 10, 36, 62, 106, 109, 161.
—— Hall, 1, 110.
—— Park, 84, 106 ; level of Thick Coal, 131.
—— Wood, 126–7.
Hinksford, 152–3, 161 ; boring, 183; gravels, 153.
Hippopotamus major, 162.
Hoards Park, 110, 169.
———————— Eardington gravels, 154, 171.
———————— Farm, 154.
———————— Gravel, 154.
Hob and Jack, 36.
Hoccum, 115.
Hodge Hill, 63.
———————— Fault, 83.
Holbeache, 13, 26, 29, 128 ; stream section, 28–9.
—— Brook, 161.
—— House, 127.
HOLLAND, L., 127.
Hollies, The, 97.
Holloway End, 114, 161.
Hollycott, 74.
Holoptychius, 24.
Holt, The, Quatt, 110, 155.
Holy Austin Rock, rock house in, 110.
—— Cross, Clent, 115.
Homer Hill, Cradley, 60.
Honey-comb sandstone, 115.
Hook Farm, 74.
Hop-drying, 178.
Hopstone, 116–7.
Horizontal movement, 130.
Horsford Brook, 131.
Horsleyhills Farm, 25, 103, 112.
HUGHES, H. W., 136.
HULL, E., 105, 116.
Huntsfield Cottage, 78.
—— Farm. 78.
Hurcott Wood, 128, 154.
Hurst Hill, 9, 123 ; pericline of, 14.
Hyde, The, 110, 166, 175.

Ice contact slope, 158.
—— front, retreat of, 154, 159.
—— -sheet, Irish Sea, 6 ; limit of, 6 ; retreat stages of, 173.

Igneous activity, Staffordian, 135.
—— rock, in Carboniferous, 48.
—— rocks, 4, 118, 134–7 ; alkaline suite of, 135 ; decomposed, in espleys, 61 ; in Carboniferous, 48 ; in S. Staffs. Coalfield, 135.
Illey Brook, 62.
———————— Limestone, 62–3.
Ingram Lane, 91.
———————— boring, 67, 73–4, 91 ; Highley Group in, 91 ; Keele Group in, 91 ; *Spirorbis* limestones in, 74.
Inliers, Silurian, in Coal Measures, 8.
Irish Sea, 151 ; glaciation, 170 ; ice, 155 ; ice-sheet, 148, 160, 170–1 ; ice-sheet, retreat stages of, 172.
Iron House, 168.
Ironbridge Gorge, 148, 150, 169–70, 174–5.
—— watershed, 171.
Ironstone, 178 ; cakes, Earnwood, 71 ; cakes of, 179.
Ironstones, 41, 45 ; above Thick Coal, 37 ; above Yard Coal, 49 ; below New Mine Coal, 37 ; in Chorley Covert, 49 ; of S. Staffs, 36.
Ischnacanthus, 24.
—— Zone, 12.
Island Pool, near Cookley, 114.
Iverley, near Stourbridge, 181.
—— House Farm, 128, 150.

Jacobs Ladder, 101, 103, 113.
JAMES AND PHILIP POUNDS, Lye, 192.
Jewstone Black Coal, 47.
JOHNSTON, M. S., 14.
JONES AND ATTWELL, Stourbridge, 190.
JONES, DANIEL, 40, 45, 47, 55, 57, 67, 78, 179.
JUKES, J. B., 38, 59, 61, 64, 103, 134 ; Red Coal Measure Clays of, 58.

Kame ; near Oreton, 171 ; terminal or recessional, 158.
Kame-like mounds, 157 ; of Highgate Common, 158.
Kames, 158.
Kates Hill, 2, 135.
KAY, H., 33, 58, 63–4.
Keele Beds., 32–3, 64, 73–5, 80, 91, 103, 110, 124, 126–7, 129.
—— Group, 3, 62, 69, 79, 81–4, 86–7, 89, 92, 95, 97–8, 101, 105, 110, 129 ; basal sandstone of, 86 ; base of, 81 ; building stone, 182 ; calcareous beds in, 181 ; Claverley, 85 ; clays of,

H 2

180 ; Clent Breccia on, 93 ; distribution of, 84 ; Eardington Pit, 90 ; flora of, 33 ; fossil plants of, 84 ; lithology, 81 ; S. Staffs. Coalfield, 84 ; thickness of, 84 ; water from, 183 ; west of Bridgnorth, 90 ; west of R. Severn, 89.
Keele Marl, in Knowlsands brick pit, 90.
—— marls, worked for bricks, Highley, 91.
Kenilworth area, 96.
—— Breccias, 94.
Keuper basement beds, near Ludstone, 117.
—— Marl, 6, 105, 109.
—— Sandstone, 124, 127, 133.
—— Series, 3.
Kidderminster, 1, 153 ; Terrace, 153–4, 163, 170–1.
KIDSTON, R., 33, 37, 39–40, 53–4, 57, 59, 64–5, 70–1, 84, 88.
KING BROS., Cradley, 192.
KING, W. WICKHAM, 7–13, 17, 19–20, 22–26, 29, 79, 82–3, 92–3, 95, 97, 99, 101–3, 111, 120, 123–4, 132, 136–7, 139, 193.
King's Barn, 187.
Kingsford, 97, 103, 154, 163.
Kingsnordley, 85, 97–8.
Kingswinford, 1, 36, 62, 96–7, 101, 106, 111, 127, 131, 180–1 ; Bunter near, 124 ; gravel ridge, 151–2, 172–3 ; sands and gravels, 153, 173.
Kinlet, 32, 41–2, 44–5, 48–50, 79–80, 180, 182 ; basalt of, 48, 144, 146 ; borings and workings near, 146 ; igneous rocks of, 134 ; marine band at, 42.
—— basalt, erosion of, 144 ; map of, 141.
—— Beds, 50, 52, 54, 138 ; at Highley, 120 ; at Shatterford, 120.
—— boring (1929), 41, 42, 44, 51 ; details of, 197–201 ; igneous rock in, 144, 147 ; section of coals, 47 ; section of part of, 145.
—— Church, quarry S.W. of, 144.
—— Colliery, 72 ; coals worked at, 52.
—— Group, 3, 6, 32–3, 40, 42, 44, 47–50, 52, 56–7, 68, 73–4, 131, 179 ; absent at Bayton and Mamble, 69 ; base of, 42 ; basement beds of, 44 ; coals of, 178 ; Compton, 57 ; grits in upper part of, 56 ; ironstone in, 179 ; sequence of, 41 ; Shatterford 55 ; thickness, 56 ; uppermost beds of, 6 ; west of R. Severn, 48.
—— Hall, 71, 143 ; thin coal near, 50.
—————— Fault, 50–1, 72, 91, 100, 129–131, 143–4.

Kinlet Park, 50–1, 71, 130.
—— Pit, 69 ; section of, 66.
—— Shaft, 44, 52, 55, 66, 68, 71–2.
Kinver, 1, 106, 110–11, 133, 153, 175, 181.
—— boring, 183.
—— church, 112.
—— Edge, 1, 107, 110, 112, 168.
Knowle Hill, 49, 142–4 ; thin coal at, 50.
Knowlesands, 110, 154, 159, 180 ; brick pit at, 90.
—— Gravel, 155.
Koninckophyllum ?, 99.

Lake deposits, of Morville Heath, 159.
—— District rocks, 151.
—— House, 88–9.
Lamellibranch, 51 ; Zones, 38.
Lamellibranchs, 197 ; non-marine, 37, 41.
Lanarkian, 39.
Langley Farm, 156 ; eskers near, 156.
LANKESTER, E. RAY, 31.
LAPWORTH, CHARLES, 62, 70, 127 ; Espley and Brick Clay Goup of, 58.
—— Club, 106.
Lays, outlier near the, 63.
Le Botwood, 75.
Lea Castle, Wolverley, 192.
—— Farm, 163.
Leasowes, 22.
Leaton Hall, kame-like ridges near, 174.
Ledbury Beds, 20.
—— fish band, 8–10, 17.
—— Group, 8–9, 12.
Ledopsis barroisi, 19.
LEES, E., 55.
Leigh House Farm, 101.
Leperditia, 25.
Lepidodendron, 198.
—— *aculeatum*, 53.
—— *sp.*, 54.
Leptaena rhomboidalis, 15, 196.
LEWIS, W. J., 9–10, 17, 19–20, 92, 136–7.
Lickey Hill, 122 ; quartzite of, 92.
Liévin, 7.
Lime, 71, 76, 180.
—— burning, 13.
Limekiln, remains of, 87, 88.
Limestone, near Ray's Farm, 73 ; of *Spirorbis* type, 155 ; with ostracods, 194.
—— balls, Shatterford Pit, 78.
—— nodules, in Worral's Brook, 79.
Limonite, 25.
Lingula, 195, 197 ; above Stinking Coal, 38 ; Claverley, 40 ; Smestow boring, 39.

Lingula band, Claverley, 42.
—— bed, in Claverley boring, 62.
—— *cornea*, 19-20.
—— *lewisi*, 19.
—— *minima*, 19.
—— *mytiloides*, 39, 42, 51, 54 ; Claverley, 40 ; Baggeridge, 38.
—— *sp. nov.*, 51.
Linley Anticline, 68, 120.
—— axis, 121.
—— Brook, 70, 173.
Linopteris obliqua, 74.
Lithostrotion (*Diphyphyllum*) *sp.*, 9.
Little or Two-Foot Coal, 35.
—— London, limestone at, 88.
—————— Brook, 87.
—————— Coppice, 87.
—————— Fields, 137.
Little Scotland, 74.
—— Welsh Glaciation, 175.
Little Wenlock basalt, 48 ; Lower Carboniferous age of, 135.
—— Woodlands, 53.
Littlegain, 181.
Littlegains Farm, 25–6, 128, 141.
Llandeilo, 7.
Llandovery sandstone, 92 ; fragments, 107 ; in espleys, 61 ; pebbles in Enville conglomerates, 92.
Lloyd House Fault, 127, 131 ; throw of, 131.
Lobby, The, 23.
Lodge Coppice, 52.
—— Farm, 19, 110, 164, 176.
—————— reservoir, 20.
Lonchopteris rugosa, 56.
Londonderry Coppice, spring in, 91 183.
Long Coppice, 25.
—— Cover, 100–1.
—— Mountain Syncline, 122.
Longmyndian conglomerates, 107.
—— rocks, 92, 101 ; pebbles of, 112.
Longmynd-Wrekin Anticline, 118.
Low Town, Bridgnorth, 164, 176.
Lowe Farm, 88.
Lower Barns Farm, 57, 112, 128.
—— Baveney, 23.
—— Beobridge, 98, 129, 163.
—— Birch Farm, 56.
—— Carboniferous fossils, in Calcareous Conglomerate Group, 99.
—— —— rocks, 4, 32, 120.
—— Clent, 165-6.
—— Coal Measures, Claverley, 39.
—— Danesford Terrace, 175–6.
—— Hagley, 115.
—— Harcourt, 50.
—— Heathen Coal, 36, 137.
—— Keuper Sandstone, 3, 6, 105, 108–9, 114–5, 117 ; Carboniferous limestone pebbles in, 117.
Lower Lias, fragments in drift, 156.
—— Limestone, of Wren's Nest, 14.
—— Ludlow Shales, 4, 9, 14, 20.
—— Mottled Sandstone, 3, 89, 103–113, 128 ; aeolian origin of, 106 ; conglomeratic base on Keele Beds, 110 ; in Knowlsands brick pit, 90 ; lithology of, 105 ; pot-holed surface of, 111, 164.
—— New Red Sandstone, of Murchison, 95.
—— Old Red Sandstone, 57, 129 ; Trimpley, 12.
—— Penn, 109, 114–5, 127, 155 ; Keuper Sandstone outlier of, 156 ; water supply of, 184.
—— Severn, Headwaters of, 169 ; pre-Glacial, 170.
Ludgbridge Brook, 10.
Ludlow Beds, inliers of, 11 ; of Linley Brook, 20.
—— Bone Bed, 4, 7–10, 17, 19–20, 22.
—— District, 9.
—— rocks, 10.
—— Series, 4, 7, 9.
Ludstone, 115, 158 ; Keuper basement beds near, 117.
—— Hall, 116.
Lusbridge (Ludgbridge) Brook, 20.
Lustre-mottling, 25, 31.
Lutley, 10.
—— Valley, 63.
Lydiates, 80–1, 103, 110.
Lye, 1, 20, 22 ; borehole at the, 182 ; water supply, 184.
—— Cemetry, 20.
—— Farm, 159.
—— Hall, 100–1.

MACKINTOSH, D., 156.
Mad Brook, 170.
Madeley, 69.
—— Heath, 151.
Magnesian Limestone Series, 95.
Main Coal, of Bayton and Mamble, 66–9, 78.
—— or Third terrace, 176.
—— Sulphur Coal, 57, 66–7, 70, 78 ; of Broseley and Coalbrookdale, 68 ; of Kinlet Shaft, 67 ; of Broseley, 67, 79–80.
—— Syncline, 123.
—— Terrace, 150, 152–5, 160, 162–5, 169, 174–5 ; feature of, 165 ; gradient-curve, 161 ; gravel of, 174–6 ; pebble

of, 165 ; of Severn, 148 ; of Smestow Brook, 166, 168 ; of Stour, 166.
Malvern, 122.
—— and Abberley axis, 121.
—— axis, 118.
Marnble, 33, 66, 69–70, 78–9, 126, 178.
Mammalian remains, 162.
Man Brook, 24–5 ; fish beds of, 25.
—— Wood, 25.
Marginal channel, 160.
Marine band, 42, 51, 197 ; at Kinlet, 41.
—— bands, 38, 41 ; of Claverley, 42 ; correlation of, 134.
—— bed, near Eymore Farm, 55.
——horizon, of Hamstead and Sandwell Park, 59.
—— horizons, in Transition or Upper Coal Measures, 59.
Mariopteris nervosa, 56.
—— *sp.*, 54, 74.
MARK PALFREY & CO., 190.
Marlbrook, 90.
MARSH AND BAXTER'S BACON FACTORY, 190.
MARSHALL, C. E., 48, 135, 137–8, 143.
MARTIN, F. W., 151, 156.
Mary Moors, 13, 26, 29, 31, 128 ; section near, 30.
Mass House, 142–4 ; thin coal near, 50.
——————— Quarry, 142.
MATHEWS, W., 64.
MATLEY, C. A., 108.
May House, 83, 87–9.
Meadowley Hill, escarpment of, 22.
Mealy Grey Coal, 35–6.
Meaton, 77.
Megalichthys sp., 51.
Mercian Highlands, 122.
Mere Farm, 158, 169–70.
——————— saddle, 174.
Meristina sp., 196.
Michelinia sp., 99.
—— *tenuisepta*, 99.
Microconchus carbonarius, 75.
Middle Coal Measures, 33, 38, 42, 45, 51, 59, 64, 126 ; basement beds of, 37 ; Coalbrookdale, 121 ; comparative sections of, 43 ; Kinlet Group, 40–57 ; Smestow, 39, 132 ; S. Staffs., 35–40, 51 ; thickness of, 35.
—— (Productive) Coal Measures, 41.
—— (Yorkian) Coal Measures, 126.
Middleton Scriven, 11, 23, 44 ; outliers near, 48.
MIDLAND COUNTIES DAIRY CO., 187.
Mill Pool, 100.
Millet-seed sand, in Lower Keuper Sandstone, 116.
—— sandstone, 106, 109–10.
Millstone Grit, 44.

Mineral Products, 178–9.
Mine-water, for industry, 183.
Mires Farm, coal crop near, 50.
MITCHELL, G. H., 38.
Mocktree Shales, 9, 20.
Modiolaris Zone, 38.
Modiolopsis cf. *complanata*, 29.
—— *complanata* var. *trimpleyensis*, 24.
—— *nilssoni*, 24.
—— cf. *nilssoni*, 19.
Molyneux Marine Band, of S. Derbyshire, 38.
Monograptus flemingii, 14.
Mons Hill, 10.
Moor Hall, 151.
—— House, 87–8, 165.
Mor Brook, 2, 22, 74–5, 90, 169 ; gorge of, 160 ; water power from, 179.
——————— valley, 164 ; boulder clay in, 160.
Morainic gravel, at Rhodes Farm, 159.
Morfe House Farm, 158, 173 ; kame near, 172.
Morfevalley, 117, 159, 174 ; gap at, 164.
—— Farm, 117.
Morville, 1, 20, 22, 148 ; gravels at, 173.
—— Heath, 159.
Mose, 158 ; boulder concentration at, 159, 173.
Moulding sand, 108, 114, 181 ; at Holloway End, 114.
Mount Pleasant, 152, 173.
Moxley, drift-filled channel of, 155.
Munster's Hill, 139.
MURCHISON, R. I., 7–8, 20, 40, 66, 73, 75, 93, 95, 156.

Nagersfield, 10.
Nanny's Rock, 110.
NASH, ISAAC, AND SONS, 191.
—— Elm, 79.
——————— Wood, 86, 88.
—— End, 86, 88–9.
——————— Sandstone, 88–9.
National Trust, 1.
Nautiloid Fauna, 38.
Ned's Garden, 49.
Nethercott, 23.
Netherend, 36.
Netherton, 9–10, 17, 72, 123, 130, 144 ; Tiphouse Fault near, 52.
—— Anticline, 10–11, 17, 123.
—— axis, 20, 61, 123.
Neuropteris, 52, 198–200.
—— *ovata*, 74.
—— *rarinervis*, 70.
—— cf. *rarinervis*, 74.
—— *scheuchzeri*, 74.

Neuropteris sp., 52, 54, 56.
New England, 72, 91, 130.
—— Mine Coal, 35-9, 179.
—————— Fireclay, No. 1, 35, 39; No. 2, 35; No. 3, 35.
———————— Ironstone, 35, 37-8.
———————— marine bed, S. Staffs., 40.
———————— and Pennystone marine band, 39.
———————— Rock, 37.
New Pool Quarry, 53; specimens from, 70.
New Series maps, 7.
New Wood, 127, 150.
Newcastle Group, 70.
Newcastle-under-Lyme Group of N. Staffs., 62.
Newhouse Farm, 180.
Newport, 124.
Newton, 113.
Nodular Beds, of Wren's Nest, 14.
Nomans Green, 69, 97, 101.
Non-sequences, 33.
Nordley, 20, 22.
North Welsh ice, 148.
—— Wood, 12-3, 57, 128.
—————— Worcester Brewery, 190.
Norton Covert, 111, 114; gravel pits in, 150.
Nortonsend Farm, 50-1, 71.

Oak bark, 185.
—— Farm, 10; (Shut End), 179.
Old Hill, 61.
—————— Marl, 33, 58-9.
—————————— Group, 32, 41-2, 61; lithological characters, 59.
Old Priory Farm, 14.
—— Red Sandstone, 4, 44, 49, 57, 68-9, 73-5, 126, 154; Deuxhill Fault in, 131; in Highley boring, 44; in gravel at Rhodes Farm, 159; in Town Mill boring, 45; at Schoolhouse Lane, 53; of Trimpley, 120.
—— Red Sandstone, Lower, 4; Middle, of Scotland, 4; Upper, 4, 8; used for building, 181; water from, 182.
—— Series maps, 7.
—— Swinford, 11, 96, 114.
—— Willey Furnace, Broseley, 49.
Oldbury, 38, 89-90, 110.
—— Church, 90.
—— Marl, 58.
Older drift, 170-1, 173.
Oldfield, 11, 22-3.
Oldhill, 37, 180.

Olivine basalt, Tansley Hill, 136.
—— basalts, 134.
—— dolerite, Claverley, 39.
—— dolerites, 134.
Onchus cf. *tenuistriatus*, 19.
—— *sp.*, 19.
—— spine, 17.
Opencast mining, 49.
Orbiculoidea, 197.
—— cf. *nitida*, 42, 51.
—— *rugata*, 18.
—— *sp.*, 196.
Ordovician, fossils in Bunter quartzite pebbles, 108; rocks of Berwyn Hills, 158; volcanic rocks of North Wales, 151.
Oreton, 109, 116, 156, 171; outliers near, 116.
Orthis (Bilobites) biloba, 15.
—— *(Dalmanella)* cf. *lunata*, 18.
—— *(Parmorthis) elegantula*, 15.
—— *(Skenidioides) lewisi*, 15.
Orthoceras cf. *bullatum*, 15.
—— *sp.*, 18.
Orthonota sp., 15.
Ostracods, 70.
Ostracoderm, 22, 53.
—— Fishes, 8.
Ounsdale, 114, 156.
Outlier, Primrose Hill, 127; Round Hill, 127.
Outliers, of Lower Keuper Sandstone, 109.
Ovalis Zone, 38.
Overend, Cradley, 179.
Overlap, 75; of Highley Group on to Old Red Sandstone, 126; within Highley Group, 68-9.
Overstep, 75, 93; of Clent Breccia by Bunter, 122; of Highley Group, 68-9; by Keele Group, 69.
Overwood, 11, 23.
—— Farm, 23.
Ox, 162.

*P*achysporangium, 19.
Pachytheca, 11, 19, 22-3.
—— *sp.*, 22.
Palaeoneilo cf. *laevirostris*, 40.
Palaeoniscid scale, 40.
Panpudding Hill, 164.
Park Attwood, 12-3, 25-6; amphitheatre, 13; cross-fault, 13, 25-6.
—— —— Fault, 128-9.
Park Farm, 80-1, 164.
—— Hill, 15.
Parkatt Wood, 128.
Parkfield, 10.

Patshull Fault, 114, 117, 129.
Pattingham Fault, 85, 89, 97–8, 114, 117, 124, 129–30, 134.
Pattingham-Kinlet Hall Fault, 130, 133.
Peat, 2.
Pebble Beds, 3, 110, 113.
—— Spreads, 168–9.
Pebbles, of basalt, 161 ; of Bunter Pebble Beds, 111–2 ; of Carboniferous limestone, 117 ; of Carboniferous limestone and chert, 107 ; of Coal Measure Sandstone, 161 ; of Longmyndian rocks, 107 ; of Main Terrace, 164 ; of Uriconian rocks, 107 ; of Welsh Border rocks, 155 ; wind-etched, 163 ; of Wrekin-Longmynd rocks, 155.
Pecopteris, 200.
Pedmore, 1, 62, 84, 96, 115, 150, 165–6.
—— Quarry, 115.
Pellet rock, 86, 88–9.
Pendlestone Rock, 110, 113.
Penecontemporaneous erosion, 69.
Penk valley, 174.
Penn, 1, 155.
—— Brickworks, 111.
—— Common, 101, 106–7, 111, 156, 171.
—— Fields, 155, 174.
Pennystone, 37.
—— Ironstone, 35 ; Coalbrookdale, 42 ; S. Staffs., 42.
—— marine band or bed, Coalbrookdale, 42 ; S. Staffs., 40.
—— New Mine Marine Band of Coalbrookdale, 38.
Pensax, 70.
Pensnett, 37, 61–2, 182.
—— Basin, 123.
Periclinal folds, 4.
Pericline, of Castle Hill, Dudley, 123 ; of Hurst Hill, 123 ; of Wren's Nest, 123.
Periclines, 9, 14.
Permian, 34, 82. 95.
—— coal, 103.
—— conglomerates, 103.
—— ice-age, of Ramsay, 93.
—— rocks, 33.
—— System, 34.
Pernopecten (*Syncyclonema*) *carboniferum*, 40.
Perryhouse Dingle, 85 ; limestone in Keele beds of, 99.
PHEMISTER, J., 137 ; on basalt, of Witnell's End, 139–40 ; on Barrow Hill basalt, 137 ; on Kinlet basalt, 144 ; on Tansley Hill basalt, 136.
Phialaspis, 23.
—— *symondsi*, 8.

Philley Brook, 102, 166.
Photographs, Survey, 202–6.
Physiography, 2.
Pickard's Farm, 87, 129.
Pidgeonhouse Farm, 57, 80, 103.
Pins Ironstone, 35.
Pipe amygdales, in Shatterford basalt, 138, 140.
Plant remains, above Brooch Coal, 37 ; of Ten Foot Measures, 37.
PLANT, THOMAS & Co., 191.
Platyschisma helicites bed, 19.
Platysomids, 70.
Plectodonta transversalis, var. *lata*, 196.
Pleistocene shells, in drift, 156.
Pleuroplax attheyi, 51.
—— *rankinei*, 51.
Pleurotomaria cf. *lloydi*, 15.
Plym Hall, 131.
Poacordites gentilis, 52.
POCOCK, R. W., 40, 48, 121, 134–6, 138, 142, 197.
Polyzoan, Kinlet Colliery, 42.
Pontesbury, 75.
Pool Hall, 156, 174.
POOLE, H. D., 38.
Poolhouse Farm, 100, 102.
Poor Robin Ironstone, 35.
Popehouse Farm, 86.
Portway Hall, Dudley, 42.
Postens Plain, 52–3.
Post-Avonian; folding, 120, 126 ; movements, 118.
—— -Clent Beds ; movements, 122.
—— -Coal Measures ; movement, 128.
—— -glacial ; deposits, 2, 175–7 ; pluvial phase, 166.
—— -Staffordian ; movements, 121.
—— -Triassic.; movements, 118, 122.
Pot-holed surface, of Lower Mottled Sandstone, 106.
Potseething Farm, 154.
—— Spring, 90.
Potters Loade Ferry, 176.
Pottery, 179 ; coarse, clays for, 37, 179 ; Shatterford, 56.
—— clays, 179.
Pound Green, 45, 52.
—— —— Common, 77.
Power House (Power Station) Terrace, 175–6.
Pre-Carboniferous ; Coal Measures on, 57 ; floor, 10 ; movement, 118, 128.
Pre-Coal Measures : denudation, 120 ; movement, 118, 120.
Prees Syncline, 122.
Pre-Glacial surface, form of, 169.
Pre-Halesowen movements, 64.
Prescott, 24.

Prestwood, 114, 162 ; boring, 108, 183.
—— House, 112.
—— Park, 153.
—— Pumping Station, 112.
Pre-Triassic : folds, 118 ; movements, 122.
Primrose Hill, 127.
PRINGLE, J., 14, 39–40, 88, 112, 151, 158, 196.
Priory, The, 14, 176.
Productive Coal Measures, 41, 58 ; west of Western Boundary Fault, 131.
—— Measures, 3, 6, 59, 61 ; Claverley, 39 ; S. Staffs., 33 ; Wyre Forest, 32.
Productus, 197 ; *P. sp.*, 39.
—— *carbonarius*, 51.
—— (*'Pustula'*) *piscariae*, 42.
—————— *rimberti*, 42, 51, 54.
PRUVOST, P., 7.
Psammosteus anglicus, 8.
—— Limestones, 8, 11–3, 22, 24–6, 29, 128 ; beds above, 26 ; burnt for lime, 180 ; springs at base of, 183.
Pteraspis, 11, 23 ; The Ostracoderm, 31.
—— *crouchii*, 31.
—— *leathensis*, 11, 22–3.
—— *rostrata*, 11.
—— *sp.*, 23.
Pterinea sp., 15.
Pterygotus, 19.
Punch Bowl Inn, 90.
Pyroclastic material, in esplevs, 61.

Quarry Bank, 61.
——, The, 115.
Quartz grains, in *Psammosteus* Limestones, 26.
Quartzite fragments, of Lickey (Cambrian) type, in espleys, 61.
Quatford, 110, 174, 176.
—— Ferry, 177.
Quatt, 89, 92, 97, 100–1, 110, 129–30, 155, 163–4 ; gravels, 174.
—— Farm, 89, 129.

Race, 81.
Racecourse Farm, 75.
Radstockian, of Kidston, 33.
Raggits quarry, 143 ; basalt in, 130.
RAMSAY, A. C., 93.
RAW, F., 83, 87, 135.
Rays Farm, 72–3.
Rea, River, 23.
Recent deposits, 2.
Red Beds, 47–8 ; above *Lingula* band, Claverley, 42 ; above marine band, Kinlet, 42 ; connected with igneous activity, 47 ; in Highley boring, 45 ; in Kinlet boring, 45, 51 ; of Kinlet Group, 51–3.
Red Clay Group (Etruria Marl), 71.
—— Coal-Measure Clays, 59.
—— Downtonian, 4, 8–11, 17, 19, 20, 22.
Redcliff, 168.
Refractory clays, 179.
Retreat stages, of Irish Sea ice-sheet, 172.
Rhabdoderma, 69.
Rhaetic, fragments in drift, 156.
Rhinoceros tichorhinus, 162.
Rhipidomella michelini, 51.
Rhodes Farm, 22, 159.
RHODES, J., 202.
Rhynchonella flags, 9.
Rhyolite, green, N. Welsh, 151.
RICHARDSON, L., 182–3.
Rider Coal, 49, 73 ; of Bayton and Mamble, 78.
Ridge Sand Mine, 110.
Ridgehill Wood, 106, 111, 169 ; escarpment, 172–3.
Ridgestone Rock, 113.
Rindleford, 176 ; borings at, 184.
—— Mill, 113, 163.
River terraces, 2 ; earlier, 153 ; higher, 2 ; later, 176.
Road metal, 182 ; possible reserves at Kinlet, 182.
ROBERTS, G. E., 55, 78, 88, 141, 179.
ROBERTSON, T., 7, 14, 22, 47–8, 68, 111, 114, 116, 120, 135, 144, 155–8, 163, 174.
Rock Houses, near Bridgnorth and Kinver Edge, 105.
Romsley, 97–8, 100, 104 ; near Halesowen, 94 ; Shropshire, 95.
—— Fault, 84–7, 97–9, 104, 129, 134, 142.
—— Group, of Arber, 94.
—— Hill, 148, 155.
Roof Coal, 78.
Rookery (Corbyn's Hall), 60.
Rough Hills White Ironstone, 35.
—— Park Wood, 80–1, 86.
—— rock, or espley, 61.
Roughton, 113, 115.
Round Hill, 22, 127, 159.
—————— Covert, 81.
—————— Pits, 127.
Roundabout Coppice, 113.
Rowley Hill, 171.
—— Regis, 133, 146 ; basalt of, 48, 136 ; basalt and dolerite of, 135 ; road metal of, 182.
Rubble Coal, 36.

Rudge, 114, 116, 129.
—— Heath, 158, 163.
Rumford Hill, 112.
Russell's Hall Fault, 37, 123, 137.

Saline Spring, at Saltwells, 182.
—— water from Downtonian, 182.
Salopian Permian, of Hull, 94.
Saltwells, 8–10.
—— Colliery, 10.
Samaropsis sp., 52.
Sand, 181 ; for pig-iron beds, 181.
—— and gravel, 155 ; glacial, 2, 6.
Sandford Hall, 187.
Sandstone, millet seed, 106 ; yellow, of Upper Old Red type, 24.
Sandyfields, 97.
Sanitary ware, 179.
Scaphaspis rectus, 31.
Schellwienella crenistria, 112.
Schoolhouse Lane, boring, 53.
Scot's Farm, 73.
SCOTT, A., 146.
Scottish rocks, 151.
Seckley, 48.
—— Cliff, 54–5.
Seckley Wood, 54 ; brick clays, 180 ; fireclays at, 179.
Section, Claverley to Smestow, 125.
Sedgley, 4, 8–10, 14–15, 17, 96, 120, 123, 126, 131, 148 ; bricks, 180 ; water supply, 183–4.
—— Hall Farm, 10.
—— (Aymestry) Limestone, 4, 9, 15 ; burnt for lime, 180.
—— Park, 171.
Seisdon, 124, 156–7, 161, 174 ; water supply, 183–4.
—— Mill, 157.
Semi-refractory clays, 179.
Senni Beds, 24.
Serpulites longissimus, 15, 18.
—— *sp.*, 51.
Seven Feet Banbury Marine Band, 38.
—— —— Marine Band of Warwickshire, 38.
Severn, River, 1, 4, 72, 77, 79, 81, 129–30, 164, 177 ; barge traffic on, 179 ; ' lower,' 150 ; rejuvenation of, 150 ; terraces of, 154 ; terrace gravels, 181.
—— basin, erosion in, 171.
—— Hall, 164, 176–7.
—— Lodge, 52–3, 71–2, 130, 165, 178.
—— Farm, 66, 76.
—— valley, 2, 110, 175–6.
Severnfield Cottages, 77.

Shatterford, 54–6, 66, 77–80, 84–5, 138, 140–1, 178, 182 ; barren Kinlet beds at, 134 ; brick clays, 180 ; fireclays at, 179 ; igneous rocks of, 134 ; pottery, 56 ; springs near, 183.
—— basalt, 56, 129, 138 ; in Eymore Wood, 55 ; section near Witnell's End, 139 ; southernmost exposure, 55 ; terminated by cross-fault, 80 ; ' dyke,' 138.
—— Deep Pit, 79 ; lower part of Highley Group in, 78.
—— sinking, 134 ; horizon of basalt in, 138.
Shavers End, 10.
Sheepwalks, 104.
SHERLOCK, R. L., 107.
Shifnal, 113.
Shipley, 114, 116, 174.
Shirlett Hill, ganister at, 44.
SHOTTON, F. W., 94, 96, 106, 122.
Shovel Bank, 116.
Shrewsbury, 175 ; map, 7.
Shropshire, 8, 11.
—— Farm, 88–9.
Shunesley, 23.
Shut End Fault, 123, 138.
Shutley, 10–1, 23, 61–2, 179–80.
Sidbury, 1, 11, 22.
—— Dingle, 48.
Sigillaria, 199.
—— *brardi* var. *denudata*, 88.
Silica sand, 181.
Silurian and Old Red Sandstone, 7–31.
——, in Smestow borehole, 39, 132, 196 ; of Dudley and Sedgley, 14 ; pebbles in Enville conglomerates, 92 ; upper limit, 7 ; water from, 182.
—— rocks, 4, 7, 123 ; beneath Coal Measures, 10 ; Claverley, 39 ; folding of, 120 ; Hayes, Lye, 20 ; in Albynes boring, 76.
—— (Downtonian) rocks, 68.
—— (Lower Ludlow) rocks, 126.
—— shales, Claverley, 39.
—— System, 7, 20.
Similis-Pulchra Zone, 38.
Singing Coal, 35.
Six Ashes, 98.
Skeets, 77.
—— Cottages, 76.
—— Farm, limestone at, 67.
SLATER, I. L., 9, 19–20.
Slickensided surfaces, 130.
Skickensides, 131 ; horizontal, 143.
Smestow, 65, 133, 161, 171, 181 ; Etruria Marl in boring at, 62 ; Silurian in boring at, 8.
—— borehole, 39, 65, 120, 124, 126, 132 ; details of, 193–6 ; Enville

Fault in, 128 ; Keele Group in, 84 ; notes on, 195–6 ; record of, 133.
Smestow Brook, 2, 112, 152–3, 156–7, 161–2, 166, 169, 174–5, 177 ; lower terrace of, 148.
—— —— valley, 172–4.
SMITH, STANLEY, 99.
Smith Coal, 45 ; Clee Hill, 178.
Solid formations, 3, 6.
South Staffordshire, 8, 33, 41–2, 66, 71 ; coal seams of, 178.
—— —— Coalfield, 32, 39, 42, 58, 118, 124, 155 ; basalt fragments from, 160 ; extensions of, 131 ; floor of, 4 ; igneous rocks of, 134 ; ironstones of, 178 ; marine bands in, 38 ; pre-Coal Measures folding in, 120 ; Silurian rocks in, 8 ; structure 123 ; uplift of, 122 ; western side of, 65.
—— —— Coal Measures, folding of, 122.
—— —— fireclays of, 179 ; northern part of, 43 ; southern part of, 43.
—— —— Waterworks, 112, 188–9.
—— —— —— Company, 183.
Southall Bank, 49.
—— Bank Brook, 49.
—— House, 50, 180.
Spadeley Rough, 90.
Sphenophyllum, 198.
—— *emarginatum*, 70, 74.
—— *sp*., 54.
Sphenopteris dilatata, 53.
Spirifer ?, 99.
—— *bisulcatus*, 151.
Spirorbis, 67, 76.
—— *pusillus*, 69, 83, 87.
—— Limestone, Bank Farm, 67.
—— limestone, 62, 67–9, 74–80, 82, 85–8, 99, 154, 194–5 ; of Bayton and Mamble, 67 ; blocks in Cantern Brook, 90 ; Claverley, 85 ; in Borle Brook, 73 ; in Calcareous Conglomerate Group, 99 ; in Cantern Brook, 89 ; in boring north of Hampton, 91 ; near Hempton Mill, 91 ; at Ingram Lane, 91 ; in Keele Beds, 130 ; in Keele Beds, S. Staffs., 82 ; at Smestow, 85 ; in Spadeley Rough, 90 ; of Tasley, 90.
—— limestones, 3, 65–6, 71 ; in Halesowen Group, 81 ; in Highley Group, 181 ; in Keele Group, 81 ; burnt for lime, 64, 180 ; for road metal, 180 ; of S. Staffs., 81–2.
Spittle Brook, 161, 166, 168.
Springs, at base of Halesowen sandstones, 64.
—— Mire, 137.

Stafford Syncline, 118, 122–3 ; Eastern limb, 123 ; Western limb, 124.
Staffordian, 40, 42 ; movement, 121, 126 ; times, 120.
—— Group, of Kidston, 33, 59.
—— Series, 53.
Staffordshire blue bricks, 59.
Stambermill, 10.
STAMP, L. D., 7.
Stanley, 23, 87–8, 165, 178 ; railway cutting, 72.
—— Hall, 110, 164, 176.
—— Seam, 73.
Stapenhill Farm, 127, 162.
—— Fault, 109–10, 114–6, 124, 127, 131, 133, 150 ; Coal Measures west of, 132 ; in Smestow borehole, 132.
Starts Green, 56, 81, 139.
Station Fault, 52, 54, 71–2, 76–7, 87, 129–30.
Stephanian, 33, 95.
Stewponey Hotel, 150, 153.
Stigmaria, 198–200.
Stinking Coal, 35–6, 38–9 ; *Lingula* above, 38.
—— Marine Band, of Leicestershire, 38.
STOBBS, J. T., 38 ; collection, 42.
Stottesdon, 1, 11, 23–4, 122.
Stour, Basin, 153.
—— River, 2, 110–1, 114, 127, 161, 163, 166, 173–5, 177 ; anomalous course of, 173 ; highest terrace of, 148 ; lower terrace of, 148 ; terrace gravels of, 181.
—— valley, 2, 153, 171, 175.
Stourbridge, 1, 2, 36–7, 95, 101, 106, 108–9, 114–5, 124, 127, 131–2, 179, 181 ; private boreholes in, 184 ; water supply, 183–4.
—— and District Water Board, 184.
—— gasworks, 161, 191.
—— Junction, 115 ; Station, 114.
—— Old Mine Fireclay, 6, 35, 39, 179.
—— Syncline, 109, 124.
—— Union Workhouse, 189.
—— Waterworks, 111, 189, 191.
Stourton, 110, 150, 153.
—— Field, 112.
—— Hall, 112.
Straits Farm borehole, S. Staffs., 85.
—— Green, 123.
——, The, 84.
Stratford Brook, 116, 177.
STRAW, S. H., 7.
Striated fragments, in trappoid breccia, 93.
Strophonella euglypha, 14–5.
—— *sp*., 196.
Structure, of Dudley and Bridgnorth area, 119.

STUBBLEFIELD, C. J., 14, 38, 42, 69, 99.
Sub-Brooch Marine Band, 40.
Sub-glacial stream, 152.
Sulphur Coal, 35–7, 39, 73.
—— (Brock Hall) Coal, 68.
—— (Stinking) Coal, 179.
—— coals, 6, 70.
—— Coal Group, 33, 40, 57, 65, 68, 178 ; plants of, 70.
—————— Series, 71.
Summerhill, 152, 173.
Sun cracks, 83.
Superficial deposits, 6 ; water from, 184.
—— Formations, 2.
Sutton, 73–4, 89–91.
—— Mill, 117.
Sweet Coal, Baveneywood, 51.
—————————— Group, 32–3, 40, 68, 178.
Swindon, 151–3, 173 ; water supply, 184.
—— church, 161.
Swynnerton dyke, 134–5.
Symon Fault, of Coalbrookdale, 69, 96, 126.
Syncline, 12, 45, 121 ; Birmingham, 118 ; Cheshire or Prees, 118 ; Clee Hill, 120 ; of Coseley, 123 ; Sedgley, 15, 123 ; Stafford, 118 ; Stourbridge, 115.
Syringopora cf. *serpens*, 196.
—— *sp.*, 99.

Tack, borehole at the, 111.
Taele, frozen sub-soil, 160, 166.
—— gravels, 168.
Tame, River, 2 ; source of, 2.
Tame-Avon watershed, 151.
Tanning, 185.
Tansley Hill, 61, 136–7, 179 ; shafts, 37–8.
Tasley, 75, 154, 159, 178, 180–1 ; limestone, 67.
Tectonic history, 118.
Tedstill, 74.
Temeside Beds, 10, 20 ; of Linley Brook, 20.
—— Group, 4, 7, 9.
—— shales, 4, 7, 9–11, 17, 19–20.
TEMPLEMAN, A., 51.
Ten Foot Ironstone, 35.
—— Yard Coal, 6.
Tertiary basalts of Scotland, 134–5.
Tettenhall-Aldersley gap, 177.
—— gap, 174.
Thalweg, of Stour, 161.
Thatchers Wood, 90.
Theralite family, 146.

Thick Coal, 6, 35–8, 40, 42, 127, 131–3, 138, 179 ; basalt in, 137 ; in Claverley borehole, 133 ; depth at Wassel Grove, 64 ; depth near Smestow, 132 ; deterioration southwards, 36 ; ironstones below the, 178 ; measures above, 38 ; measures below, 38 ; relation of dolerite to, 135 ; rise in, 64 ; sections of measures above, 60 ; Smestow boring, 39 ; split into Top, Middle and Bottom coals, 36.
—————— Rock, 37.
—— Sandstone, of Bayton and Mamble, 67, 69.
Third or Main Terrace, 160.
Three-quarter Coal, 47.
Thyestes (*Auchenaspis*) *egertoni*, 22.
—————— *sp.*, 22.
—————— Group, 8.
Tiddle Brook, 73.
Tile clays, 179.
Tiles, 179.
Tilestones, 7.
Tinker's Castle, 111.
Tiphouse, 52, 71.
—— Cottages, 51, 144.
—— Farm, 71–2.
—— Fault, 50–2, 71–2, 77, 91, 130, 144.
Tipton, 1, 36, 59, 61 ; water supply, 183.
—— Moat, 35.
Titterstone Clee, 45, 118.
—— Clee Hill Coalfield, 32, 39.
TOMLINSON, M. E., 151.
Top Coal, Coalbrookdale, 42.
Tournaisian sea, of the south-west province, 120.
Town Mill Boring, 45.
Transition Coal Measures, 59.
Trappoid breccia, 6, 93 ; origin of, 96.
Trescott, 156, 161, 174.
Trias, 105–17 ; conglomeratic base of, 90 ; type area of English, 105 ; water from, 183.
Triassic rocks, 3, 105, 124, 127.
Trimpley, 2, 4, 13, 29, 31, 55, 57, 81, 97, 101, 103 122, 128.
—— anticline, 8; 56, 66, 80–1, 84, 86, 94, 121–2, 124, 126, 128–9, 139 ; Caledonian trend of, 122 ; pitch of, 124.
—— axis, 55, 122.
—— Fish-bed, 24.
—— Fish Zone, 12, 24.
—— inlier, 12–13, 24, 120, 124, 128 ; lime-burning in, 180 ; water from sandstones in, 183.
TRUEMAN, A. E., 38.
Trysull, 128, 156–7, 161, 171, 174–5 ; building sand at, 181 ; water supply, 184.

Tuckhill, 86, 98, 158, 173.
Turlshill House, 10, 15.
Turner's Hill, 8–10, 17, 123 ; anticline, 11.
TURNEY, W. J., & CO., 190.
Turritella, 156.
Twenty Yard Rock, 37.
Twin Quarry, near Oakham, 135.
Two Foot Coal, 41, 47, 49–50, 53, 59, 61 ; of Highley, 51, 199.
—— ——,or Little Coal, 35.

Uffmoor Wood, 63–4, 84.
Unconformity, 3–4, 40, 42, 44, 48, 57–8, 64 ; base of Breccia Group, 95 ; base of Bunter, 96 ; base of Clent Beds, 121 ; base of Clent Breccia, 83 ; base of Clent Group, 6, 34, 96 ; base of Enville Group, 86 ; base of Halesowen Group, 6, 33, 63–4, 96 ; base of Highley Group, 6, 77, 96 ; base of Keele Beds, 86 ; base of Upper Old Red Sandstone, 8 ; base of Triassic rocks, 6 ; between Etruria Marl and Halesowen Group, 121 ; between Kinlet and Highley Groups, 121, 126 ; between Trias and Clent Beds, 122 ; between Upper and Lower Old Red Sandstone, 118 ; (the " Symon Fault "), 121.
Unconformities, 33, 68 ; within Highley Group, 69.
Underton, 22.
Uplands, 90, 160.
Upper Arley, 1, 66, 76–7, .84, 129–30, 165, 176–7, 181–2 ; springs, 183.
Upper Carboniferous, earth movements in, 96.
—— —— Rocks, 32.
Upper Coal Measures, 44, 52, 124, 127 ; floor of, 126 ; Enville Beds, 91–104 ; Etruria Marl Group, 58–62 ; Halesowen Group, 62–5 ; Highley Group, 65–81 ; Keele Group, 81–91.
Upper Danesford Terrace, 175–6.
—— Farmcote, 115, 117, 169.
—— Forge, Eardington, 90.
—— Gornal, 10.
—— Limestone, of Wren's Nest, 14.
—— Linley Brook, 159.
—— Ludlow beds, in shafts, 10.
—— Ludlow Shales, 4, 9, 15, 17, 19.
—— Ludstone, 116, 163.
—— Mottled Sandstone, 3, 108–9, 113–7, 128, 181.
—— Old Red Sandstone, 11, 122.
—— Overton, 23.

Upper Penn, 92–3, 96, 111, 123, 155–6, 171.
—— Permian, 102.
—— Sulphur Coal, 59, 61.
Uriconian rocks, 59, 61.
—— type pebbles, 112.
—— volcanic rocks, 101.
URRY, W. D., on age of Clee Hill basalt, 135.

Valley train, 148.
Vein quartz, in espleys, 61.
VERNON, R. D., 94.
Volcanic rocks, North Welsh, 157.

Wadeley, 23, 48.
—— Farm, 70.
Wallheath, 152, 161, 173 ; building sand near, 181.
Walls, The, 116.
Wallsbatch, 22.
Walltown, 24.
Walton, 24, 131.
—— Farm, 24.
—— Hill, 166.
—— Hills, 96.
WARING, GEO., 38.
Warren's Hall Colliery, 37–8, 136.
Warshill Farm, 101.
Wassel Grove, 11.
—— —— Coal, 62–3 ; clays with, 65.
—— —— Pits, 10 ; Etruria Marl in, 64.
—— —— sinking, 132.
Water Supply, 178, 182–3 ; Bridgnorth, 90 ; Highley, 91 ; wells and boreholes for, 186–92.
Watershed, Humber—Bristol Channel, 2.
Waterstones, 109.
WATSON, D. M. S., 51, 53, 69.
WATTS, W. W., 70, 134.
WEDD, C. B., 122.
Wednesbury Oak, No. 8 Pit, 59–60.
Wellsbach, 11.
Welsh Borderland, 7.
Welsh ice-sheet, 170, 173.
—— Re-advance (or ' Little Welsh Glaciation'), 175.
Wenlock Limestone, 4, 8–9, 14 ; boulders of, .in Coal Measures sandstones of Chawn Hill, 11 ; fragments of, in Clent Breccia of Doctor's Hill, 11 ; lithological sub-divisions, 14 ; of Shropshire, 14 ; underground workings in, 9 ; used in building Dudley Castle, 181.

Wenlock Series, 4, 7–8.
—— Shales, 4, 8–9, 14 ; in Smestow boring, 39.
West Midlands, 151.
Western Boundary Fault, 17, 37, 62–3, 84, 96, 120, 122–4, 126–7, 131–2, 161.
'Western Drift', 151.
Westhope Hill, quartz in *Psammosteus* Limestones of, 26.
Westphalian, 39 ; of de Lapparent, 33 ; of Kidston, 33, 59.
—— Series, 33.
Westwood, 159.
WHITE, E. I., 31.
White Cross, 103, 158, 174 ; borehole, 102.
—— Ironstone, 35, 37–9.
—— Well spring, 98.
Whitehaven Sandstone, Cumberland, marine horizon in, 59.
WHITEHEAD, T. H., 10, 17, 20, 37, 40, 42, 59, 63, 82, 94, 108, 120–1, 131, 135, 175, 193.
Whitestone Ironstone, 137–8.
Whittington, 113.
—— Common, 150, 169.
Wilderness, 81.
WILLIAMS, DAVID, 151.
WILLIAMSON, W. O., 61.
WILLS, L. J., 107, 118, 121–2, 148, 150–1, 153–6, 158–60, 162, 164, 166, 169–71, 173–6, 196.
Windmill End, 123.
Winnal Farm, 51, 71.
Winwood Heath, 192.
Winwoods, 70, 77, 181 ; fauna of limestone at, 69 ; limestone near, 67.
Witley, 64.
—— Colliery, 64 ; fossil wood at, 64.
Witnells End, 24, 56, 79–80, 86, 138–40, 180.
Wodehouse, 110–11, 127.
Wollaston, 1, 107, 109, 111, 114, 116, 150, 181.
Wollescote, 20, 62, 123 ; water supply, 184.
—— Fault, 123.
—— Hall, 63.
Wolverhampton, 1, 2, 10, 155, 171, 174 ; boulder clay of, 155 ; map, 7.
—— Corporation Waterworks, 184, 186–7.

Wolverley, 106, 108, 111, 113, 122, 154, 163, 175.
—— Waterworks, 192.
Wom Brook, 161.
Wombourne, 39, 109–111, 114–15, 127, 133, 148, 155–7, 161, 181 ; water supply, 184.
Wombourne Church, 114, 116, 156.
Woodhouse Farm, 67, 76, 129–30.
Woodfield, 115.
—— Grange, 156.
Woodlands, 23, 74, 131.
Woodsetton, 14.
WOODWARD, A. SMITH, 83.
Wooton, 84–5, 89, 129, 163 ; boulders near, 173 ; gravels, 164, 174.
—— Dingle, 85.
Worcester Terrace, 175.
Worcestershire, 8, 12.
Wordesley, 109, 111 ; borehole, 127 ; Bunter near, 124 ; water supply, 184.
Worfe, River, 2, 113, 115, 169, 174, 177 ; gorge of, 163; mouth of, 164, 176.
—— Bridge, 113, 163, 176.
—— Valley, 158, 163, 169–71, 174–5 ; ice-stream in, 171.
Worfield, 115, 163, 170, 174, 176–7.
Worralls Brook, 77–9.
Wounsdale, gravel ridge near, 174.
Wren's Nest, 8, 9, 123 ; lime-burning on, 180 ; pericline of, 14.
Wrickton Forge, 49.
Wychbury Fault, 83.
—— Hill, 96, 101, 165.
Wyken, 176–7, 186.
Wyre Forest, 33, 71, 126.
—— —— Coalfield, 39–40, 67, 179 ; Etruria Marl in, 121 ; section across, 46.

Yard Coal, 49 ; of Coalbrookdale, 42.
Yorkian, 39, 42, 52, 57, 59 ; plants, 138 ; times, 120 ; of Watts, 33.
—— Series, 39.
—— strata, variation in thickness of, 126.

Zaphrentis ?, 99.
—— *delanouei* ?, 99.
—— *konincki* ?, 99.